APPLIED MATHEMATICS

P J F Horril

Head of the Mathematics Department,
Nottingham High School

Also by the same author

Pure Mathematics Books 1 and 2
by L. Bostock and S. T. Chandler

Longman

Revised by the same author

Pure Mathematics Books 1 and 2
by J. K. Backhouse and S. P. T. Houldsworth

Longman Group UK Limited
Longman House, Burnt Mill, Harlow, Essex CM20 2JE, England and
Associated Companies throughout the world.

First published 1989

ISBN 0 582 35575 3

Set in 10/12pt Times (Lasercomp)
Produced by Longman Group (FE) Ltd
Printed in Hong Kong

Contents

Revision Exercise 3

Chapter 14   Forces in equilibrium

Chapter 15   The moment of a force

Chapter 16   Centre of gravity

Chapter 17   Miscellaneous statics

Preface

This book is intended for students preparing for the Applied Mathematics part of A-level Mathematics, AS-level Mathematics and similar examinations.

At one time, this subject suffered from a tendency, on the part of examiners and textbook writers, to make excessive use of algebraic and trigonometrical manipulation; undoubtedly this put many students off the subject. More recently examiners have tried to set more straightforward questions, and this book is written with this welcome trend in mind.

It is expected that most readers will be taking a corresponding course in pure mathematics, and, in particular will be developing their knowledge of calculus. At various points in this book reference is made to *Pure Mathematics Books 1 and 2*, by Backhouse and Houldsworth and revised by the present author, which is published by the Longman Group. Any other similar text book should be suitable. It is also assumed that most readers will have an electronic calculator.

In the early chapters we study the relationship between displacement, velocity and acceleration i.e. kinematics; this is followed (Chapters 8–13) by some applications of Newton's laws of motion, in other words, dynamics. After this (Chapters 14–17), we examine forces in equilibrium i.e. statics, and finally return to dynamics in Chapter 18. Sadly, pressure of time has squeezed 'moments of inertia' out of many A-level Mathematics syllabuses, but it is included here because many potential engineers will find it useful, and hopefully many readers will want to pursue the subject beyond the confines of examination requirements. Vector methods are introduced at an early stage (Chapter 5) and subsequently they are used whenever they make a constructive contribution to the subject.

Many of the misconceptions which students develop in this subject stem from failing to fully understand that 'acceleration' is a vector quantity; this in turn leads to misunderstanding the relationship between forces and the resulting acceleration, (see *Studies in Mechanics Learning*, published by the Centre for Studies in Science and Mathematics Education at the University of Leeds). With this in mind, this book is arranged so that the ideas of vectors in general, and velocity and acceleration in particular, are fully discussed before embarking on Newton's laws of motion. Statics is dealt with later. However, some teachers may prefer to start this topic at an earlier stage; if so they will find it quite easy to adapt the order of these chapters to their own liking.

How to use this book

Each chapter is divided into convenient sections, each dealing with a single concept, or a group of related concepts. These are punctuated by individual questions (marked **Qu.** in the text), which enable the reader to check whether he or she has understood the preceding work. Students working on their own, without a regular teacher, should try to do all of these questions. They are frequently used to encourage the student to consider alternative methods to those expounded in the text.

At intervals, there are sets of exercises for the student to consolidate the work to date. Most chapters end with a Miscellaneous Exercise which tests understanding of the whole chapter. Many of the questions in these exercises are from recent A-level examinations. Some are from examinations in Additional Mathematics: these form a useful 'half-way house' before tackling the more difficult A-level questions, and are marked with an asterisk. There are Revision Exercises after Chapters 5, 9, 13 and 17. It is suggested that some of the later questions in the Miscellaneous Exercises are also reserved for revision purposes.

The standard abbreviations used by examination boards are employed in this book. Readers should be aware of the convention that algebraic symbols are always printed in *italic*, but abbreviations of units always appear in normal type; for example, a distance of 'x metres' is printed x m.

Accuracy

Answers are provided in the end of the book. Where a calculator has been used, the answer is normally rounded off to three significant figures (but teachers should encourage students to consider whether the data and simplifying assumptions warrant such a degree of accuracy). Where a numerical value of g, the acceleration due to gravity, is required, $9.81 \, \text{m s}^{-2}$ is normally used, but in some cases, following the practice of many examining boards, $10 \, \text{m s}^{-2}$ is used. (Readers should always follow the instructions in the particular exercise.) This is plainly a very convenient number, even if it is not very accurate. Also, following the usual convention, terms like π and $\sqrt{2}$ are left in answers where this is appropriate, i.e. in questions of an algebraic character.

Notation in diagrams

The references are to the chapters where the notation first appears. Units are frequently omitted from diagrams in the interests of clarity.

VECTORS (Chapter 5)

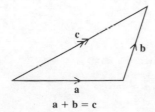

$a + b = c$

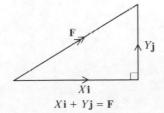

$Xi + Yj = F$

VELOCITY (Chapter 2)

5 m s^{-1} v

ACCELERATION (Chapter 2)

2 m s^{-2} a

FORCE (Chapter 8)

T

mg

IMPULSE (Chapter 11)

I

It would not be possible to write a book like this without a great deal of help and support from other people. I am deeply indebted to my friend John Pitts who has read the manuscript, made many helpful suggestions and worked through the questions. Those who have met John will know he has immense experience as an examiner and teacher of mathematics; his help has been invaluable and I am most grateful for it. Some of the text and exercises have also been tried by colleagues and pupils at Nottingham High School and this has also been much appreciated. I would particularly like to mention John Wood, who has made many useful suggestions. Even so, any faults or errors which remain are my responsibility.

My thanks are also due to Andrew Ransom and all his colleagues at Longman House for their encouragement and expert advice.

Last, but not least, I would like to thank my wife Pamela for her endless patience over household chores neglected while I have been occupied with this project.

P. J. F. HORRIL
Nottingham
1989

Acknowledgements

We are grateful to the following for permission to reproduce photographs:

Allsport, pages 26 (photo Tony Duffy), 120 (photo Vandystadt), 174, 372 (photo Vandystadt); Camera Press/Deutsche Presse-Agentur GmbH, page 230 (photo Harmut Reeh); Camera Press/Interfoto MTI, Hungary, page 156 (photo Petrovits); Trevor Clifford, page 292; Harlow Council, Leisure Services Dept, page 78; The Hutchison Library, page 320; London Transport Museum, page 344; Martin Mulcahy, page 266; Salter Industrial Measurement Ltd, page 200; Science Photo Library, pages xii (photo NASA), 100 (photo Dr Harold Edgerton); Sporting Pictures (UK), page 40; West Air Photography, page 6.

Cover: Detail from *Les Disques dans la Ville*, Fernand Léger, 1920. Photographie Musée national d'Art moderne, Centre Georges Pompidou, Paris.

We are grateful to the following examining bodies for permission to reproduce questions from past examination papers:

Associated Examining Board (AEB); University of Cambridge Local Examinations Syndicate (C); Joint Matriculation Board (JMB); University of London, School Examinations Department (L); Oxford and Cambridge Schools Examination Board (O & C) and University of Oxford Delegacy of Local Examinations (O).

Questions from the above bodies are indicated by the letters shown in parentheses. The letters MEI (Mathematics in Education and Industry) and SMP (Schools Mathematics Project) indicate questions from syllabuses set by the Oxford and Cambridge Schools Examination Board. Those questions marked with an asterisk are taken from Additional Mathematics papers.

Newtonian Mechanics

Chapter 1

Introduction to Mechanics

1.1 Newtonian Mechanics

The term 'Applied Mathematics' has acquired several meanings in recent years, but in this book we shall only be concerned with its traditional meaning, namely, Newtonian Mechanics.

Newtonian Mechanics is chiefly concerned with the mass of an object, the forces acting on it, and the resulting motion. As the name suggests, we are indebted to Sir Isaac Newton (1642–1727) for his brilliant exposition of the subject in his *Principia* (1687). He in turn owed a debt to the great experimentalists like Galileo (1564–1642), Huygens (1629–1695) and Kepler (1571–1630), whose work on astronomy provided the motivation and empirical evidence from which Newton derived his mathematical theories.

The mathematics evolved by Newton, in turn, provided the techniques which enabled the scientists and engineers of the 19th century to develop the technology which gave Europe its Industrial Revolution; most of the benefits of science, which we enjoy today, depend in some way or other on Newton's work. Indeed, some of the most recent and spectacular scientific achievements, namely the use of Earth satellites and interplanetary flight, are directly related to Newton's work on orbital motion. It is one of the remarkable features of Newtonian Mechanics that, although it was derived from experiments on Earth with fairly primitive apparatus, it has been successful in explaining the motion of the solar system.

However, in the 20th century it was realised that Newtonian Mechanics does have some limitations: it is not suitable for investigating phenomena which occur in the most distant parts of the universe, nor is it suitable for interpreting very small scale, i.e. subatomic, phenomena. For these branches of science, mathematicians like Einstein (1879–1955) and Schrödinger (1887–1961) have invented the subjects of relativity and quantum Mechanics. Nevertheless, for the study of physical phenomena on the terrestrial, or even the planetary scale, Newtonian Mechanics is still the right approach. It also has the advantage, which should not be despised, that the mathematical skills required are within the grasp of a competent A-level student, whereas relativistic and quantum Mechanics are much more demanding subjects. So from the educational point of view, Newtonian

Mechanics is the natural starting point for the aspiring scientist or applied mathematician.

1.2 Newton's model

In making his mathematical model of the universe, Newton made a number of very plausible assumptions: in particular he assumed that it is possible to have a fixed set of (3-dimensional) axes and that it is possible for observers in different positions to synchronise their clocks and hence agree on the time and place where an event occurs. These assumptions however have to be questioned when dealing with very large scale astronomical events and very small, subatomic, events.

1.3 Units

The basic quantities which are used in the study of Mechanics are mass, length and time. Most, but not all, scientists these days use the Système International d'Unités (SI units), that is kilograms for mass, metres for length and seconds for time, together with appropriate multiples and submultiples of these. However there are many situations, navigation for instance, where it is still the normal practice to use Imperial units, especially for distances and speeds. So Imperial units have been used occasionally in this book, but the reader is not expected to learn the conversion factors either between Imperial units, or between SI and Imperial units.

1.4 Assumptions and simplifications

A motor car is a very large and complicated piece of machinery, with thousands of moving parts; its passengers are also massive objects which are capable of infinitely variable motion. It would not be possible for a car designer, even using the most powerful computer, to take into account all the moving parts of the vehicle and its contents. Plainly it is necessary to simplify the problem. In fact, useful deductions about the speed and acceleration of the vehicle can be made by disregarding most of the complications and focussing one's attention on the total mass of the vehicle and its passengers, and the forces acting on it, e.g. the driving force and the air-resistance.

The skill of a good applied mathematician frequently depends on the ability to select the approximations and simplifications which reduce the mathematical problems to manageable proportions, while still producing results which are useful and realistic.

One very common simplification is to regard the object being studied as a *particle*, that is, an object of finite mass, whose size is negligible. At first sight it may seem ridiculous to regard say, a space rocket whose mass is several thousand tonnes, as a 'particle', but if the rocket is being tracked on radar from a space station, thousands of miles away, it may be a perfectly reasonable assumption to make, since its dimensions are extremely small compared with the distances involved. In fact, it can be shown that the centre of mass of a large object behaves like a particle whose mass equals the total mass of the object; so mathematical techniques, learnt in the course of studying the motion of a particle, are not wasted when the time comes to study the motion of a large object.

1.5 Accuracy

One very important consideration in all engineering and scientific calculations, is the effect of errors in the data. The reader should be aware that it is always possible, at least in theory, to *count* a finite number of objects with total accuracy; however, it is never possible to make a *measurement* with total accuracy. The design of the measuring instrument and the precision of the scales from which the readings are taken, will always introduce unavoidable errors.

However, in a textbook, the reader should not become too obsessed with the problems of accuracy, lest this should detract from the chief purpose of the textbook, namely the mastery of the new concepts being studied. One particular approximation occurs throughout the subject: the value of the acceleration due to gravity. In fact this quantity varies, slightly, from place to place on the Earth's surface, although it has an official 'standard' value of $9.806\,65\,\mathrm{m\,s^{-2}}$. In this book $9.81\,\mathrm{m\,s^{-2}}$ will normally be used, although in several places it will be taken to be $10\,\mathrm{m\,s^{-2}}$, since this clearly simplifies many calculations.

The reader should, however, not lose sight of the fact that, in a real calculation, the combined effect of the various errors and approximations needs to be analysed before a particular degree of accuracy can be claimed for the results.

Exercise 1

1 Consider the two statements:
   (a) DISTANCE = SPEED $\times$ TIME
   (b) COOKING TIME = 20 minutes to the pound plus an extra 20 minutes.
   Find four more rules which, like statement (a), are true for any consistent set of units, and four rules like statement (b) which are only true for a particular set of units.
2 Two quantities x and y are measured and found to be 5.1 and 4.9 respectively. It is known that each measurement is subject to a maximum error of 0.1. Find the greatest and least possible values of the following quantities:
   (a) $x + y$ (b) $x - y$ (c) xy (d) $x : y$ (e) $y \div x$
   (f) $\dfrac{1}{x + y}$ (g) $\dfrac{1}{x - y}$
3 Under suitable conditions a large massive object may be regarded as a particle. Describe three such situations.
4 Find out about the Cd factor (the drag factor) and the effect it has on the design of motor cars. Describe briefly the importance of the following factors to the purchaser of a new car:
   (a) the price   (b) the fuel consumption   (c) the power
   (d) the top speed   (e) the acceleration   (f) the Cd factor
   (g) the weight   (h) the length   (i) the insurance class.

5 Why would it be difficult for observers in different parts of the universe to synchronise their clocks?

6 Describe some of the simplifications which might be made when considering the motion of

(a) a child on a swing;

(b) washing rotating in a spin-drier;

(d) a heavy roller being used to roll a cricket pitch;

(d) a sports car travelling at high speed in a straight line;

(e) a parachutist   (i) in 'free-fall',   (ii) with the parachute deployed.

In questions 7–10 write a brief essay upon:

7 The life and work of Galileo.

8 The life and work of Newton.

9 Kepler's laws of motion.

10 Why the study of astronomy has had such a profound effect on the development of mathematics.

Chapter 2

Motion in a straight line

2.1 Introduction

In this chapter we shall be studying the motion of an object moving along a straight line. In some cases the speed will be constant, but in others it may vary. Some important terms will be introduced, especially displacement, velocity and acceleration.

Positive and negative numbers will be used in the same way as they are used when drawing graphs in pure mathematics, and it will be assumed that the reader is familiar with the idea of the gradient of a graph.

2.2 The displacement–time graph

As in any technical subject, it is important that the terms we use are clearly defined. The word 'displacement', in this context, has a stronger meaning than 'distance', as the first two examples illustrate.

Example 1

A train moves at constant speed along a track which runs North–South. It passes a signal box at noon, travelling due North at 60 km/h. After 2 hours it stops and remains stationary for 1 hour. Then it travels South at 80 km/h for 3 hours. (Ignore the effects of it slowing down and speeding up.)
(a) How far North does it travel?
(b) How far South does it travel?
(c) At what times is it 40 km North of the signal box?
(d) Find the total distance travelled.
Draw a graph showing its distance North of the signal box (on the vertical axis) plotted against time (on the horizontal axis).

Motion in a straight line

The distance travelled at constant speed is given by speed × time. Hence

(a) distance travelled North = 60 × 2 km
                        = 120 km
(b) distance travelled South = 80 × 3 km
                        = 240 km
(c) The time taken to travel a given distance at constant speed is given by distance ÷ speed. Hence the time taken to reach the point 40 km North of the signal box on the outward journey
                        = 40 ÷ 60 hours
                        = 40 minutes

On the return journey the train has to travel 80 km to reach the required point. This takes one hour. Hence the train is at the point 40 km North of the signal box at 12.40 hours and again at 16.00 hours. (See broken lines in the diagram.)

(d) The total distance travelled = 120 + 240 km
                        = 360 km

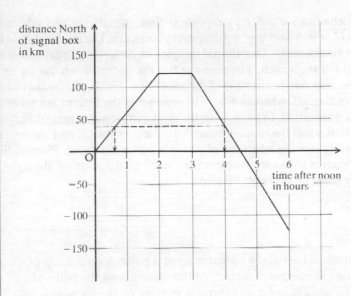

Figure 2.1

There are several points to note in this example. Parts (a) and (c) are concerned with the position of the train relative to the signal box, whereas part (d) is only concerned with the total distance. Plainly, questions like (a) and (c) are important if we need to know where the train is. Part (d) would be important if we were interested in the amount of fuel it needed.

The position of an object relative to some fixed point is called its **displacement**; note that both the distance and the direction of the object must be given. In the graph, Figure 2.1, negative numbers are used to describe displacements South of the signal box.

It should be noted that while the train is travelling North its graph is a straight line with gradient 60, i.e. for each unit of 1 hour on the time scale, there is an increase of 60 km on the distance scale. In other words the gradient on the displacement–time graph gives the speed of the train. What is more, the gradient on the part of the graph representing the return journey is − 80 (a negative sign is always used for a line which slopes downward as we read from left to right). This again corresponds to the speed on the return journey; the negative sign indicates that the train is moving South.

When we wish to describe the motion of an object, giving both its speed and direction, we use the word **velocity**. For motion in a straight line, the direction can be indicated by a positive or negative sign, provided the sign convention is made clear from the outset; in Example 1, displacements and velocities North are positive, while those South are negative. In the next example, the velocity is variable, both in magnitude and direction. Displacements above the initial point and upward velocities are positive; displacements below the initial point and downward velocities are negative.

Throughout Mechanics we use t to represent time, usually in seconds, and phrases like '$t = 0$', '$t = 5$' and '$t = -10$' are very common. The reader should be clear what such phrases mean. One should imagine an observer at a car race who times the cars with a stopwatch. The time $t = 0$ is the instant when the watch is started; it may, or may not, be the start of the race. Time $t = 5$ is the instant when five seconds have elapsed, whereas $t = -10$ represents the instant ten seconds before the watch was started. In particular, the reader must not assume that $t = 0$ refers to the instant when the motion 'starts'. In the race mentioned above, the cars may be already moving when the observer's watch is started. Phrases like 'the *initial* displacement/velocity/acceleration' refer to the state of the system when $t = 0$; they may, or may not, be zero.

Example 2

A stone is projected vertically upwards from a point A on the top of a tall building. The height, h metres, of the stone above the point of projection, t seconds after it was thrown, is given by the formula

$$h = 40t - 5t^2$$

As it falls back to Earth, it misses the edge of the building and continues to fall vertically. Find the height above A when $t = 0, 1, 2, 3, \ldots, 10$.

Draw the graph of h against t.

Show that $80 - 5(t - 4)^2 = 40t - 5t^2$ and deduce that the highest point on the stone's path is 80 m above A.

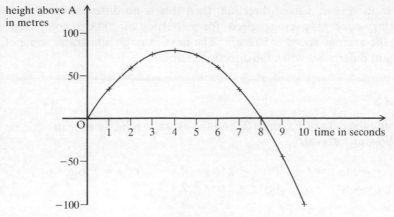

Figure 2.2

The values of h are shown in the table below:

t	0	1	2	3	4	5	6	7	8	9	10
h	0	35	60	75	80	75	60	35	0	-45	-100

$$80 - 5(t - 4)^2 = 80 - 5(t^2 - 8t + 16)$$
$$= 80 - 5t^2 + 40t - 80$$
$$= 40t - 5t^2$$

Hence the height of the stone can be expressed in the form

$$h = 80 - 5(t - 4)^2$$

Since the term $(t - 4)$ is squared, it cannot be negative, so h is always less than or equal to 80; it is only equal to 80 when $t = 4$. Consequently the highest point is 80 m above the point of projection and it is reached after 4 seconds.

Note the following points in Example 2.
(a) For any given height less than 80 m above A, there are two possible times, one when the stone is rising, and one when it is falling.
(b) When $t = 8$, $h = 0$. This means the displacement is zero, i.e. the stone is level with its initial position.
(c) For $t > 8$, h is negative. This indicates that the stone is *below* its initial point A; e.g. when $t = 10$, the stone is 100 m below A.

2.3 Average velocity

The average velocity over an interval of time is defined as follows:

$$\text{average velocity} = \frac{\text{change in displacement}}{\text{time taken}}$$

If the motion is in a fixed direction, then this is no different from the more elementary concept, average speed; for a journey of 200 km, completed in 4 hours, the average speed is 50 km/h. The next example illustrates some of the important differences when the direction changes.

Example 3

Using the data of Example 2, find the average velocity over the following intervals:

(a) $t = 0$ to $t = 4$ (b) $t = 2$ to $t = 4$ (c) $t = 2$ to $t = 6$
(d) $t = 4$ to $t = 6$ (e) $t = 2$ to $t = 2.1$

(a) Average velocity over the 4 second interval from $t = 0$ to $t = 4$

$$= \frac{80 - 0}{4 - 0}$$

$$= 20$$

The average velocity is 20 m/s upwards.

(b) Average velocity over the 2 second interval from $t = 2$ to $t = 4$

$$= \frac{80 - 60}{4 - 2}$$

$$= 10$$

The average velocity is 10 m/s upwards.

(c) Average velocity over the 4 second interval from $t = 2$ to $t = 6$

$$= \frac{60 - 60}{6 - 2}$$

$$= 0$$

The average velocity is zero.
(Note that this follows from the fact that the *change in the displacement* is zero.)

(d) Average velocity over the 2 second interval from $t = 4$ to $t = 6$

$$= \frac{60 - 80}{6 - 4}$$

$$= -10$$

The average velocity is 10 m/s *downwards*.

(e) For this part we need the original formula in Example 2, i.e.

$$h = 40t - 5t^2$$

When $t = 2.1$,

$$h = 40 \times 2.1 - 5 \times 2.1^2$$
$$= 84 - 22.05$$
$$= 61.95$$

Hence the average velocity over the 0.1 seconds from $t = 2$ to $t = 2.1$

$$= \frac{61.95 - 60}{2.1 - 2}$$

$$= \frac{1.95}{0.1}$$

$$= 19.5$$

The average velocity is 19.5 m/s upwards.

2.4 Velocity at an instant

The concept of average velocity over a large interval of time has only limited usefulness; much more useful is the idea of the velocity at an instant. In a motor car, we can look at the speedometer to read the speed at a particular moment. If we wish to calculate the value of the velocity at an instant, we start by investigating the average velocity over a very small interval of time.

Let us return to the problem of the stone in Examples 2 and 3. In Example 3 part (e), we saw that the average velocity over an interval of 0.1 seconds was 19.5 m/s. We will now reduce the interval to 0.01 s.

When $t = 2.01$,

$$h = 40 \times 2.01 - 5 \times 2.01^2$$
$$= 80.4 - 20.2005$$
$$= 60.1995$$

$$\text{Average velocity} = \frac{60.1995 - 60}{2.01 - 2}$$

$$= \frac{0.1995}{0.01}$$

$$= 19.95 \text{ m/s}$$

The reader should now work through Questions 1 and 2.

Qu. 1 For the data in Example 2, calculate the average velocity over the interval $t = 2$ to $t = 2.001$.

Qu. 2 Repeat Qu. 1 for $t = 2$ to $2 + h$. Verify that this agrees with the results above. What can you deduce about the average velocity as h tends to zero?

Putting these results together, it seems reasonable to deduce that as the time interval nears zero, the average velocity approaches $20 \, \text{m/s}$.

2.5 Instantaneous velocity and the displacement–time graph

It is instructive to consider the effect of the process described in §2.4 on the displacement–time graph.

Suppose the graph in Figure 2.3 represents the displacement–time graph of a moving object, and that points P and Q represent the displacements at two moments which are separated by a time interval, h.

$$\text{displacement at time } t = MP$$
$$\text{displacement at time } (t + h) = NQ$$
$$\text{change in displacement} = NQ - MP$$
$$= NQ - NR$$
$$= RQ$$
$$\text{time interval} = MN = PR$$
$$\therefore \quad \text{average velocity} = \frac{RQ}{PR}$$

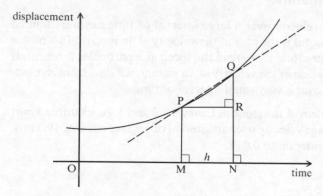

Figure 2.3

Now, this ratio is the gradient of the chord through P and Q, so we have a very useful connection between the displacement–time graph and velocity, namely

> the average velocity is numerically equal to the gradient of the corresponding chord on the displacement–time graph

Furthermore, when the time interval h becomes smaller and smaller, the line through P and Q merges into the tangent at the point P, hence

> the instantaneous velocity is numerically equal to the gradient of the corresponding tangent on the displacement–time graph

The graph in Figure 2.4 is the displacement–time graph for the stone in Example 2 with the tangent at (2,60) drawn.

From the graph, we see that the gradient of the tangent at (2,60) is $\frac{80}{4} = 20$. Consequently the instantaneous velocity when $t = 2$ is 20 m/s. This agrees with the conclusion we reached in §2.4.

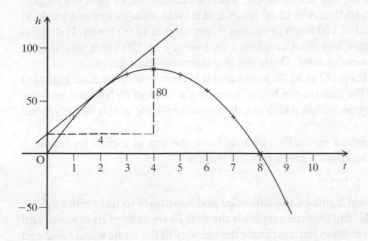

Figure 2.4

Summary

(a) A **displacement** defines the position of a variable point relative to a fixed point; both the distance and the direction must be given. For motion in a straight line, the direction can be indicated by a positive or negative sign.

(b) The **velocity** of a point is the rate at which its displacement is changing; both its speed and direction of motion must be given.

(c) The **gradient** of the tangent at a point on the displacement–time graph is numerically equal to the instantaneous velocity.

Note on abbreviations

The reader's attention is drawn to the following conventions.

(a) When a letter is used in algebra to represent a number, it is always printed in *italic type*.

(b) Abbreviations of units are always printed in normal type (called roman type), for instance, the reader will see 'length = 5 cm', or 'length = x cm'.

(c) The letter s is the abbreviation for 'seconds', it is *never* used to make the plural form of an abbreviation (do not write 5 centimetres as 5 cms).

(d) The abbreviation for 'metres per second' and similar units may be written either as m/s or as $\mathrm{m\,s^{-1}}$. (In the next section the reader will meet acceleration; the units for this can be written as $\mathrm{m/s^2}$ or as $\mathrm{m\,s^{-2}}$.)

Exercise 2a

1 Buses leave a city centre at 20 minute intervals, and travel non-stop along a route to a point 10 km from the centre, at 20 km/h. A jogger leaves the centre at the same time as a bus and follows the same route at a steady 8 km/h. Draw a displacement–time graph and from it estimate the times in minutes and the distances from the city centre in km, when successive buses pass the jogger.

2 A train leaves a station A at 12.00 hours and travels, along a straight track, at a constant speed of 100 km/h to station B, arriving at 13.00 hours. It stops at B for 30 minutes and then continues its journey at 120 km/h, arriving at station C, 45 minutes later. Draw the displacement–time graph.

Another train leaves C at 12.30 hours and travels non-stop to A, at constant speed, passing the first train while it is waiting at station B. Indicate on your diagram the region within which the displacement–time graph of the second train lies.

3 A stone is projected vertically upwards from the top of a cliff. Its height, h metres, after t seconds is given by the formula

$$h = 30t - 5t^2$$

On the way down it misses the cliff-edge and continues to fall vertically. Draw the displacement–time graph for the first 10 seconds of its motion, and by drawing suitable tangents estimate the velocity of the stone when $t = 2$ and when $t = 5$.

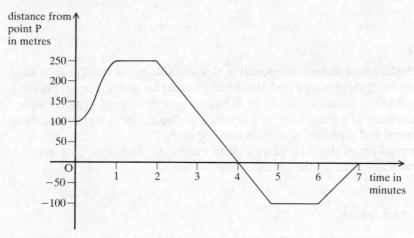

Figure 2.5

4 The diagram (Figure 2.5) represents the displacement of a shunting engine moving along a straight track. Describe its motion in words.

5 A spring is suspended from a fixed point A and hangs vertically with a heavy bob B at the other end. When the spring is at rest $AB = 50$ cm. The bob is displaced and it then oscillates in a vertical line about its resting position (point O). After t seconds its height, x cm, above O is given by

$$x = 5 \sin t$$

Using a calculator *in radian mode* complete the following table (giving the values of *x* correct to 2 decimal places).

t	0	0.5	1.0	1.5	2.0	2.5	3.0	3.5	4.0	4.5	5.0	5.5	6.0	6.5	7.0	7.5	8.0
x	0	2.40						0.71			−4.89		−1.40				

Draw the displacement–time graph and from it estimate:
(a) the times when the bob is at O;
(b) the times when the bob is 2 cm *above* O;
(c) the times when the bob is 2 cm *below* O;
(d) the velocity when $t = 1$;
(e) the velocity when $t = 4$.

6 A stone is dropped from a high building and after t seconds it has fallen x metres, where $x = 5t^2$. Draw its displacement–time graph, and, by drawing a suitable tangent, estimate its velocity after 3 seconds.

7 Calculate the average velocity of the stone in question 6, over the following intervals:
(a) $t = 3$ to $t = 3.1$ (b) $t = 3$ to $t = 3.01$ (c) $t = 3$ to $t = 3 + h$
Deduce the instantaneous velocity when $t = 3$. Compare this answer with the answer to question 6.

8 Copy the displacement–time graph for Example 2 (p. 8) and by drawing a tangent at (3,75) estimate the instantaneous velocity when $t = 3$. Calculate the average velocity over the interval $t = 3$ to $t = 3 + h$ and by letting h tend to zero, deduce the instantaneous velocity when $t = 3$.

9 Draw a displacement–time graph for a stone which is projected vertically so that its height h metres, after t seconds, is given by

$$h = 50t - 5t^2$$

Estimate the instantaneous velocity when $t = 7$ and compare this with the value obtained from considering the average velocity over the interval $t = 7$ to $t = 7 + h$.

10 A particle oscillates about a point O and its displacement, x cm, from O, after t seconds, is given by $x = 10 \sin t$ (where t is measured in radians). Using a calculator, *in radian mode*, compile a table of values of x for $t = 0$ to 8. Draw the tangent at the origin and hence estimate the velocity of the particle as it passes through O.
Calculate the average velocity of the particle over the following intervals, making sure your calculator is still in radian mode, and recording your results correct to 2 decimal places:
(a) $t = 0$ to $t = 1$ (b) $t = 0$ to $t = 0.5$
(c) $t = 0$ to $t = 0.1$ (d) $t = 0$ to $t = 0.01$
What do these results suggest about the instantaneous velocity when $t = 0$?

2.6 Acceleration and retardation

When a car is advertised, it is a common practice to give the rate at which it will accelerate: a statement such as '0 to 90 km/h in 10 seconds' indicates that the car can increase its speed at an average rate of 9 km/h per second. In mathematics and science it is inconvenient to have a mixture of units, and so the speed in this example should be converted from km/h to m/s, i.e.

$$90 \, \text{km/h} = 90\,000 \, \text{m/h}$$
$$= \frac{90\,000}{3600} \, \text{m/s}$$
$$= 25 \, \text{m/s}$$

We can then write the acceleration as 2.5 m/s per second. Although the expression 'm/s per second' conveys clearly what is meant by acceleration, the units are usually abbreviated to m/s^2 (read 'metres per second squared'); as an alternative m s^{-2} is often used.

The average acceleration of an object moving in a straight line is defined as follows:

$$\text{average acceleration} = \frac{\text{final velocity} - \text{initial velocity}}{\text{time taken}}$$

Qu. 3 A car accelerates in a straight line from 60 km/h to 90 km/h in 5 seconds; find its average acceleration in m/s^2.

Qu. 4 A cyclist accelerates in a straight line from 0 to 12.5 m/s in 18.5 seconds; find the average acceleration, giving the answer correct to two significant figures.

When an object slows down, its retardation could be defined in a similar way; for example if the speed of a car decreases from 30 m/s to 10 m/s in 5 seconds, we could describe this as a retardation of 4 m/s². However in mathematics it is more satisfactory to use the definition of acceleration above and write:

$$\text{average acceleration} = \frac{10 - 30}{5} \, \text{m/s}^2$$
$$= -4 \, \text{m/s}^2$$

Earlier in the chapter, we made a distinction between *distance* (magnitude only) and *displacement* (magnitude and direction), and a similar distinction was made between speed and velocity. While we are considering motion in a straight line, the direction can be specified by a clearly defined sign convention (but we shall reconsider this issue when we examine motion in more than one dimension). Since acceleration has been defined in terms of velocity (not just speed), this too will have an associated direction. Unfortunately there is no suitable alternative word to use for acceleration when we mean the magnitude only.

The next example illustrates the idea of acceleration as a directed quantity.

Example 4

A car moves in a straight line. Find its average acceleration when the velocity makes the following changes over a period of 10 seconds:

(a) initial velocity = $+10\,\text{m/s}$, final velocity = $+25\,\text{m/s}$
(b) initial velocity = $+10\,\text{m/s}$, final velocity = $+5\,\text{m/s}$
(c) initial velocity = $-50\,\text{m/s}$, final velocity = $0\,\text{m/s}$
(d) initial velocity = $-10\,\text{m/s}$, final velocity = $-20\,\text{m/s}$

By definition,

$$\text{average acceleration} = \frac{\text{final velocity} - \text{initial velocity}}{\text{time taken}}$$

(a) Average acceleration $= \dfrac{(+25) - (+10)}{10}\,\text{m/s}^2$

$$= \frac{+15}{10}\,\text{m/s}^2$$

$$= +1.5\,\text{m/s}^2$$

(b) Average acceleration $= \dfrac{(+5) - (+10)}{10}\,\text{m/s}^2$

$$= \frac{-5}{10}\,\text{m/s}^2$$

$$= -0.5\,\text{m/s}^2$$

(In everyday speech we would regard this as a retardation of $0.5\,\text{m/s}^2$)

(c) Average acceleration $= \dfrac{0 - (-50)}{10}\,\text{m/s}^2$

$$= \frac{+50}{10}\,\text{m/s}^2$$

$$= +5\,\text{m/s}^2$$

(d) Average acceleration $= \dfrac{(-20) - (-10)}{10}\,\text{m/s}^2$

$$= -1\,\text{m/s}^2$$

These results are shown schematically in Figure 2.6. In each case, A is the initial position and B the final one, (but the displacement is *not* drawn to scale). The sign convention is that motion to the *right* is positive (note the arrows pointing to the *left* in parts (c) and (d)). The single-headed arrows indicate the velocities and the double-headed arrows represent the accelerations.

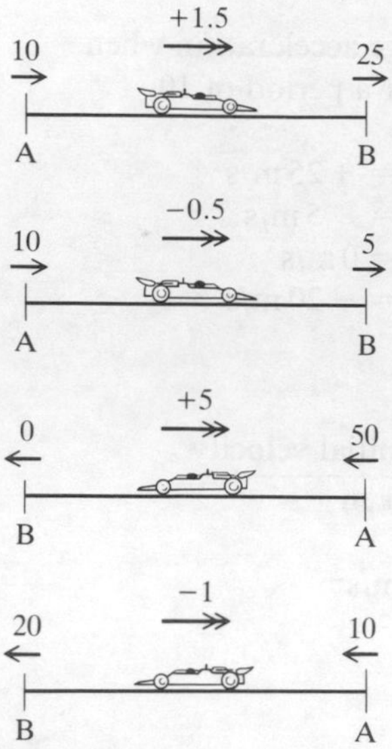

	Speed	Initial velocity	Final velocity	Acceleration
	increasing	+10	+25	+1.5
	decreasing	+10	+5	−0.5
	decreasing	−50	0	+5
	increasing	−10	−20	−1

Figure 2.6

In the next example, we consider a slightly less abstract situation. From the start it is implied that any upward displacement, velocity or acceleration is positive. The acceleration due to gravity (9.8 m/s² towards the Earth's surface) is expressed as the directed number −9.8.

Example 5

A ball is projected upwards with an initial velocity of 20 m/s. The magnitude of the acceleration due to gravity is 9.8 m/s²; write down an expression for the velocity of the ball after t seconds. Hence find
(a) the velocity when $t = 2$;
(b) the velocity when $t = 3$;
(c) the time when the velocity is zero;
(d) the time when the velocity is 20 m/s downwards.

The situation described in the question is represented in Figure 2.7.

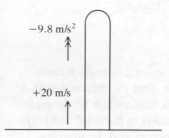

Figure 2.7

The initial velocity is $+20\,\text{m/s}$ and the acceleration is $-9.8\,\text{m/s}^2$. Hence the change in the velocity in t seconds is $-9.8t\,\text{m/s}$, so the velocity $v\,\text{m/s}$ after t seconds, is given by

$$v = 20 - 9.8t$$

(a) When $t = 2$, $v = 20 - 9.8 \times 2 = 0.4$.
   Hence the velocity after 2 seconds is $0.4\,\text{m/s}$ upwards.
(b) When $t = 3$, $v = 20 - 9.8 \times 3 = -9.4$.
   Hence the velocity after 3 seconds is $9.4\,\text{m/s}$ downwards.
(c) The velocity is zero when

$$20 - 9.8t = 0$$
$$9.8t = 20$$
$$t = 20 \div 9.8$$
$$= 2.04 \quad \text{(correct to 3 significant figures)}$$

Hence the velocity is zero after 2.04 seconds (at this moment the ball is at the highest point of its trajectory).

(d) If the velocity is $20\,\text{m/s}$ downwards then $v = -20$. Hence

$$20 - 9.8t = -20$$
$$9.8t = 40$$
$$t = 40 \div 9.8$$
$$= 4.08 \quad \text{(correct to 3 significant figures)}$$

Hence the velocity is $20\,\text{m/s}$ downwards after 4.08 seconds.

(The reader should note that, in this example, the arithmetic has been done to an accuracy of 3 significant figures. However, the validity of this degree of accuracy will depend upon the accuracy of the data.)

Applying the argument in §2.4 to the velocity–time graph we can deduce that

> the average acceleration is numerically equal to the gradient of the corresponding chord on the velocity–time graph

and also that

> the acceleration at an instant is numerically equal to the gradient of the corresponding tangent on the velocity–time graph

Qu. 5 A particle P moves in a straight line and its velocity $v\,\text{m/s}$ after t seconds, is given by $v = 16\sqrt{t}$. Draw the velocity–time graph, and from it estimate
(a) the average acceleration over the first 4 seconds;
(b) the instantaneous acceleration, when $t = 25$.

2.7 The velocity–time graph

Example 6

A car accelerates at 2 m/s^2 for 10 s. It then travels at constant speed for 20 s, but then makes an emergency stop with a retardation of 8 m/s^2. Calculate the maximum speed and the time taken in the emergency stop. Draw the velocity–time graph.

The maximum speed $= (2 \times 10) \text{ m/s}$
$\qquad\qquad\qquad\quad = 20 \text{ m/s}$

Time taken stopping $= (20 \div 8)$ seconds
$\qquad\qquad\qquad\qquad = 2.5$ seconds

The velocity–time graph is shown in Figure 2.8. (Notice that since acceleration and retardation are constant, the graphs for these stages are straight lines, with gradients 2 and -8 respectively; for the middle stage the graph is a straight line parallel to the axis, i.e. its gradient is zero.)

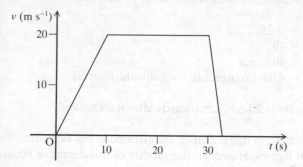

Figure 2.8

Qu. 6 Describe in words the motion represented by the velocity–time graphs in Figure 2.9.

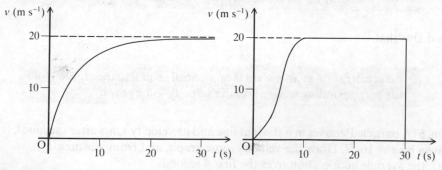

Figure 2.9a   **Figure 2.9b**

2.8 The area under the velocity–time graph

The graph in Figure 2.10 represents the motion of an object which moves with a constant velocity of 10 m/s for 20 s; the distance moved by this object is 200 m.

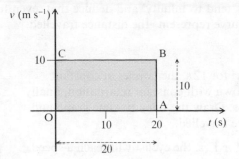

Figure 2.10

Notice that the area under the graph, that is, the area of the rectangle OABC, is 200 units. (The actual area will, of course, depend on the scale of the graph, but here we use the dimensions indicated on the two axes, i.e. 20 by 10.) In other words the area under the graph is numerically equal to the distance travelled. Clearly the validity of this statement does not depend on the particular speed and time in this illustration; the same argument could be applied to any *constant* speed and any time. We shall now show that the argument can be extended to include the case when the speed is *variable*.

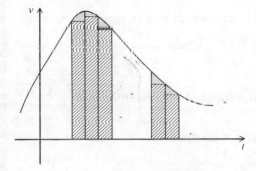

Figure 2.11

Figure 2.11 represents the motion of an object moving with variable speed. We can estimate the area under the graph by considering the area enclosed by a large number of very narrow rectangles. Although there will be a slight error, (the area enclosed by the curved boundary and the top of the 'steps') we can make this error small by taking a very large number of rectangles. Also, instead of regarding the velocity as gradually changing, we can regard it as being constant over each small interval of time (represented by the width of each rectangle). Again we can make the error which this introduces as small as we please by taking the width of the rectangles to be sufficiently small. To each small rectangle, we can apply the argument used above to show that the area inside the rectangle represents the distance travelled.

We can then, at least in our imagination, let the width of the rectangles 'tend to zero' and the number of rectangles 'tend to infinity' and deduce that, even for variable speed, the area under the curve represents the distance travelled.†

Example 7

A cyclist accelerates at $0.75\,\mathrm{m/s^2}$ for $12\,\mathrm{s}$, then cycles at constant speed for $30\,\mathrm{s}$ and then slows down with constant retardation, finally coming to rest after a further $8\,\mathrm{s}$. Draw the velocity–time graph and from it deduce the total distance travelled.

After accelerating at $0.75\,\mathrm{m/s^2}$ for $12\,\mathrm{s}$, the cyclist will reach a speed of $9\,\mathrm{m/s}$. While accelerating and slowing down the motion is represented by straight-line graphs.

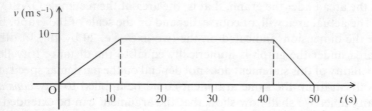

Figure 2.12

In the first stage, the distance is represented by a triangular region:

$$\begin{aligned}\text{area of the triangle} &= \tfrac{1}{2} \times \text{base} \times \text{height}\\ &= \tfrac{1}{2} \times 12 \times 9\\ &= 54\end{aligned}$$

In the second stage, the distance travelled is represented by a rectangular region:

$$\begin{aligned}\text{area of the rectangle} &= \text{base} \times \text{height}\\ &= 30 \times 9\\ &= 270\end{aligned}$$

The final stage is represented by a triangular region:

$$\begin{aligned}\text{area of the triangle} &= \tfrac{1}{2} \times 8 \times 9\\ &= 36\end{aligned}$$

The total area under the velocity–time graph is $(54 + 270 + 36)$ units $= 360$ units. Hence the total distance travelled by the cyclist is $360\,\mathrm{m}$.

---

†For a fuller discussion of this point the reader should refer to 'the area under a curve' in a pure mathematics text-book, e.g. *Pure Mathematics Book 1* by J. K. Backhouse and S. P. T. Houldsworth, revised by P. J. F. Horril, published by Longman.

Summary

(a) **Velocity** has magnitude (its speed) and direction. For motion in a straight line the direction can be indicated by a positive or negative sign.
(b) The **acceleration** is the rate at which the velocity changes. Retardation is recorded as a negative acceleration.
(c) The **gradient** of the tangent at a point on the velocity–time graph is numerically equal to the acceleration at that instant.
(d) The **area** under the velocity–time graph is numerically equal to the distance travelled.

Exercise 2b

1 Express the following accelerations in m/s^2:
  (a) 7.2 km/h per second   (b) 44 km/h per minute
  (c) 32 ft/s per second (1 ft = 0.3048 m)
2 Find the average acceleration, in m/s^2:
  (a) 0 to 20 km/h in 12 s   (b) 30 km/h to 60 km/h in 5 s
  (c) 50 km/h to 10 km/h in 4 s
3 A car accelerates at a constant rate of $2 m/s^2$. If it starts from rest, how fast, in km/h, will it be moving after 5 s?
4 The maximum retardation of a car with good brakes is
  (a) $8 m/s^2$, on a dry road
  (b) $4 m/s^2$, on a wet road
  (c) $1 m/s^2$, on an icy road
  Find the time which this car will take to stop under each of these conditions, from a speed of 100 km/h.
5 Using the data of question 4, find the distance travelled in each case. (Readers may care to reflect on the fact that 100 km/h is approximately 60 mph and a typical medium-sized car is about 4.5 m long!)
6 A motorcyclist accelerates at a rate of $a m/s^2$ for 10 s, then travels at constant speed for 20 s and then stops with a retardation of $2a m/s^2$. Given that the total distance travelled is 550 m, find the value of a and the maximum speed.
7 A jet aeroplane can take off when its speed reaches 72 m/s. The runway must be sufficiently long to allow for the take-off to be aborted in an emergency. Given that the plane accelerates at a constant rate for 15 s to reach its take-off speed and that in an emergency its retardation is $4 m/s^2$, calculate the length of runway required.
8 Using the diagrams in Qu. 6 (see Figure 2.9, p. 20), estimate the distance travelled in each case.
9 A parachutist opens his parachute when his speed has reached 12 m/s. From the moment when it is fully deployed his downward speed is given in the following table:

t (s)	0	0.2	0.4	0.6	0.8	1.0	1.5	2.0	3.0
v (m/s)	12	9.5	8.0	7.0	6.3	5.7	5.0	4.6	4.2

Draw the velocity–time graph and from it estimate the retardation when   (a) $t = 0.5$,   (b) $t = 1$,   (c) $t = 2$.
Describe the motion of the parachutist in words.

10 Using the velocity–time graph for question 9, estimate the distance the parachutist falls in the 3 s following the deployment of the parachute.

Exercise 2c (Miscellaneous)

1 Express the following speeds in ms^{-1}, correct to significant figures:
(a) 50 km/h   (b) 44 km/h   (c) 88 ft/s (1 ft = 0.3048 m)
(d) 100 mph (1 mile = 1.609 34 km)

2 A car journey from A to B is completed at an average speed of 60 km/h, and the return journey from B to A is completed with an average speed of 90 km/h. What is the average speed for the round trip? (The answer is not 75 km/h.)

3 A cyclist sets out from a point A and rides at a steady speed of 24 km/h for 60 minutes. He then stops for 30 minutes before cycling back along the same route at 18 km/h. At the moment when he left point A, a jogger set out from the same point, running the same route, with a steady speed of 12 km/h. Draw a displacement–time graph and use it to find the time and the distance from A when they pass each other.

4 A stone is dropped from the top of a high cliff. After t s, its distance x m below the top of the cliff is given by the following table:

t (s)	0	1	2	3	4	5
x (m)	0	5	20	44	78	122

Calculate its average velocity over the intervals   (a) $t = 0$ to 4,   (b) $t = 2$ to 3.
Draw the displacement–time graph and from it estimate (c) the instantaneous velocity when $t = 3$.

5 An object is projected vertically upwards from the top of a high building, and its height, h m, after t s, is given by

$$h = 35t - 5t^2$$

As it falls back, it misses the point of projection and continues falling vertically. Find
(a) the times when it is 50 m above the point of projection;
(b) the time when it is 40 m below the point of projection.

6 A point P moves in a straight line and its displacement x m from a fixed point O after t s is given by

$$x = 10t - t^2$$

Find the times when   (a) $x = 0$,   (b) $x = 16$,   (c) $x = -24$.
By writing the displacement in the form $x = a - (b - t)^2$, show that the greatest (positive) displacement is 25 m.

Draw the displacement–time graph and use it to estimate the instantaneous velocity when   (d) $t = 2$,   (e) $t = 6$,   (f) $t = 10$.

In questions 7 and 8 no units are mentioned. Any consistent set of units can be used, e.g. distance in metres, time in seconds, speed in metres per second and acceleration in metres per second per second; or again, miles, hours, miles per hour and miles per hour per hour, respectively.

7 A point moves along a straight line and its displacement from a fixed point O is given by $x = 4t^2$.
Calculate the average velocity over the following intervals:
(a) $t = 0$ to $t = 5$ (b) $t = 2$ to $t = 3$ (c) $t = 2$ to $t = 2.1$
Find, in terms of h, the average velocity over the interval $t = 2$ to $t = 2 + h$.
By letting h tend to 0, deduce the instantaneous velocity when $t = 2$.

8 Using the data of question 7, find, in terms of T and h, the average velocity over the interval $t = T$ to $t = T + h$. Deduce an expression in terms of T for the instantaneous velocity when $t = T$.

9 Find the average acceleration in m/s² for the following changes in velocity:
(a) from $+40$ km/h to $+100$ km/h in 10 s
(b) from $+60$ km/h to rest in 5 s
(c) from $+20$ m/s to -10 m/s in 10 s
(d) from -10 m/s to $+10$ m/s in 5 s

10 Find the velocity after 5 s in the following cases:
(a) initial velocity 10 m/s, acceleration 2 m/s²
(b) initial velocity 20 m/s, acceleration -2 m/s²
(c) initial velocity -5 m/s, acceleration 4 m/s²
(d) initial velocity 5 m/s, acceleration -10 m/s²

11 A car accelerates uniformly† from 0 to 72 km/h in 10 s. Find the distance travelled.

12 As a car accelerates uniformly from 20 m/s to 80 m/s it travels 400 m. Find the time taken and the acceleration.

†'Uniformly' means 'at a constant rate'.

Vertical motion under gravity

Chapter 3

The constant acceleration formulae

3.1 Introduction

In this chapter, we shall confine our attention to motion in a straight line *with constant acceleration*. (The term 'uniform' acceleration is often used to mean the same thing.)

We shall develop two of the important characteristics of the velocity–time graph, which were discussed in the previous chapter, namely
(a) the gradient represents the acceleration;
(b) the area under the graph represents the displacement.

There are five related quantities which repeatedly occur in this context, and these are represented by standard abbreviations; they are

the initial velocity, u

the final velocity, v

the time taken, t

the acceleration, a

the displacement, s

[The expression 'initial velocity' sometimes worries students who are new to this subject. It is advisable to think in terms of an observer with a stopwatch; the initial velocity u is the velocity at the moment when the stopwatch is started (i.e. when $t = 0$). The 'final' velocity v is the velocity t seconds later, and s is the displacement at this instant.]

It is a common practice, especially in algebraic problems, to omit the units. This may be done when the result is valid for any consistent set of units, e.g. the distance in metres, the time in seconds and the speed in metres per second; or, distance in miles, time in hours and speed in miles per hour.

3.2 The constant acceleration formulae

First, we consider a numerical example.

Example 1

A car accelerates uniformly from $10\,\mathrm{m\,s}^{-1}$ to $30\,\mathrm{m\,s}^{-1}$ in 5 seconds. Calculate (a) its acceleration, (b) the distance travelled.

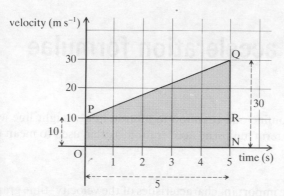

Figure 3.1

(a) The gradient of $PQ = RQ/ON$
$$= (30 - 10)/5$$
$$= 4$$
$\therefore$ the acceleration is $4\,\mathrm{m\,s}^{-2}$.

(b) The area of the trapezium ONQP
$$= \tfrac{1}{2}(OP + NQ)ON$$
$$= \tfrac{1}{2} \times (10 + 30) \times 5$$
$$= \tfrac{1}{2} \times 40 \times 5$$
$$= 100$$
$\therefore$ the distance travelled is $100\,\mathrm{m}$.

Using Example 1 as our guide, we shall now develop five formulae, each of which relates four of the quantities u, v, t, a and s (see §3.1). Refer to Figure 3.2.

The gradient of $PQ = RQ/ON$
$$= (NQ - OP)/ON$$
$$= (v - u)/t$$
$\therefore$ the acceleration, $a = (v - u)/t$.

Multiplying both sides of this equation by t, we obtain

$$at = v - u$$

Hence

$$v = u + at$$

$\cdot$ $\cdot$ $\cdot$ (1)

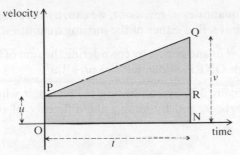

Figure 3.2

The area of the trapezium ONQP

$$= \tfrac{1}{2}(OP + NQ)ON$$
$$= \tfrac{1}{2}(u + v)t$$

∴   the displacement, s, is given by

$$s = \tfrac{1}{2}(u + v)t \qquad \qquad \qquad \text{. . . . (2)}$$

Using equation (1), we substitute $v = u + at$, in equation (2).

$$s = \tfrac{1}{2}[u + (u + at)]t$$
$$= \tfrac{1}{2}(2u + at)t$$
$$= \tfrac{1}{2}(2ut + at^2)$$

∴ $$s = ut + \tfrac{1}{2}at^2 \qquad \qquad \qquad \text{. . . (3)}$$

On the other hand, if we use equation (1) to eliminate u from equation (2), we obtain

$$s = \tfrac{1}{2}[(v - at) + v]t$$
$$= \tfrac{1}{2}(2v - at)t$$

$$s = vt - \tfrac{1}{2}at^2 \qquad \qquad \qquad \text{. . . . (4)}$$

Finally, we rearrange equations (1) and (2), as follows:

$$v - u = at \qquad \text{(from (1))}$$
$$v + u = \frac{2s}{t} \qquad \text{(from (2))}$$

and, multiplying these together to eliminate t, we obtain

$$(v - u)(v + u) = at \times \frac{2s}{t}$$

∴ $$v^2 - u^2 = 2as$$

$$v^2 = u^2 + 2as \qquad \qquad \qquad \text{. . . . (5)}$$

Notice that, given any three of the quantities u, v, s, a or t, we can, by choosing a suitable equation from (1) to (5) above, find either of the missing quantities.

Qu. 1 In Figure 3.2, show that $RQ = at$ and hence, by considering the sum of the areas of triangle PRQ and rectangle ONRP, show that $s = ut + \frac{1}{2}at^2$.

Qu. 2 Derive equation (4) from Figure 3.2, by completing the rectangle which has O, N and Q as three of its vertices, and by finding the difference of two suitable areas.

Example 2

A stone falls with constant acceleration $9.8 \, \text{m s}^{-2}$. Find the distance it falls as its speed increases from $20 \, \text{m s}^{-1}$ to $50 \, \text{m s}^{-1}$.

With the usual notation, $u = 20$, $v = 50$ and $a = 9.8$; we wish to find s. The only equation which uses these four quantities (and excludes t) is equation (5),

$$v^2 = u^2 + 2as$$

Substituting for u, v and a, we obtain

$$50^2 = 20^2 + (2 \times 9.8 \times s)$$
$$2500 = 400 + 19.6s$$
$$19.6s = 2100$$
$$s = 2100/19.6$$
$$= 107 \quad \text{(correct to 3 significant figures)}$$

Hence the stone falls 107 m.

The reader should note that when stating that the answer is correct to 3 significant figures, there are three important factors to consider. Firstly, has the arithmetic been completed to this degree of accuracy? The answer in this example is 'yes'. Secondly, does the accuracy of the data warrant this degree of accuracy? Thirdly, do the assumptions which have been made (in this case, that the acceleration is constant), invalidate the accuracy of the result? In a real problem, the latter factors would require further investigation, but we shall not tackle these complications here.

Example 3

A stone is projected vertically upwards, and it is subject to an acceleration, directed downwards, of $9.8 \, \text{m s}^{-2}$. It is observed that it reaches its highest point after $4 \, \text{s}$. Calculate the greatest height.

First we note that the stone reaches the highest point when its velocity is (momentarily) zero, i.e. $v = 0$, and we are given that this point is reached when $t = 4$. The acceleration is directed *downwards*; that is, while the stone travels upwards it is slowing down, so we put $a = -9.8$. The equation which includes v, t, a and s (which we need to find) is equation (4):

$$s = vt - \tfrac{1}{2}at^2$$

Substituting the data, we have

$$s = 0 \times 4 - \tfrac{1}{2} \times (-9.8) \times 4^2$$
$$= +4.9 \times 16$$
$$= 78.4$$

∴   the greatest height reached is 78.4 m.

Example 4

A particle moves along the x-axis. It is projected from the origin O, with an initial velocity of $+10 \, \text{m s}^{-1}$ and it moves with a constant acceleration of $-2 \, \text{m s}^{-2}$. Taking the units of the x-axis to be metres, calculate:
(a) the times when it is at the point $x = +16$;
(b) its position after 12 s;
(c) its velocity after 12 s.

In this question, $u = +10$, and $a = -2$.
(a) Substituting the data into $s = ut + \tfrac{1}{2}at^2$, we obtain

$$+16 = 10t - t^2$$

Hence

$$t^2 - 10t + 16 = 0$$
$$(t - 8)(t - 2) = 0$$
$$∴ \qquad\qquad t = 2 \text{ or } 8$$

Therefore the particle is at $x = +16$ after 2 seconds and again after 8 seconds.
(b) Given that $t = 12$ and using

$$s = ut + \tfrac{1}{2}at^2$$

we obtain

$$s = 10 \times 12 - \tfrac{1}{2} \times 2 \times 12^2$$
$$= 120 - 144$$
$$= -24$$

After 12 seconds, the particle is at $x = -24$, i.e. 24 m to the left of the origin.

(c) Given that $t = 12$ and using

$$v = u + at$$

we obtain

$$v = 10 - 2 \times 12$$
$$= 10 - 24$$
$$= -14$$

After 12 seconds, the particle is moving at $14 \, \text{m s}^{-1}$, in the *negative* direction, i.e. to the left.

Example 5

In a greyhound race, the dogs chase a mechanical hare along a straight track which is 100 m long; the hare travels at a constant speed of $10 \, \text{m s}^{-1}$. The dogs are released from their starting boxes T seconds after the hare has passed them. It is estimated that the fastest dog accelerates for 4 s at $5 \, \text{m s}^{-2}$, and thereafter runs at constant speed. The dog must not catch the hare before the finishing line. Show that $T > 3$.

The hare takes 10 s to travel 100 m, and since it is given a 'start' of T seconds, the dogs will have been running for $(10 - T)$ seconds when the hare reaches the finishing line.

The distance travelled by the fastest dog while it is accelerating, can be obtained from the formula

$$s = ut + \tfrac{1}{2}at^2$$

With $u = 0$, $a = 5$ and $t = 4$, that is

$$s = \tfrac{1}{2} \times 5 \times 16$$
$$= 40$$

So this dog travels 40 m while it is accelerating. By the end of this stage, its velocity is v, where

$$v = 0 + 5 \times 4$$
$$= 20$$

The maximum speed reached by the dog is $20 \, \text{m s}^{-1}$. It now travels at this speed for the remaining $(6 - T)$ seconds; in this time it will cover $20(6 - T)$ metres. Hence the total distance covered by the dog, at the moment when the hare reaches the finishing line is $[40 + 20(6 - T)] \, \text{m}$.

Since the dog must not catch the hare before the finishing line, this distance must be less than 100 m, i.e.

$$40 + 20(6 - T) < 100$$
$$\therefore \quad 40 + 120 - 20T < 100$$
$$160 - 20T < 100$$
$$20T > 60$$
$$\therefore \quad T > 3$$

Summary

There are five constant acceleration formulae; each one relates four of the quantities u, v, t, a and s. The letter which is omitted is shown in brackets:

(1) $\qquad v = u + at \qquad\qquad (s)$

(2) $\qquad s = \frac{1}{2}(u + v)t \qquad\quad (a)$

(3) $\qquad s = ut + \frac{1}{2}at^2 \qquad\quad (v)$

(4) $\qquad s = vt - \frac{1}{2}at^2 \qquad\quad (u)$

(5) $\qquad v^2 = u^2 + 2as \qquad\quad (t)$

Readers are reminded that these equations can *only* be used when the acceleration is constant.

Exercise 3a

Throughout this exercise, the letters u, v, s, a and t will be used in the usual way (see §3.1).
In questions 1–5, values are assigned to three of the letters; find the value of the letters in brackets.

1 $u = 5$, $v = 15$, $a = 2.5$; (t).
2 $s = 20$, $t = 4$, $a = 0.5$; (u).
3 $u = 20$, $s = 75$, $a = -2$; (v).
4 $u = 40$, $v = 10$, $t = 3$; (s).
5 $s = 100$, $v = 20$, $t = 10$; (a).
6 Express a in terms of v, u and t.
7 Express t in terms of s, u and v.
8 Express a in terms of s, u and t.
9 Express s in terms of v, u and a.
10 Express u in terms of s, a and t.
11 A jet aircraft accelerates uniformly from rest to its take-off speed of 270 km/h, in a distance of 1500 m. Calculate its acceleration in m s^{-2} and the time taken in seconds.
12 A car brakes with a retardation of 2.5 m s^{-2}. Calculate the distance it requires to stop, if its initial speed is   (a) 20 m s^{-1},   (b) 40 m s^{-1}.
13 A car, which is initially travelling at 20 m s^{-1}, accelerates at 2 m s^{-2}. What speed will it reach after 10 s?

14 A car skidding in a straight line on ice, has a retardation of $1\,\mathrm{m\,s^{-2}}$. In 10 s, it travels 150 m, before hitting an obstacle. Find the speed of the car on impact.

15 A motorbike accelerates from 0 to 90 km/h in 10 s. During this time it travels 150 m. Show that the acceleration cannot be uniform.

16 A car accelerates uniformly from rest for 15 s, it then continues at constant speed for 60 s, then it slows down with constant retardation, coming to rest in 20 s. Given that the total distance is 2325 m, find
(a) the constant speed　　　(b) the acceleration　　　(c) the retardation.

17 A missile is projected vertically upwards, with an initial velocity of $100\,\mathrm{m\,s^{-1}}$; it is subject to a constant downward acceleration of $10\,\mathrm{m\,s^{-2}}$. Find
(a) the times when it is 320 m above the point of projection;
(b) the times when its speed is $40\,\mathrm{m\,s^{-1}}$.

18 An aircraft starts from rest and accelerates at a rate of $2\,\mathrm{m\,s^{-2}}$. After 30 s, it develops an engine fault and its acceleration falls instantly to $1.5\,\mathrm{m\,s^{-2}}$. After 4 s at this rate of acceleration, the pilot decides to abort the take-off and brings the aircraft to rest with a retardation of $3\,\mathrm{m\,s^{-2}}$. Calculate the total length of runway used.

19 Two cars are travelling, one behind the other, at a steady speed of $25\,\mathrm{m\,s^{-1}}$, with a gap of 25 m between them. The driver of the first car, seeing danger ahead, brakes with a retardation of $5\,\mathrm{m\,s^{-2}}$. The driver of the second car takes 2 s to react to the brake lights of the car in front, and then does an 'emergency stop' with a retardation of $7.5\,\mathrm{m\,s^{-2}}$. Show that a collision is inevitable and find the time which elapses between the moment when the first driver applies the brakes and the collision.

20 A sports car starts from rest and accelerates at $3\,\mathrm{m\,s^{-2}}$ for 10 s and then travels at constant speed. Five seconds after it starts a racing car follows it, accelerating at $4\,\mathrm{m\,s^{-2}}$ for 10 s and then travelling at constant speed. Find the time (measured from the moment the sports car starts), which elapses before the racing car catches up with the sports car, and calculate the distance they have both travelled at this moment.

3.3 Vertical motion under gravity

If a feather and a golf ball were dropped simultaneously and allowed to fall freely, we would expect the feather to fall more slowly than the golf ball. However, there are two factors which affect the motion, (a) the pull of gravity and (b) the air-resistance. If we could separate these two factors, say by conducting the experiment in a vacuum, then a very different picture would emerge: the two objects would fall at the same rate. Over three hundred years ago, Galileo Galilei (1564–1642) made the remarkable discovery that, in a given location, the pull of gravity causes all objects, regardless of size or mass, to fall with the same acceleration. The difference in the rates of descent of the feather and the golf ball is due to the markedly greater effect of air-resistance on the motion of the feather.

Taking into account the effect of air-resistance requires mathematical techniques which are beyond the scope of this book, so we shall confine ourselves to problems in which air-resistance is neglected. Although this means our results will only be an approximation to the real situation, this approximation can be quite good provided the object is fairly heavy and the speed fairly small.

Neglecting air resistance, vertical motion under gravity becomes a special case of motion in a straight line with constant acceleration. With modern techniques, it is possible to measure this acceleration quite accurately. In the United Kingdom it is about $9.81 \, \text{m s}^{-2}$. (However, although it is constant at a particular location, it does vary slightly with latitude.) The acceleration due to gravity is usually denoted by the letter g. In text-book exercises and examination questions it is frequently convenient to take g to be $10 \, \text{m s}^{-2}$, since this simplifies the arithmetic.

Example 6

A stone is projected vertically upwards with an initial velocity of $20 \, \text{m s}^{-1}$. Find its vertical velocity when its height above the point of projection is $10 \, \text{m}$. (Take g to be $9.81 \, \text{m s}^{-2}$.)

In the usual notation, $u = 20$, $a = -9.81$, $s = 10$, and we wish to find v. Hence using

$$v^2 = u^2 + 2as$$

$$\begin{aligned} v^2 &= 20^2 + 2 \times (-9.8) \times 10 \\ &= 400 - 196.2 \\ &= 203.8 \end{aligned}$$

$$\therefore \quad v = \pm 14.3 \quad \text{(correct to 3 significant figures)}$$

The velocity, when the height is $10 \, \text{m}$, is $\pm 14.3 \, \text{m s}^{-1}$.

[Note that the speed of $14.3 \, \text{m s}^{-1}$ occurs twice: once on the way up ($+$ sign) and again on the way down ($-$ sign).]

Example 7

A projectile is fired vertically upwards with an initial velocity u. After an interval of T seconds (and while the first projectile is still in the air) a second projectile is fired vertically upwards, also with initial velocity u. Find the time and the height at which they meet in terms of u, g and T.

First, we must find the time when the projectiles meet. Using the formula

$$s = ut + \tfrac{1}{2}at^2$$

with $a = -g$ and $s = h$, we obtain

$$h = ut - \tfrac{1}{2}gt^2 \qquad \qquad \qquad \text{. . . (1)}$$

When the projectiles meet, the second one has been in the air for $t - T$ seconds (where $t \geqslant T$), so for the second projectile

$$h = u(t - T) - \tfrac{1}{2}g(t - T)^2 \qquad \qquad \text{. . . (2)}$$

Equating the two expressions for h gives

$$ut - \tfrac{1}{2}gt^2 = u(t - T) - \tfrac{1}{2}g(t - T)^2$$
$$\therefore \quad ut - \tfrac{1}{2}gt^2 = ut - uT - \tfrac{1}{2}g(t^2 - 2tT + T^2)$$
$$ut - \tfrac{1}{2}gt^2 = ut - uT - \tfrac{1}{2}gt^2 + gtT - \tfrac{1}{2}gT^2$$
$$0 = -uT + gtT - \tfrac{1}{2}gT^2$$
$$gtT = uT + \tfrac{1}{2}gT^2$$
$$gt = u + \tfrac{1}{2}gT$$
$$\therefore \quad t = \frac{u}{g} + \frac{T}{2}$$

Therefore the projectiles meet after $\left(\dfrac{u}{g} + \dfrac{T}{2}\right)$ seconds.

Substituting this into equation (1) gives

$$h = u\left(\frac{u}{g} + \frac{T}{2}\right) - \frac{g}{2}\left(\frac{u}{g} + \frac{T}{2}\right)^2$$

$$= \frac{u^2}{g} + \frac{uT}{2} - \frac{g}{2}\left(\frac{u^2}{g^2} + \frac{uT}{g} + \frac{T^2}{4}\right)$$

$$= \frac{u^2}{g} + \frac{uT}{2} - \frac{u^2}{2g} - \frac{uT}{2} - \frac{gT^2}{8}$$

$$= \frac{u^2}{2g} - \frac{gT^2}{8}$$

Therefore the projectiles meet at a height of $\left(\dfrac{u^2}{2g} - \dfrac{gT^2}{8}\right)$.

Exercise 3b

In questions 1 to 7, take g to be $10\,\mathrm{m\,s^{-2}}$.

1 A girl steps off a diving board into a swimming pool, which is 10 m above the level of the surface of the water, and falls vertically. Find the time which elapses before she enters the water and find her speed at this instant.

2 A coin is thrown vertically downwards into an empty well which is 44 m deep. Find the initial velocity, given that the coin reaches the bottom of the well in 2 seconds.

3 A ball is projected vertically upwards with an initial velocity of $17.5\,\mathrm{m\,s^{-1}}$. Find the time between the instants when it is at a height of 7.5 m above the point of projection.

4 A ball is released from rest from a height of 5 m above a horizontal surface. It falls vertically onto the surface and it continues bouncing in a vertical line. Each time it strikes the horizontal surface, its speed is halved. Find the time

taken for the first 'bounce' (i.e. from the moment it strikes the plane for the first time to the moment it strikes the plane for the second time).

5 If the ball in question 4 continues bouncing indefinitely, what time will elapse from the moment the ball strikes the plane for the first time to the moment when the bouncing ceases. [Hint: the sum of the infinite Geometric Progression $1 + x + x^2 + x^3 + \ldots$, where $|x| < 1$, is $1/(1 - x)$.]

6 A hot air balloon rests, ready for take-off, with the basket just touching the ground. It then rises vertically with an acceleration of $0.5\,\mathrm{m\,s^{-2}}$. After $20\,\mathrm{s}$ a stone is dropped from the balloon. Find the time taken for the stone to reach the ground.

7 A tennis player throws the ball vertically into the air. It rises to its highest point, and as it descends, the player strikes it when it is $1\,\mathrm{m}$ above the level from which it was thrown. Given that the time which elapses between throwing the ball and striking it is $2\,\mathrm{s}$, find the greatest height it reaches above the point of projection.

8 A stone is released from rest and allowed to fall vertically. After T seconds, another stone is projected vertically downwards from the same point, with an initial velocity u. This stone overtakes the first after a further T seconds. Show that $u = 3gT/2$.

9 A particle is projected vertically upwards from a point O. After t seconds (and before the first particle has reached its highest point) a second particle is projected vertically upwards from O, with the same initial velocity, and they collide after a further t seconds. Show that $t = 2T/3$, where T seconds is the time taken by the first particle to reach its highest point.

10 A ball is projected vertically upwards with an initial velocity u. Find, in terms of u and g, the times when it is at a height of $4u^2/(9g)$.

Exercise 3c (Miscellaneous)

1 A train accelerates uniformly from rest and travels $\frac{1}{2}\,\mathrm{km}$ in the first $20\,\mathrm{s}$. Find its acceleration in $\mathrm{m\,s^{-2}}$.

2 A car, which is initially travelling at constant speed, starts to accelerate uniformly at $2\,\mathrm{m\,s^{-2}}$. After $6\,\mathrm{s}$, it reaches a speed of $120\,\mathrm{km/h}$. Find the distance it travels while it is accelerating, and its original speed.

3 A particle is projected vertically with an initial velocity of $u\,\mathrm{m\,s^{-1}}$. When it reaches a height of $200\,\mathrm{m}$, its speed is $15\,\mathrm{m\,s^{-1}}$. Taking g to be $10\,\mathrm{m\,s^{-2}}$, calculate the value of u, and find the greatest height reached by the particle.

4 An aircraft lands at $120\,\mathrm{m\,s^{-1}}$ on a runway whose total length is $3\,\mathrm{km}$. The aircraft touches down $500\,\mathrm{m}$ from one end of the runway and slows down with a uniform retardation of $4\,\mathrm{m\,s^{-2}}$. How far is it from the other end of the runway when it comes to rest?

5 A particle moves with gradually increasing acceleration. Show, by means of a diagram or otherwise, that its average velocity is less than the mean of its initial and final velocities.

6 The diagram below represents the velocity–time graphs of two runners who are using the same running track, and starting from the same point.

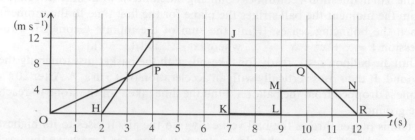

Figure 3.3

The graph through O, P, Q and R is that of runner A, and the graph through H, I, J, K, L, M and N is that of runner B.

Find the distance which each runner has covered when

(a) $t = 4$, (b) $t = 7$, (c) $t = 12$.

7 Describe (in words) the motion of the runners in question 6.

8 A car accelerates uniformly from rest. It accelerates at $1.5\,\mathrm{m\,s}^{-2}$ for 6 s, and then at $2.5\,\mathrm{m\,s}^{-2}$ for 4 s. Calculate

(a) the distance travelled;

(b) the average speed over the 10 s period;

(c) the average acceleration over this period.

9 A particle is projected vertically upwards with an initial speed of $80\,\mathrm{m\,s}^{-1}$. After 4 s, a second particle is projected vertically upwards from the same point, with an initial speed of $160\,\mathrm{m\,s}^{-1}$. Find the further time which elapses before they meet and calculate the height above the point of projection at this instant. (Take g to be $10\,\mathrm{m\,s}^{-2}$.)

10 A train normally travels along a certain track at 180 km/h. On a certain day, it is obliged to slow down to 36 km/h while it passes engineering work on 1 km of the track. It slows down with uniform retardation over a distance of 900 m, and, after passing the restricted part of the track, it accelerates uniformly at $2\,\mathrm{m\,s}^{-2}$, until it reaches its normal speed. Calculate the time lost.

11 Two particles, X and Y, are moving in the same direction on parallel horizontal tracks.

At a certain point O, the particle X, travelling with a speed of 16 m/s and retarding uniformly at 6 m/s², overtakes Y, which is travelling at 8 m/s and accelerating uniformly at 2 m/s². Calculate

(a) the distance of Y from O when the velocities of X and Y are equal;

(b) the velocity of X, when Y overtakes X.

*(C)

12 A particle X is projected vertically upwards from the ground with a velocity of 80 m/s. Calculate the maximum height reached by X. (Take g to be $10\,\mathrm{m\,s}^{-2}$.)

A particle Y is held at a height of 300 m above the ground. At the moment when X has dropped 80 m from its maximum height, Y is projected downwards with a velocity of v m/s. The particles reach the ground at the same time. Calculate the value of v.

*(C)

13 A train which is travelling at a constant speed of 36 m/s is required to slow down in order to pass through a station. It does so by decelerating uniformly at 0.2 m/s² until its speed is reduced to 12 m/s; it maintains a speed of 12 m/s for 1.8 km and then accelerates uniformly at 0.1 m/s² until it reaches its cruising speed of 36 m/s once more.

Find, in minutes, for how long the speed of the train is less than 36 m/s, and show that in this time it travels 10.44 km.

*(SMP)

14 A skier is sliding out of control down a straight, steady slope and is accelerating uniformly at 1.5 m/s².

(a) While he covers a particular stretch of 25 m his speed is doubled. Find by calculation how fast he is going at the end of the 25 m.

(b) If he started from rest (with the same acceleration), calculate

   (i)   the time he would take to cover 25 m;

   (ii) the speed he would have attained in that time.

*(SMP)

15 A lift travels vertically a distance of 22 m from rest at the basement to rest at the top floor. Initially the lift moves with constant acceleration x m/s² for a distance of 5 m; it continues with a constant speed u m/s for 14 m; it is then brought to rest at the top floor by a constant retardation y m/s². State, in terms of u only, the times taken by the lift to cover the three stages of the journey.

If the total time that the lift is moving is 6 seconds, calculate

(a) the value of u;

(b) the values of x and y.

(AEB; part)

Velocity and acceleration

Chapter 4

Displacement, velocity, acceleration (using calculus)

4.1 Introduction

In this chapter we shall develop the concepts introduced in the earlier chapters by applying to them the techniques of calculus. It is assumed that, by this stage, the reader has made a start on the study of calculus, and can now differentiate and integrate polynomials and simple trigonometrical functions. If this is not the case, the chapter may be postponed until these basic ideas have been studied.

In the context of applied mathematics, the independent variable (that is, the variable whose values are represented by points on the *horizontal* axis) is usually t, the time, and in this chapter the dependent variable (the one represented by points on the *vertical* axis) will be either x, the displacement, or v, the velocity.

4.2 The displacement–time graph

From Chapter 2, we know that the gradient of a **chord** on a displacement–time graph represents the *average* velocity over the time interval concerned. If the displacement x is given as a function of t, the time, and if δx is an increment in x corresponding to an increment δt in the value of t, then the gradient of the chord PQ (see Figure 4.1) is $\dfrac{\delta x}{\delta t}$, and this gives the average velocity over the interval t to $t + \delta t$.

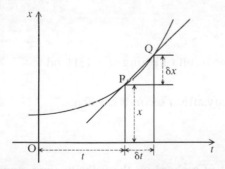

Figure 4.1

When δt tends to zero, the line which passes through P and Q becomes the tangent at the point P, and $\dfrac{\delta x}{\delta t}$ becomes $\dfrac{dx}{dt}$, the derivative of x with respect to t. In other words, if x is expressed as a function of t, then

> the instantaneous velocity at time t is $\dfrac{dx}{dt}$

Example 1

A particle moves along a straight line and its displacement x metres, from the origin, after t seconds is given by $x = 4.9t^2$. Find, in terms of t,

(a) the average velocity over the interval t to $(t + \delta t)$;
(b) the instantaneous velocity.

(a) We are given that

$$x = 4.9t^2 \qquad\qquad\qquad \ldots \ . \ (1)$$

so, when t increases to $(t + \delta t)$, the displacement becomes

$$x + \delta x = 4.9(t + \delta t)^2$$
$$= 4.9t^2 + 9.8t(\delta t) + 4.9(\delta t)^2 \qquad \ldots \ . \ (2)$$

Subtracting equation (1) from equation (2) gives

$$\delta x = 9.8t(\delta t) + 4.9(\delta t)^2$$

Hence

$$\frac{\delta x}{\delta t} = 9.8t + 4.9\delta t \qquad\qquad \ldots \ . \ (3)$$

Hence, the average velocity over the interval t to $t + \delta t$ is $(9.8t + 4.9\delta t)\,\mathrm{m\,s^{-1}}$.

(b) To find the instantaneous velocity, we differentiate the expression for x in terms of t, i.e.

$$x = 4.9t^2$$

$$\frac{dx}{dt} = 9.8t$$

(Note that this expression is the result of letting δt in (3) tend to zero.)

Hence the instantaneous velocity after t seconds is $9.8t\,\mathrm{m\,s^{-1}}$.

4.3 The velocity–time graph

In Chapter 2 we saw that the gradient of a **chord** on the velocity–time graph represents the *average* acceleration, and that the gradient of a **tangent** at a point gives the *instantaneous* acceleration. Consequently, if the velocity v is given as a function of t, we can find the instantaneous acceleration by differentiating v with respect to t; in other words

the instantaneous acceleration at time t is $\dfrac{dv}{dt}$

Example 2

The velocity, $v \, \mathrm{m\,s^{-1}}$, of a point moving in a straight line, after t seconds, is given by $v = 10t + 3t^2$. Find the acceleration
(a) when $t = 0$, (b) when $t = 4$.

$$v = 10t + 3t^2$$

$$\frac{dv}{dt} = 10 + 6t$$

(a) When $t = 0$, $\dfrac{dv}{dt} = 10$, i.e. the acceleration is $10 \, \mathrm{m\,s^{-2}}$.

(b) When $t = 4$, $\dfrac{dv}{dt} = 34$, i.e. the acceleration is $34 \, \mathrm{m\,s^{-2}}$.

Notice that we can combine the ideas explained above: if we are given x as a function of t, we can deduce both the velocity and the acceleration, for example, if we are given that $x = 4.9t^2$, then the velocity $v = \dfrac{dx}{dt} = 9.8t$, and, differentiating again, the acceleration $= \dfrac{dv}{dt} = 9.8$.

Example 3

A particle is moving along the x-axis, and its displacement from the origin in metres after t s, is given by

$$x = 3 \sin t + 4 \cos t$$

Find its displacement, velocity and acceleration when (a) $t = 0$,
(b) $t = \frac{1}{2}\pi$.

We are given that

$$x = 3 \sin t + 4 \cos t$$

Differentiating this, we obtain the velocity, $v \, \mathrm{m\,s^{-1}}$:

$$v = \frac{dx}{dt} = 3 \cos t - 4 \sin t$$

and the acceleration, $a\,\mathrm{m\,s^{-2}}$, is given by

$$a = \frac{dv}{dt} = -3\sin t - 4\cos t$$

(a) When $t = 0$,

$$x = \quad 3\sin 0 + 4\cos 0 = +4$$
$$v = \quad 3\cos 0 - 4\sin 0 = +3$$
$$a = -3\sin 0 - 4\cos 0 = -4$$

Therefore, when $t = 0$,
the displacement is $+4\,\mathrm{m}$,
the velocity is $+3\,\mathrm{m\,s^{-1}}$
and the acceleration is $-4\,\mathrm{m\,s^{-2}}$
(i.e. the particle is slowing down with a retardation of $4\,\mathrm{m\,s^{-2}}$.)

(b) When $t = \frac{1}{2}\pi$,

$$x = 3\sin\left(\tfrac{1}{2}\pi\right) + 4\cos\left(\tfrac{1}{2}\pi\right) = +3$$
$$v = 3\cos\left(\tfrac{1}{2}\pi\right) - 4\sin\left(\tfrac{1}{2}\pi\right) = -4$$
$$a = -3\sin\left(\tfrac{1}{2}\pi\right) - 4\cos\left(\tfrac{1}{2}\pi\right) = -3$$

Therefore, when $t = \frac{1}{2}\pi$,
the displacement is $+3\,\mathrm{m}$ (the particle is on the positive
axis and 3 m from O),
the velocity is $-4\,\mathrm{m\,s^{-1}}$ (the particle is moving towards O
at $4\,\mathrm{m\,s^{-1}}$),
the acceleration is $-3\,\mathrm{m\,s^{-2}}$ (the particle is accelerating
towards O, i.e. its *speed* is increasing)

Exercise 4a

*Throughout this exercise, x represents the displacement from the origin, in metres,
of a particle P which is moving in a straight line; the velocity after t seconds is
$v\,\mathrm{m\,s^{-1}}$. In each question, either x or v is given in terms of t.*

1 $x = 10t - 5t^2$; find the velocity when (a) $t = 0$, (b) $t = 1$, (c) $t = 2$.
2 $x = \frac{1}{10}t^3$; find the velocity when (a) $t = 10$, (b) $t = 12$. Hence find the *average*
   acceleration over the interval $t = 10$ to $t = 12$.
3 $x = 20t - 5t^2$; find the value of t when the velocity is zero and find the displacement at this instant.
4 $v = 10 + 6t^2 - t^3$; find the acceleration when (a) $t = 0$, (b) $t = 2$, (c) $t = 4$.
5 $v = t^2 - 7t + 10$; find the acceleration at the times when the velocity is zero.
6 $x = 20t + 5t^2$; show that the acceleration is constant. Find the velocity when
   $t = 0$.
7 $x = 5 + 2t + t^3$; find the velocity and the acceleration when $t = 2$.
8 $x = \sin 3t$; find the velocity and the acceleration when (a) $t = 0$, (b) $t = \frac{1}{6}\pi$,
   (c) $t = 2$.
9 $x = \sqrt{t}$; find the velocity and the acceleration when $t = 25$.
10 $x = \sin t + \cos t$; find the smallest positive value of t for which
   (a) the velocity is zero, (b) the acceleration is zero.
   Find also the displacements at these times.

4.4 Integrating the velocity and acceleration functions

In the preceding section we saw that, if the displacement is expressed as a function of time, then, by differentiating this expression we can find the velocity, and differentiating the velocity gives the acceleration. Consequently, if we know the **acceleration** as a function of time, then by reversing the process, in other words integrating, we can find the velocity, and integrating this gives the displacement.

However there is an extra complication: when we differentiate a **constant**, it becomes zero, so when we integrate we must be careful to allow for this. Such 'constants of integration' remain as unknown quantities, unless we are provided with further information about the motion. The next two examples illustrate this.

Example 4

A particle is moving in a straight line and its acceleration after
t seconds is $(20 - 6t)\,\mathrm{m\,s^{-2}}$. Find its velocity in terms of t, given that
its initial velocity is $5\,\mathrm{m\,s^{-1}}$, and hence find its displacement
x metres, in terms of t, given that when $t = 2$, $x = 50$.

Let the velocity after t seconds be $v\,\mathrm{m\,s^{-1}}$. Then its acceleration is
the derivative of v with respect to t, i.e. $\dfrac{dv}{dt}$, so

$$\frac{dv}{dt} = 20 - 6t$$

Integrating this, gives

$$v = 20t - 3t^2 + A \quad \text{where } A \text{ is a constant.}$$

We are given that when $t = 0$, $v = 5$,

$$\therefore \quad 5 = 20 \times 0 - 3 \times 0^2 + A$$
$$\therefore \quad A = 5$$

Hence the velocity after t seconds is

$$v = 20t - 3t^2 + 5$$

We now write v as $\dfrac{dx}{dt}$:

$$\frac{dx}{dt} = 20t - 3t^2 + 5$$

and integrating this gives

$$x = 10t^2 - t^3 + 5t + B \quad \text{where } B \text{ is a constant.}$$

When $t = 2$, $x = 50$, so

$$50 = 10 \times 2^2 - 2^3 + 5 \times 2 + B$$
$$50 = 40 - 8 + 10 + B$$
$$\therefore \quad B = 8$$

Hence

$$x = 10t^2 - t^3 + 5t + 8$$

Example 5

A particle moves in a straight line and its acceleration after t seconds is $12t \, \text{m s}^{-2}$. Find an expression for its displacement, x metres, after t seconds, in terms of t.

Given that when $t = 0$, $x = 0$ and that when $t = 2$, $x = 20$, find the displacement and the velocity when $t = 5$.

Let the velocity after t seconds be $v \, \text{m s}^{-1}$. Then

$$\frac{\mathrm{d}v}{\mathrm{d}t} = 12t$$

Integrating,

$$v = 6t^2 + A$$

and integrating again,

$$x = 2t^3 + At + B, \quad \text{where } A \text{ and } B \text{ are constants.}$$

We are given that when $t = 0$, $x = 0$, therefore

$$0 = 2 \times 0^3 + A \times 0 + B$$
$$\therefore \quad B = 0$$

Hence
$$x = 2t^3 + At$$

We are also given that, when $t = 2$, $x = 20$. Therefore

$$20 = 2 \times 2^3 + A \times 2$$
$$20 = 16 + 2A$$
$$2A = 4$$
$$\therefore \quad A = 2$$

Hence after t seconds the displacement and velocity are given by

$$x = 2t^3 + 2t$$
and $$v = 6t^2 + 2$$
respectively.

Hence, when $t = 5$,

$$x = 2 \times 5^3 + 2 \times 5$$
$$= 2 \times 125 + 10$$
$$= 260$$

and $$v = 6 \times 5^2 + 2$$
$$= 6 \times 25 + 2$$
$$= 152$$

Hence, when $t = 5$, the displacement is $260 \, \text{m}$, and the velocity is $152 \, \text{m s}^{-1}$.

In the next example, the velocity is related to time by a table of numerical data; the area under the velocity–time graph then has to be estimated by a numerical method, in this case the **trapezium rule**.

Example 6

A car moves along a straight line, and its speed $v\,\text{m}\,\text{s}^{-1}$ is recorded at 5 s intervals. The data is recorded below:

t	0	5	10	15	20	25	30
v	0	5	12	18	20	19	15

Use the data to estimate the distance travelled.

A velocity–time graph illustrating the data is shown below:

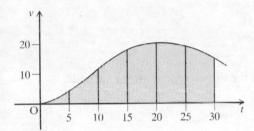

Figure 4.2

The distance travelled is represented by the area under the graph
To estimate this, we divide the region into 6 parts and use the trapezium rule, that is, we estimate the area of each part by applying the formula for calculating the area A of a trapezium, namely

$$A = \tfrac{1}{2}(a + b)h$$

where a and b are the lengths of the parallel sides, and h is the distance between them.

Distance travelled

$$\approx \tfrac{1}{2}(0 + 5) \times 5 + \tfrac{1}{2}(5 + 12) \times 5 + \tfrac{1}{2}(12 + 18) \times 5 +$$
$$+ \tfrac{1}{2}(18 + 20) \times 5 + \tfrac{1}{2}(20 + 19) \times 5 + \tfrac{1}{2}(19 + 15) \times 5$$
$$= 2.5 \times (5 + 17 + 30 + 38 + 39 + 34)$$
$$= 2.5 \times 163$$
$$= 407.5$$

The distance travelled, correct to 2 significant figures, is 410 m.

(Readers who know Simpson's Rule should use it and compare the answer with the one above.)

4.5 The constant acceleration formulae (using calculus)

In the previous chapter, we proved the 'constant acceleration formulae' by referring to the appropriate straight-line graph. We can also prove these formulae by using calculus methods.

Let the (constant) acceleration be denoted by a, then, in the usual notation, the velocity v after t seconds is related to a by the differential equation

$$\frac{dv}{dt} = a$$

Integrating this gives

$v = at + c$ where c is a constant.

But when $t = 0$, $v = u$ (the initial velocity), hence

$u = a \times 0 + c$

$\therefore \quad c = u$

Hence

$v = u + at$

(This is equation (1) on p. 28.)

Now we write v as $\dfrac{ds}{dt}$, where s is the displacement after t seconds:

$$\frac{ds}{dt} = u + at$$

and integrating this gives

$s = ut + \frac{1}{2}at^2 + k$ where k is a constant.

However, if $s = 0$ when $t = 0$, then $k = 0$. Hence

$s = ut + \frac{1}{2}at^2$

(and this is equation (3) on p. 29).

The remaining equations can now be found algebraically, as before. However there is an important alternative method which can be used here and elsewhere in applied mathematics: using the **chain rule**†, we write

$$\frac{dv}{dt} = \frac{dv}{ds} \times \frac{ds}{dt}$$ where s is the displacement.

But $\dfrac{ds}{dt} = v$, so the acceleration may be written as follows:

$$\frac{dv}{dt} = v\frac{dv}{ds}$$

---

† If necessary the reader should refer to a suitable textbook on Pure Mathematics e.g. *Pure Mathematics Book I* by J. K. Backhouse, S. P. T. Houldsworth, revised by P. J. F. Horril, published by Longman.

Note also that the RHS is the derivative of $\frac{1}{2}v^2$, with respect to s, so

$$\frac{dv}{dt} = \frac{d}{ds}(\tfrac{1}{2}v^2)$$

If the acceleration is constant ($=a$), then we can write

$$\frac{dv}{dt} = a$$

and hence

$$\frac{d}{ds}(\tfrac{1}{2}v^2) = a$$

and, integrating with respect to s, we have

$\frac{1}{2}v^2 = as + c$ where c is a constant.

But, when $t = 0$, $s = 0$ and v is equal to u, hence

$\frac{1}{2}u^2 = a \times 0 + c$

$\therefore \quad c = \frac{1}{2}u^2$

Hence $\frac{1}{2}v^2 = as + \frac{1}{2}u^2$

i.e. $v^2 = u^2 + 2as$

(This is equation (5) on p. 29.)

The substitution of $v\dfrac{dv}{ds}$ for $\dfrac{dv}{dt}$ should be regarded as a standard move in any problem where the acceleration is expressed in terms of the *displacement*.

Example 7

A particle moves along the positive x-axis and its acceleration is $6x$ m s^{-2}, where x is its displacement from the origin O, in metres. Given that the initial velocity is 10 m s^{-1}, find its velocity in terms of x.

Let the velocity be v m s^{-1} after t s, then

$$\frac{dv}{dt} = 6x$$

Note that this equation employs *three* variables, v, x and t; we cannot proceed with integration until the number of variables has been reduced to *two*. This is achieved by replacing the acceleration, $\dfrac{dv}{dt}$ by $\dfrac{d}{dx}(\tfrac{1}{2}v^2)$ or $v\dfrac{dv}{dx}$.

Hence

$$v\frac{dv}{dx} = 6x$$

Integrating with respect to x gives

$\frac{1}{2}v^2 = 3x^2 + c$ where c is a constant.

Putting $x = 0$, $v = 10$, gives

$\frac{1}{2} \times 10^2 = 3 \times 0^2 + c$

Hence

$$c = 50$$

$\therefore \quad \frac{1}{2}v^2 = 3x^2 + 50$

$\therefore \quad v^2 = 6x^2 + 100$

$\therefore \quad v = \pm\sqrt{(6x^2 + 100)}$

However, since the initial velocity and the acceleration are both positive, the subsequent velocity is also positive, so we take the positive square root, i.e.

$$v = \sqrt{(6x^2 + 100)}$$

Example 8

A particle moves along a straight line and when its distance from a fixed point O is x metres, its retardation is $(5 + \frac{1}{2}x)\,\mathrm{m\,s^{-2}}$. Given that its velocity is $20\,\mathrm{m\,s^{-1}}$ when $x = 0$, express its velocity v in terms of x.

Find also the value of x when the particle first comes to rest.

We are given that the retardation is $(5 + \frac{1}{2}x)$; writing this as $-\dfrac{dv}{dt}$, we have

$$\frac{dv}{dt} = -(5 + \tfrac{1}{2}x)$$

and replacing $\dfrac{dv}{dt}$ by $v\dfrac{dv}{dx}$ gives

$$v\frac{dv}{dx} = -(5 + \tfrac{1}{2}x)$$

Integrating this with respect to x, it becomes

$$\tfrac{1}{2}v^2 = -5x - \tfrac{1}{4}x^2 + A$$

However, when $x = 0$, $v = 20$, so

$$200 = A$$

Therefore

$$\tfrac{1}{2}v^2 = 200 - 5x - \tfrac{1}{4}x^2$$
$$v^2 = 400 - 10x - \tfrac{1}{2}x^2$$

Hence

$$v = \pm\sqrt{(400 - 10x - \tfrac{1}{2}x^2)}$$

The particle comes to rest when

$$400 - 10x - \tfrac{1}{2}x^2 = 0$$
$$x^2 + 20x - 800 = 0$$
$$\therefore \quad (x + 40)(x - 20) = 0$$
$$\therefore \quad x = -40 \text{ or } +20$$

However, the particle is initially moving in the positive direction, so it will first come to rest when x is positive, i.e. when $x = 20$.

Exercise 4b

Throughout this exercise, x represents the displacement from the origin, in metres, of a point P which is moving in a straight line; the velocity after t seconds is v m s^{-1}, *and the corresponding acceleration is a* m s^{-2}. *In each question, either v or a is given.*

1 $v = 6t + 3t^2$; find x in terms of t, given that when $t = 0$, $x = 10$. Hence find the displacement of P when $t = 4$.

2 $v = 6\sqrt{t}$; find x when $t = 4$, given that when $t = 1$, $x = 12$.

3 $v = 12 + 6t^2$; find the distance travelled by P in
(a) the first second   (b) the fifth second   (c) the nth second

4 $v = 5\cos 4t$; given that when $t = 0$, $x = 0$, show that $a = -16x$. Find the maximum displacement of P.

5 $a = 6t$; find v and x in terms of t, given that when $t = 0$, $v = 10$ and $x = 0$.

6 $a = 30\sqrt{t}$; given that when $t = 0$, $x = 4$, and that when $t = 4$, $x = 100$, find x in terms of t.

7 $a = 2v$; given that when $x = 0$, $v = 5$, find v in terms of x. Hence find the displacement when $v = 25$.

8 $a = 4x^3$; given that when $x = 0$, $v = 2$, express v in terms of x. Hence find x when $v = 6$.

9 A particle moves in a straight line so that

$$\frac{dv}{dt} = -4x$$

Given that $v = 0$ when $x = 5$, express v in terms of x. Hence find the velocity when $x = 3$. Describe the motion in words.

10 A particle moves in a straight line so that

$$\frac{dv}{dt} = -\frac{8}{x^2}$$

Given that when $x = 1$, $v = 5$, find v in terms of x, and show that when x tends to infinity, v tends to 3.

Exercise 4c (Miscellaneous)

1 A particle is moving in a straight line and its displacement from a fixed point O, after t s, is x m. Its velocity v m s^{-1} is given by

$$v = 24t - 6t^2$$

Given that, when $t = 2$, $x = 0$, find
(a) an expression for x, in terms of t;
(b) the displacement when $t = 5$;
(c) the acceleration, in terms of t;
(d) the maximum velocity.

2 A particle is moving in a straight line and its displacement, x m, after t s, is given by

$$x = \sin 2t + \cos 2t$$

Find its velocity, v m s^{-1}, in terms of t, and find the smallest positive value of t, for which $v = 0$. Prove that for all values of t

$$-\sqrt{2} \leqslant x \leqslant +\sqrt{2}$$

3 The velocity of a sky-diver is given by

$$v = 10t - t^2, \quad 0 \leqslant t \leqslant 5$$
$$v = 25, \qquad\quad t > 5$$

Sketch the velocity–time graph. Find the distance fallen in (a) the first five seconds, (b) the first ten seconds. Describe the motion in words.

4 (In the previous chapter we regarded the acceleration due to gravity as constant, and denoted it by g. In this question we use the techniques developed in the current chapter to take into account the 'inverse-square law' of gravity, using g for the gravitational acceleration at the Earth's surface.)

The acceleration towards the centre of the Earth, due to gravity, at a distance x from the centre of the Earth, is k/x^2, where k is a constant, provided that $x \geqslant R$, where R is the radius of the Earth. At the Earth's surface the acceleration is g m s^{-2}; show that $k = gR^2$.

A projectile is fired vertically upwards from a point on the Earth's surface, and its velocity at a distance x m from the centre of the Earth is v m s^{-1}. Show that

$$v\frac{dv}{dx} = -\frac{gR^2}{x^2}$$

Given that the initial velocity is $u\,\mathrm{m\,s^{-1}}$, prove that the greatest height reached, above the surface of the Earth, is

$$\frac{u^2 R}{2gR - u^2}$$

provided $2gR > u^2$. What will happen if $u^2 \geqslant 2gR$?

Taking the radius of the Earth to be 6380 km and g to be 9.81 m s^{-2}, estimate, in km h^{-1}, the least value of the initial velocity required in order to 'escape' from Earth's gravity.

5 The velocity $v\,\mathrm{m\,s^{-1}}$ of a small object at time t seconds as it moves in a straight line is given by $v = 4t^2 - 24t + 11$. At what times is the object instantaneously at rest, and what is the acceleration of the object at these times? The object is at a certain point when $t = 0$. How far away from this point is the object 3 seconds later?

*(MEI)

6 The forward speed, $v\,\mathrm{m\,s^{-1}}$, of a particle travelling in a straight line after t seconds is given by $v = 16 - 3t^2$.
   (a) How far does it move in the first second?
   (b) After how many seconds does it start to move backwards?
   (c) Show that it is back at the starting point after 4 seconds.
   (d) What total distance (forwards and backwards) does it travel in these 4 seconds?

*(SMP)

7 A train starts from rest at station A and travels on a straight track to station B, where it stops. Between the stations the speed of the train, $v\,\mathrm{km\,h^{-1}}$, t hours after leaving A, is given by $v = 180t - 135t^2$.
   (a) Find the time taken to travel from A to B and hence calculate the distance AB.
   (b) Find the accelerations at A and B, giving your answers in km h^{-2}.
   (c) Find the maximum speed of the train.

(C: part)

8 A particle moves in a straight line. At time t its displacement from a fixed point O on the line is x, its velocity is v, and its acceleration is $k \sin \omega t$, where k and ω are constant. The displacement, velocity and acceleration are all measured in the same direction. Initially the particle is at O and is moving with velocity u. Show that

$$v = \frac{k}{\omega}(1 - \cos \omega t) + u$$

and hence find x in terms of t, k, ω and u.

Show that, if $u = 0$, the particle first comes to rest after travelling a distance $2\pi k/\omega^2$.

(JMB: part)

9 A particle moves in a straight line, so that at time t seconds its velocity is v metres per second, given by

$$v = \sin(2\pi t) + \tfrac{1}{2}$$

At time $t = 0$ the particle is at point O of the line.

(a) Obtain an expression for the displacement of the particle from O at time t, and show that, if t is positive, the particle never returns to O.

(b) During the first second of its motion, find the time interval during which the particle is moving towards O.

(C: part)

10 An athlete runs in a straight line, and his speed $v\,\mathrm{m\,s^{-1}}$ is recorded at 10 second intervals. The data is recorded below:

t	0	10	20	30	40	50	60
v	0	4	7	6	6.5	5	7

Use the trapezium rule to estimate the distance covered in one minute.

Chapter 5

Vectors in applied mathematics

5.1 Introduction

Many of the quantities which we meet in applied mathematics have both magnitude and direction, and are 'added' by the parallelogram law of addition, for example, displacement, velocity and acceleration (see §5.4). Any such quantity is called a **vector**. We shall soon be meeting further vector quantities; what is more we shall wish to consider them in the context of two or three dimensions (so far the discussion has been limited to one dimension). A quantity which has magnitude only is called a **scalar**.

In this chapter we shall consider the mathematical properties of vectors, especially those which are useful in applied mathematics.†

5.2 Displacement vectors

Displacement vectors are important in applied mathematics and they provide a straightforward introduction to vector methods.

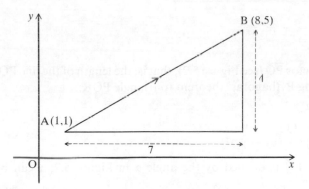

Figure 5.1

†For a further discussion of vectors, see *Pure Mathematics Books 1* and *2*, by J. K. Backhouse and S.P.T. Houldsworth, revised by P. J. F. Horril, published by Longman.

In Figure 5.1, the displacement from A to B is a combination of 7 units to the right (i.e. parallel to the x-axis and in the positive direction) and 4 units upwards (i.e. parallel to the y-axis and in the positive direction). We write this in the form

$$\overrightarrow{AB} = \begin{pmatrix} 7 \\ 4 \end{pmatrix}$$

The displacement from B to A is written

$$\overrightarrow{BA} = \begin{pmatrix} -7 \\ -4 \end{pmatrix}$$

For any displacement vector, $\overrightarrow{AB}$

$$\overrightarrow{AB} + \overrightarrow{BA} = \mathbf{0}$$

(the left-hand side indicates a displacement from A to B, and then back to A again, i.e. the resultant displacement is zero). Consequently we write

$$\overrightarrow{BA} = -\overrightarrow{AB}$$

In general, the displacement from a point P (x_1, y_1) to a point Q (x_2, y_2) is written

$$\overrightarrow{PQ} = \begin{pmatrix} x_2 - x_1 \\ y_2 - y_1 \end{pmatrix}$$

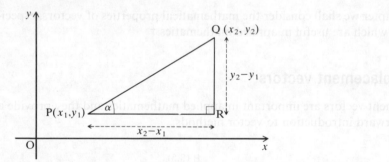

Figure 5.2

The magnitude of the vector $\overrightarrow{PQ}$ (see Figure 5.2), that is, the length of the line PQ, can be found by applying Pythagoras' theorem to triangle PQR:

$$PQ^2 = PR^2 + RQ^2$$
$$= (x_2 - x_1)^2 + (y_2 - y_1)^2$$
$$\therefore \quad PQ = \sqrt{[(x_2 - x_1)^2 + (y_2 - y_1)^2]}$$

The direction of vector $\overrightarrow{PQ}$ is defined by the angle α in Figure 5.2; α can be calculated from

$$\tan \alpha = \frac{y_2 - y_1}{x_2 - x_1} \quad \dagger$$

†However, since this calculation yields *two* possible values for α, it is always essential to examine the position of $\overrightarrow{PQ}$ on a diagram.

Note that the angle α should always be measured anticlockwise from the *positive* x-axis.

Qu. 1 For each of the following pairs of points, A and B, write down the vector $\overrightarrow{AB}$ and find (i) the magnitude of $\overrightarrow{AB}$, (ii) the angle $\overrightarrow{AB}$ makes with the positive x-axis:

(a) A (3,5)   B (7,10)   (b) A (5,4)   B (2,6)
(c) A (−5,1)   B (1, −5)   (d) A (4,2)   B (−2,2)

Qu. 2 Find the magnitude and direction of the following vectors:

(a) $\begin{pmatrix} 0.8 \\ 0.6 \end{pmatrix}$ (b) $\begin{pmatrix} -0.8 \\ 0.6 \end{pmatrix}$ (c) $\begin{pmatrix} 0.8 \\ -0.6 \end{pmatrix}$ (d) $\begin{pmatrix} 0 \\ 5 \end{pmatrix}$ (e) $\begin{pmatrix} -3 \\ 0 \end{pmatrix}$

Qu. 3 Write down expressions for the magnitude and direction of the following vectors:

(a) $\begin{pmatrix} \cos\alpha \\ \sin\alpha \end{pmatrix}$ (b) $\begin{pmatrix} \sin\alpha \\ \cos\alpha \end{pmatrix}$ (c) $\begin{pmatrix} -\cos\alpha \\ \sin\alpha \end{pmatrix}$ (d) $\begin{pmatrix} \cos\alpha \\ -\sin\alpha \end{pmatrix}$

where α is measured in radians, and $0 < \alpha < \dfrac{\pi}{2}$.

5.3 Unit vectors

Any vector whose magnitude is 1 unit (see Qu. 2 and Qu. 3 above) is called a **unit vector**. The unit vectors $\begin{pmatrix} 1 \\ 0 \end{pmatrix}$ and $\begin{pmatrix} 0 \\ 1 \end{pmatrix}$ are especially important because they are parallel to the x-axis and y-axis respectively. They are given a standard notation, namely

$$\mathbf{i} = \begin{pmatrix} 1 \\ 0 \end{pmatrix} \quad \text{and} \quad \mathbf{j} = \begin{pmatrix} 0 \\ 1 \end{pmatrix}$$

In three-dimensional work, the unit vector parallel to the z-axis is represented by **k**: the unit vectors **i**, **j** and **k** are then

$$\mathbf{i} = \begin{pmatrix} 1 \\ 0 \\ 0 \end{pmatrix} \quad \mathbf{j} = \begin{pmatrix} 0 \\ 1 \\ 0 \end{pmatrix} \quad \text{and} \quad \mathbf{k} = \begin{pmatrix} 0 \\ 0 \\ 1 \end{pmatrix}$$

Note that **i**, **j** and **k** are always printed in **bold type**; in manuscript they should always be written with a wavy line underneath, i.e. ḭ, j̰ and k̰.

Figure 5.3 shows a unit vector $\overrightarrow{OP}$ making an angle α with the x-axis.

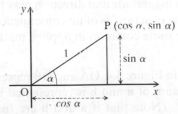

Figure 5.3

The unit vector $\overrightarrow{OP}$ can be expressed in terms of α as follows:

$$\overrightarrow{OP} = \begin{pmatrix} \cos \alpha \\ \sin \alpha \end{pmatrix}$$

The vectors $\mathbf{i}$ and $\mathbf{j}$ are special cases in which the angle α is 0 (for $\mathbf{i}$), and $\pi/2$ radians (for $\mathbf{j}$).

Notation

It is frequently convenient to represent a vector by a single letter. When this is done, the letter is always printed in **bold type** (in manuscript it should be written with a wavy line underneath). The magnitude of the vector is usually represented by the same letter, but printed in *italic type*.

For example, if $\mathbf{F} = \begin{pmatrix} 3 \\ 4 \end{pmatrix}$ then the magnitude of vector $\mathbf{F}$ is given by

$$F = \sqrt{(3^2 + 4^2)}$$

$$F = 5$$

In general, if $\mathbf{F} = \begin{pmatrix} x \\ y \end{pmatrix}$, then $F = \sqrt{(x^2 + y^2)}$.

On the other hand if the vector $\mathbf{F}$ is inclined at an angle α to the positive x-axis, then we may write

$$\mathbf{F} = \begin{pmatrix} F \cos \alpha \\ F \sin \alpha \end{pmatrix}$$

The displacement vector of a point P from the origin is called the **position vector** of P, and it is usual to use the corresponding lower case (small) letter, in this case $\mathbf{p}$, as an abbreviation for $\overrightarrow{OP}$. For example, if A is the point (3,5) then its position vector is written

$$\mathbf{a} = \begin{pmatrix} 3 \\ 5 \end{pmatrix}$$

In general the position vector of P (x, y) is written $\mathbf{p} = \begin{pmatrix} x \\ y \end{pmatrix}$.

5.4 The addition of vectors

Two displacement vectors which have the same magnitude and direction may be regarded as equal to each other; their actual 'line of action' is of no consequence. Any vector having this property (and they are quite common in applied mathematics) is called a **free vector**.

Vectors are added by the parallelogram law: if, in Figure 5.4, $\overrightarrow{OA}$ and $\overrightarrow{OB}$ represent the vectors $\mathbf{a}$ and $\mathbf{b}$ respectively, then the sum of $\mathbf{a}$ and $\mathbf{b}$ is represented in magnitude and direction by the diagonal $\overrightarrow{OR}$. (Note that if $\mathbf{a}$ and $\mathbf{b}$ are free vectors then $\overrightarrow{OA} = \overrightarrow{BR} = \mathbf{a}$, and $\overrightarrow{OB} = \overrightarrow{AR} = \mathbf{b}$.)

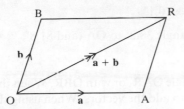

Figure 5.4

Example 1

Given that **a** and **b** are two vectors of magnitude 8 units and 5 units respectively, and that the angle between **a** and **b** is 120°, find the magnitude and direction of **a** + **b**.

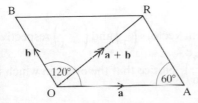

Figure 5.5

In the diagram, $\overrightarrow{OA}$ and $\overrightarrow{OB}$ represent the vectors **a** and **b**; the diagonal $\overrightarrow{OR}$ represents **a** + **b**.

Applying the cosine rule to triangle OAR, we obtain

$$QR^2 = OA^2 + AR^2 - 2 \times OA \times AR \times \cos \angle OAR$$
$$= 8^2 + 5^2 - 2 \times 8 \times 5 \times \cos 60°$$
$$= 64 + 25 - 80 \times 0.5$$
$$= 49$$
$$\therefore \quad OR = 7$$

Hence the magnitude of **a** + **b** is 7 units.

Applying the sine rule to triangle OAR, gives

$$\frac{\sin \angle AOR}{AR} = \frac{\sin \angle OAR}{OR}$$

$$\frac{\sin \angle AOR}{5} = \frac{\sin 60°}{7}$$

$$\therefore \quad \sin \angle AOR = 5 \times \frac{\sin 60°}{7}$$

$$= 0.6186 \quad \text{(correct to 4 significant figures)}$$

(If a calculator is used, this line should be omitted and all the significant figures available on the calculator retained.)

∴　∠ AOR = 38.2°, correct to the nearest 0.1°

Hence the vector **a** + **b** is inclined at an angle 38.2° to $\overrightarrow{OA}$ (and 81.8° to $\overrightarrow{OB}$).

Note that it is sufficient to work with the triangle OAR, or with OBR. When this is done we say we are using the 'triangle law' to add the vectors. When using the triangle law to add vectors (see Figure 5.6) the vectors must be joined 'head-to-tail'. The third side of the triangle represents the sum in magnitude and direction. (In this chapter the sum will be indicated by a double-headed arrow.)

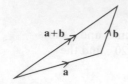

Figure 5.6

For example, if **a** and **b** are the displacement vectors $\begin{pmatrix} 5 \\ 2 \end{pmatrix}$ and $\begin{pmatrix} -2 \\ 4 \end{pmatrix}$ respectively (see Figure 5.7) then **a** + **b** is the vector $\begin{pmatrix} 3 \\ 6 \end{pmatrix}$. Notice that the order in which the vectors are added does not matter.

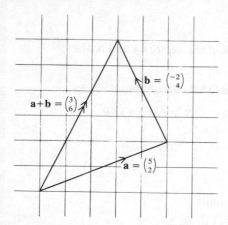

Figure 5.7

In general, if $\mathbf{a} = \begin{pmatrix} x_1 \\ y_1 \end{pmatrix}$ and $\mathbf{b} = \begin{pmatrix} x_2 \\ y_2 \end{pmatrix}$, then $\mathbf{a} + \mathbf{b} = \begin{pmatrix} x_1 + x_2 \\ y_1 + y_2 \end{pmatrix}$. This rule may be extended to any number of vectors, i.e.

$$\begin{pmatrix} x_1 \\ y_1 \end{pmatrix} + \begin{pmatrix} x_2 \\ y_2 \end{pmatrix} + \begin{pmatrix} x_3 \\ y_3 \end{pmatrix} + \cdots + \begin{pmatrix} x_n \\ y_n \end{pmatrix} = \begin{pmatrix} x_1 + x_2 + x_3 + \cdots + x_n \\ y_1 + y_2 + y_3 + \cdots + y_n \end{pmatrix}$$

For a given vector **v**, the vector which has the same magnitude as **v** but the opposite direction is written −**v** (see §5.2). If $\mathbf{v} = \begin{pmatrix} x \\ y \end{pmatrix}$ then $-\mathbf{v} = \begin{pmatrix} -x \\ -y \end{pmatrix}$. Note that $\mathbf{v} + (-\mathbf{v}) = \mathbf{0}$.

The difference of two vectors **a** and **b** is found by adding **a** and $(-\mathbf{b})$, i.e.

$$\mathbf{a} - \mathbf{b} = \mathbf{a} + (-\mathbf{b})$$

Hence, in the notation above,

$$\mathbf{a} - \mathbf{b} = \begin{pmatrix} x_1 \\ y_1 \end{pmatrix} - \begin{pmatrix} x_2 \\ y_2 \end{pmatrix}$$

$$= \begin{pmatrix} x_1 \\ y_1 \end{pmatrix} + \begin{pmatrix} -x_2 \\ -y_2 \end{pmatrix}$$

$$= \begin{pmatrix} x_1 - x_2 \\ y_1 - y_2 \end{pmatrix}$$

The difference of **a** and **b**, that is **a** − **b**, is shown in Figure 5.8.

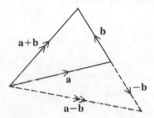

Figure 5.8

To add vectors in applied mathematics, especially if there are more than two, it is usually convenient to express them as column vectors, i.e. in the form $\begin{pmatrix} x \\ y \end{pmatrix}$. The next example illustrates the procedure.

Example 2

The magnitudes and directions of three vectors **a**, **b** and **c**, are as follows:

    a, magnitude 10 units, direction 0°
    b, magnitude 5 units, direction 60°
    c, magnitude 8 units, direction 150°

where the angles are all measured from the positive x axis.

Find the magnitude and direction of **a** + **b** + **c**.

First, we express **a**, **b** and **c** as column vectors:

$$\mathbf{a} = \begin{pmatrix} 10 \\ 0 \end{pmatrix}$$

$$\mathbf{b} = \begin{pmatrix} 5\cos 60° \\ 5\sin 60° \end{pmatrix}$$

$$\mathbf{c} = \begin{pmatrix} -8\cos 30° \\ 8\sin 30° \end{pmatrix}$$

Hence

$$\mathbf{a} + \mathbf{b} + \mathbf{c} = \begin{pmatrix} 10 + 5\cos 60° - 8\cos 30° \\ 0 + 5\sin 60° + 8\sin 30° \end{pmatrix}$$

$$= \begin{pmatrix} 5.572 \\ 8.330 \end{pmatrix} \quad †$$

The magnitude of $\mathbf{a} + \mathbf{b} + \mathbf{c}$ is

$$= \sqrt{(5.572^2 + 8.330^2)}$$
$$= \sqrt{100.4}$$
$$= 10.0 \quad \text{(correct to 3 significant figures)}$$

Hence the magnitude of $\mathbf{a} + \mathbf{b} + \mathbf{c}$ is 10.0 units.

The angle which $\mathbf{a} + \mathbf{b} + \mathbf{c}$ makes with the x-axis is θ, where

$$\tan \theta = \frac{8.330}{5.572}$$

$$= 1.495$$

$$\therefore \qquad \theta = 56.2° \quad \text{correct to the nearest } 0.1°$$

Hence vector $\mathbf{a} + \mathbf{b} + \mathbf{c}$ is inclined at 56.2° to the x-axis.

5.5 Multiplication by a scalar

Figure 5.9 shows a vector $\mathbf{F} = \begin{pmatrix} x \\ y \end{pmatrix}$

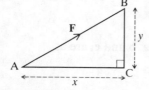

Figure 5.9

If the magnitude of $\mathbf{F}$ is multiplied by a positive factor k, but its direction is left unchanged, the triangle ABC in Figure 5.9 will be enlarged by a scale factor k. Consequently

$$k\mathbf{F} = k\begin{pmatrix} x \\ y \end{pmatrix} = \begin{pmatrix} kx \\ ky \end{pmatrix}$$

The factor k is often called a **scalar** .

† If the reader is using a calculator, then its capacity to handle the arithmetic with a high degree of accuracy should be exploited. In particular the memory function should be used to avoid 'rounding off' the intermediate results.

For example,

$$5\binom{2}{7} = \binom{10}{35}$$

Using this idea, we can write a column vector $\binom{x}{y}$ in a rather more convenient form, namely

$$\binom{x}{y} = \binom{x}{0} + \binom{0}{y}$$

$$= x\binom{1}{0} + y\binom{0}{1}$$

$$= x\mathbf{i} + y\mathbf{j}$$

where $\mathbf{i}$ and $\mathbf{j}$ are the unit vectors parallel to the x-axis and the y-axis, respectively (see §5.3).

Qu. 4 Given that $\mathbf{p} = 2\mathbf{i} + 4\mathbf{j}$ and $\mathbf{q} = 3\mathbf{i} - 6\mathbf{j}$, express in terms of $\mathbf{i}$ and $\mathbf{j}$:
(a) $\mathbf{p} + \mathbf{q}$ (b) $\mathbf{p} - \mathbf{q}$ (c) $5\mathbf{p} + 2\mathbf{q}$ (d) $\tfrac{1}{2}\mathbf{p} - \tfrac{1}{3}\mathbf{q}$

Summary

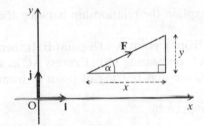

Figure 5.10

(a) $\mathbf{i} = \binom{1}{0}$, $\mathbf{j} = \binom{0}{1}$

(b) $\mathbf{F} = x\mathbf{i} + y\mathbf{j} = \binom{x}{y}$

$$= (F\cos\alpha)\mathbf{i} + (F\sin\alpha)\mathbf{j} = \binom{F\cos\alpha}{F\sin\alpha}$$

(c) $F = \sqrt{(x^2 + y^2)}$ and $\tan\alpha = y/x$

(d) If $\mathbf{F}_1 = x_1\mathbf{i} + y_1\mathbf{j}$ and $\mathbf{F}_2 = x_2\mathbf{i} + y_2\mathbf{j}$,
   then $\mathbf{F}_1 + \mathbf{F}_2 = (x_1 + x_2)\mathbf{i} + (y_1 + y_2)\mathbf{j}$
   and $\mathbf{F}_1 - \mathbf{F}_2 = (x_1 - x_2)\mathbf{i} + (y_1 - y_2)\mathbf{j}$

(e) $k\mathbf{F} = k\binom{x}{y} = \binom{kx}{ky}$, in particular $-\mathbf{F} = \binom{-x}{-y}$

Exercise 5a

The reader is reminded that when using coordinate axes, angles are measured anti-clockwise from the positive x-axis.

1 Given that $\mathbf{p} = 2\mathbf{i} + 3\mathbf{j}$ and $\mathbf{q} = 4\mathbf{i} - 5\mathbf{j}$, express in terms of $\mathbf{i}$ and $\mathbf{j}$:
   (a) $\mathbf{p} + \mathbf{q}$ (b) $\mathbf{p} - \mathbf{q}$ (c) $2\mathbf{p} + 3\mathbf{q}$ (d) $5\mathbf{p} - 2\mathbf{q}$

2 Find the magnitude and direction of the vectors in each part of question 1.

3 Given that the magnitude of a vector $\mathbf{F}$ is 10 units and that it is inclined to the (positive) x-axis at an angle of 30°, express $\mathbf{F}$ in the form $x\mathbf{i} + y\mathbf{j}$.

4 Given that a vector $\mathbf{V}$ has a magnitude of 20 units and that it is inclined to the (positive) x-axis at an angle of 120°, express $\mathbf{V}$ as a column vector.

5 Given that $\mathbf{p} = \begin{pmatrix} 0 \\ 10 \end{pmatrix}$, express $\mathbf{p}$ as the sum of two vectors $\mathbf{a}$ and $\mathbf{b}$, which are inclined at angles of 30° and 120°, respectively, to the (positive) x-axis.

6 Find scalars m and n, such that

$$m\begin{pmatrix} 3 \\ 4 \end{pmatrix} + n\begin{pmatrix} 7 \\ 10 \end{pmatrix} = \begin{pmatrix} 1 \\ 0 \end{pmatrix}$$

7 Find scalars α, β, γ (not all zero) such that

$$\alpha\begin{pmatrix} 2 \\ 5 \end{pmatrix} + \beta\begin{pmatrix} 6 \\ 7 \end{pmatrix} + \gamma\begin{pmatrix} 10 \\ 13 \end{pmatrix} = \begin{pmatrix} 0 \\ 0 \end{pmatrix}$$

Are the values α, β, γ unique? If not, explain the relationship between the possible sets of values.

8 A ship sails 50 km on a bearing of 015° from a point A to a point B. It then changes course and sails 100 km to point C on bearing 150°. Express $\overrightarrow{AC}$ as a column vector, and hence find the distance and bearing of point C from point A.

9 The position of a particle P at time t s is given by

$$\overrightarrow{OP} = (1 + 3t)\mathbf{i} + (2 + t)\mathbf{j}$$

Draw a pair of axes and mark the position of P when $t = 0, 1, 2, \ldots, 10$. Show that the path of P is a straight line and find the angle at which this line is inclined to the x-axis.

10 A particle P moves in a plane and its position at time t s is given by the vector

$$\overrightarrow{OP} = (1 + t^2)\mathbf{i} + 3t\mathbf{j}$$

Write down the vector which represents
   (a) its position when $t = 2$;
   (b) its position when $t = 5$;
   (c) its displacement over the interval $t = 2$ to $t = 5$.

5.6 Differentiating and integrating vectors

Many of the vectors which arise in applied mathematics vary with time, and the methods of calculus can be used to find, for example, the rate at which a vector is changing.

Suppose that **F** is a vector which is changing with time and that

$$\mathbf{F} = x\mathbf{i} + y\mathbf{j} \qquad \qquad \cdots \quad (1)$$

where x and y are functions of time. Suppose also that in a small time interval δt, the functions x and y increase to $x + \delta x$ and $y + \delta y$, respectively, with a corresponding increase in **F** of $\delta \mathbf{F}$, i.e.

$$\mathbf{F} + \delta \mathbf{F} = (x + \delta x)\mathbf{i} + (y + \delta y)\mathbf{j} \qquad \qquad \cdots \quad (2)$$

Then, subtracting equation (1) from equation (2), we have

$$\delta \mathbf{F} = (\delta x)\mathbf{i} + (\delta y)\mathbf{j}$$

Dividing this by δt gives the average rate of change of **F** over the time interval δt, and when δt tends to zero this becomes

$$\frac{d\mathbf{F}}{dt} = \frac{dx}{dt}\mathbf{i} + \frac{dy}{dt}\mathbf{j}$$

Example 3

Given that $\mathbf{F} = (15t + 4)\mathbf{i} - (5t^2 + 3)\mathbf{j}$, find the rate of change of **F** when $t = 4$.

$$\mathbf{F} = (15t + 4)\mathbf{i} - (5t^2 + 3)\mathbf{j}$$

Differentiate with respect to t:

$$\frac{d\mathbf{F}}{dt} = 15\mathbf{i} - 10t\mathbf{j}$$

When $t = 4$

$$\frac{d\mathbf{F}}{dt} = 15\mathbf{i} - 40\mathbf{j}$$

The rate of change of **F**, when $t = 4$, is $15\mathbf{i} - 40\mathbf{j}$.

(Note that the rate of change of **F** is a *vector* quantity; however this is not surprising, since as t changes, **F** also changes both in *magnitude and direction*.)

Not only can we differentiate a vector with respect to t, but we can also integrate, for example given that

$$\frac{d\mathbf{v}}{dt} = 20t\mathbf{i} + 3t^2\mathbf{j}$$

then, integrating

$$\mathbf{v} = 10t^2\mathbf{i} + t^3\mathbf{j} + \mathbf{c}$$

As always, it is important to remember the 'constant of integration' **c**; in this context, however, it must be a vector. This constant can only be determined if we are given some further data. If, for example, we know that when $t = 2$, $\mathbf{v} = 50\mathbf{i} + 10\mathbf{j}$, then substituting this we have

$$50\mathbf{i} + 10\mathbf{j} = 10 \times 2^2\mathbf{i} + 2^3\mathbf{j} + \mathbf{c}$$
$$50\mathbf{i} + 10\mathbf{j} = 40\mathbf{i} + 8\mathbf{j} + \mathbf{c}$$
$$\therefore \qquad \mathbf{c} = 10\mathbf{i} + 2\mathbf{j}$$

Hence

$$\mathbf{v} = 10t^2\mathbf{i} + t^3\mathbf{j} + 10\mathbf{i} + 2\mathbf{j}$$
i.e. $\quad \mathbf{v} = (10t^2 + 10)\mathbf{i} + (t^3 + 2)\mathbf{j}$

Qu. 5 Given that $\mathbf{F} = (10 \cos t)\mathbf{i} + (10 \sin t)\mathbf{j}$, find $\dfrac{d\mathbf{F}}{dt}$. Show also that $\dfrac{d^2\mathbf{F}}{dt^2} = -\mathbf{F}$.

Qu. 6 Given that $\dfrac{d\mathbf{V}}{dt} = (2\mathbf{i} + 3\mathbf{j})t$, and that $\mathbf{V} = 5\mathbf{i} + 3\mathbf{j}$ when $t = 0$, express $\mathbf{V}$ in terms of t.

5.7 The scalar product

We can now add and subtract vectors, multiply a vector by a scalar, and apply calculus methods to vectors, but so far we have not considered the possibility of multiplying a vector by another vector. Rather surprisingly, there are two distinct rules for combining a pair of vectors, both called 'multiplication': in one the result, or 'product' is a scalar quantity, i.e. it has magnitude only, and in the other, the 'product' is a vector. In this book, we shall only be concerned with the scalar product.

Definition

The scalar product of two vectors, $\mathbf{a}$ and $\mathbf{b}$, whose magnitudes are a and b respectively, and which are at an angle θ to one another (see Figure 5.11), is defined as follows:

$$\mathbf{a} \cdot \mathbf{b} = ab \cos \theta$$

(The symbol for scalar multiplication is a very distinct dot; it is a common practice to call the scalar product 'the dot product'.)

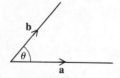

Figure 5.11

Example 4

Given that the magnitudes of vectors **a** and **b** are 6 units and 4 units, respectively, and that the angle between **a** and **b** is 60°, find the scalar product **a** . **b**.

$$\mathbf{a} . \mathbf{b} = 6 \times 4 \times \cos 60°$$
$$= 24 \times 0.5$$
$$= 12$$

(Note that no specific units have been mentioned: comment on this will be reserved until we meet particular applications of this technique.)

Scalar multiplication has several important properties, which are listed below. With the exception of (g), the proofs of these properties are fairly trivial; even so, the reader should pause and think about them. In each case, **a** and **b** represent a pair of non-zero vectors:

(a) $\mathbf{a} . \mathbf{b} = \mathbf{b} . \mathbf{a}$

(b) if **a** is perpendicular to **b**, then $\mathbf{a} . \mathbf{b} = 0$

(c) if $\mathbf{a} . \mathbf{b} = 0$, then **a** is perpendicular to **b**

(d) $\mathbf{i} . \mathbf{j} = 0$

(e) $\mathbf{i} . \mathbf{i} = \mathbf{j} . \mathbf{j} = 1$

(f) $\mathbf{a} . \mathbf{a} = a^2$

(g) $\mathbf{a} . (\mathbf{b} + \mathbf{c}) = \mathbf{a} . \mathbf{b} + \mathbf{a} . \mathbf{c}$

Notes

(i) One of the great advantages of vector methods is that they can easily be extended to use in three dimensions. In this context, the unit vectors parallel to the axes are always denoted **i**, **j** and **k** (see §5.3), and properties (d) and (e) then become:
$$\mathbf{i} . \mathbf{j} = \mathbf{j} . \mathbf{k} = \mathbf{k} . \mathbf{i} = 0 \quad \text{and} \quad \mathbf{i} . \mathbf{i} = \mathbf{j} . \mathbf{j} = \mathbf{k} . \mathbf{k} = 1$$

(ii) Property (c) should be used as the standard method for showing that a pair of vectors are perpendicular.

(iii) Property (g) is vital, since it allows us to remove brackets according to the normal rules of algebra, for example

$$5\mathbf{i} . (3\mathbf{i} + 4\mathbf{j}) = 15\mathbf{i} . \mathbf{i} + 20\mathbf{i} . \mathbf{j}$$
$$= 15 \times 1 + 20 \times 0$$
$$= 15$$

The proof of this property is quite difficult; it will be given in the next section.

Qu. 7 Given that $\mathbf{a} = 2\mathbf{i} + 3\mathbf{j}$ and $\mathbf{b} = 6\mathbf{i} - 5\mathbf{j}$, find $\mathbf{a} . \mathbf{b}$.
Qu. 8 Prove that $(10\mathbf{i} + 15\mathbf{j})$ is perpendicular to $(21\mathbf{i} - 14\mathbf{j})$.

When vectors are expressed in terms of $\mathbf{i}$ and $\mathbf{j}$, the scalar product takes a special form, which should be memorised:

if $\mathbf{a} = x_1\mathbf{i} + y_1\mathbf{j}$ and $\mathbf{b} = x_2\mathbf{i} + y_2\mathbf{j}$
then $\mathbf{a}.\mathbf{b} = (x_1\mathbf{i} + y_1\mathbf{j}).(x_2\mathbf{i} + y_2\mathbf{j})$

$$\mathbf{a}.\mathbf{b} = (x_1\mathbf{i} + y_1\mathbf{j}).(x_2\mathbf{i} + y_2\mathbf{j})$$

$\mathbf{a}.\mathbf{b} = x_1x_2 + y_1y_2$
(because $\mathbf{i}.\mathbf{i} = \mathbf{j}.\mathbf{j} = 1$, and $\mathbf{i}.\mathbf{j} = 0$).

The corresponding result in three dimensions is as follows:

if $\mathbf{a} = x_1\mathbf{i} + y_1\mathbf{j} + z_1\mathbf{k}$ and $\mathbf{b} = x_2\mathbf{i} + y_2\mathbf{j} + z_2\mathbf{k}$
then

$$\mathbf{a}.\mathbf{b} = x_1x_2 + y_1y_2 + z_1z_2$$

Example 5

Given that $\mathbf{p} = 12\mathbf{i} + 3\mathbf{j} + 4\mathbf{k}$ and $\mathbf{q} = 4\mathbf{i} + 3\mathbf{j}$, find p^2, q^2 and $\mathbf{p}.\mathbf{q}$. Hence find the angle between $\mathbf{p}$ and $\mathbf{q}$.

$$
\begin{aligned}
p^2 = \mathbf{p}.\mathbf{p} &= (12\mathbf{i} + 3\mathbf{j} + 4\mathbf{k}).(12\mathbf{i} + 3\mathbf{j} + 4\mathbf{k}) \\
&= 12 \times 12 + 3 \times 3 + 4 \times 4 \\
&= 144 + 9 + 16 \\
&= 169
\end{aligned}
$$
$\therefore \qquad p = 13$

$$
\begin{aligned}
q^2 = \mathbf{q}.\mathbf{q} &= (4\mathbf{i} + 3\mathbf{j}).(4\mathbf{i} + 3\mathbf{j}) \\
&= 4 \times 4 + 3 \times 3 \\
&= 16 + 9 \\
&= 25
\end{aligned}
$$
$\therefore \qquad q = 5$

$$
\begin{aligned}
\mathbf{p}.\mathbf{q} &= (12\mathbf{i} + 3\mathbf{j} + 4\mathbf{k}).(4\mathbf{i} + 3\mathbf{j}) \\
&= 12 \times 4 + 3 \times 3 + 4 \times 0 \\
&= 48 + 9 \\
&= 57
\end{aligned}
$$

By definition

$\mathbf{p}.\mathbf{q} = pq \cos\theta$ where θ is the angle between $\mathbf{p}$ and $\mathbf{q}$.

So, substituting the values above, we have

$57 = 13 \times 5 \cos\theta$
$\therefore \quad \cos\theta = \dfrac{57}{65}$
$\therefore \qquad \theta = 28.7°$ correct to the nearest $0.1°$

Hence the angle between the vectors $\mathbf{p}$ and $\mathbf{q}$ is $28.7°$, correct to the nearest $0.1°$.

Exercise 5b

1 Prove that the vector $4\mathbf{i} + 3\mathbf{j} + 2\mathbf{k}$ is perpendicular to the vector $4\mathbf{i} - 2\mathbf{j} - 5\mathbf{k}$.

2 Find the cosine of the angle between the vectors $12\mathbf{i} + 5\mathbf{j}$ and $3\mathbf{i} - 4\mathbf{j}$.

3 Find the value of the real number m, such that the vector $m\mathbf{i} + 4\mathbf{j}$ is perpendicular to the vector $2\mathbf{i} - 5\mathbf{j}$.

4 Write down in terms of α, the unit vector $\mathbf{u}$ which is inclined to the x-axis at an angle $\alpha°$; write down also the unit vector $\mathbf{v}$ which is inclined to the x-axis at an angle $(90 + \alpha)°$. Verify that $\mathbf{u} . \mathbf{v} = 0$.

5 Given that $\mathbf{u} = (\cos t)\mathbf{i} + (\sin t)\mathbf{j}$, find $\dfrac{d\mathbf{u}}{dt}$ and verify that it is perpendicular to $\mathbf{u}$.

6 Given that $\mathbf{r} = t^2\mathbf{i} + t^3\mathbf{j}$, find $\dfrac{d\mathbf{r}}{dt}$.

7 Given that $\mathbf{V} = (\cos 5t)\mathbf{i} + (\sin 5t)\mathbf{j}$, verify that $\dfrac{d^2\mathbf{V}}{dt^2} = -25\mathbf{V}$.

8 Given that $\dfrac{d\mathbf{p}}{dt} = 4t\mathbf{i} + 6t^2\mathbf{j}$, find an expression for $\mathbf{p}$ in terms of t. If $\mathbf{p} = 10\mathbf{j}$ when $t = 0$, find $\mathbf{p}$ in terms of t.

9 Given that $\dfrac{d^2\mathbf{r}}{dt^2} = 10\mathbf{j}$, find an expression for $\mathbf{r}$ in terms of t. Given also that, when $t = 0$, $\mathbf{r} = \mathbf{0}$ and $\dfrac{d\mathbf{r}}{dt} = 2\mathbf{i} + \mathbf{j}$, show that $\mathbf{r} = 2t\mathbf{i} + (t + 5t^2)\mathbf{j}$.

10 A unit vector $\mathbf{V}$ makes angles α, β and γ with the x-axis, the y-axis and the z-axis respectively. By considering $\mathbf{V} . \mathbf{i}$, $\mathbf{V} . \mathbf{j}$ and $\mathbf{V} . \mathbf{k}$, or otherwise, show that $\mathbf{V} = (\cos\alpha)\mathbf{i} + (\cos\beta)\mathbf{j} + (\cos\gamma)\mathbf{k}$. Hence show that

$$\cos^2\alpha + \cos^2\beta + \cos^2\gamma = 1$$

(The values of $\cos\alpha$, $\cos\beta$ and $\cos\gamma$ are called the direction cosines of the vector $\mathbf{V}$.)

5.8 The proof of the distributive law

In Figure 5.12, $\overrightarrow{OA} = \mathbf{a}$, $\overrightarrow{OB} = \mathbf{b}$, $\overrightarrow{OC} = \mathbf{c}$ and $\overrightarrow{OR} = \mathbf{b} + \mathbf{c}$. The angles AOB, AOC and AOR are denoted in the proof (see next page) by β, γ and θ, respectively.

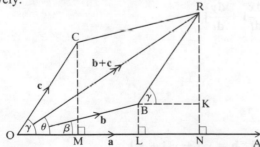

Figure 5.12

By definition

$$\mathbf{a} \cdot \mathbf{b} = OA \times OB \times \cos \beta$$
$$= OA \times OL \quad \text{(since } OL = OB \times \cos \beta \text{)}$$

Similarly

$$\mathbf{a} \cdot \mathbf{c} = OA \times OC \times \cos \gamma$$
$$= OA \times OM$$

and adding these expressions gives

$$\mathbf{a} \cdot \mathbf{b} + \mathbf{a} \cdot \mathbf{c} = OA \times OL + OA \times OM$$

Now, the quantities of the right-hand side are lengths (not vectors), so we may apply the normal rules of algebra to them, i.e.

$$OA \times OL + OA \times OM = OA \times (OL + OM)$$
$$\therefore \quad \mathbf{a} \cdot \mathbf{b} + \mathbf{a} \cdot \mathbf{c} = OA \times (OL + OM)$$

But $OM = LN$
(because $OM = OC \cos \gamma$, and $LN = BK = BR \cos \gamma = OC \cos \gamma$)

$$\mathbf{a} \cdot \mathbf{b} + \mathbf{a} \cdot \mathbf{c} = OA \times (OL + LN)$$
$$= OA \times ON$$

But $ON = OR \times \cos \theta$, therefore

$$\mathbf{a} \cdot \mathbf{b} + \mathbf{a} \cdot \mathbf{c} = OA \times OR \times \cos \theta$$

The right-hand side of this equation, however, is the scalar product $\mathbf{a} \cdot (\mathbf{b} + \mathbf{c})$. Hence

$$\mathbf{a} \cdot \mathbf{b} + \mathbf{a} \cdot \mathbf{c} = \mathbf{a} \cdot (\mathbf{b} + \mathbf{c})$$

This important rule, known as the **distributive law** for scalar multiplication, justifies the removal of brackets, according to the ordinary rules of elementary algebra, as in Example 5. (See §5.7.)

Summary

(a) If $\mathbf{F} = x\mathbf{i} + y\mathbf{j}$, then $\dfrac{d\mathbf{F}}{dt} = \dfrac{dx}{dt}\mathbf{i} + \dfrac{dy}{dt}\mathbf{j}$

(b) $\mathbf{a} \cdot \mathbf{b} = ab \cos \theta$

(c) $\mathbf{i} \cdot \mathbf{i} = \mathbf{j} \cdot \mathbf{j} = \mathbf{k} \cdot \mathbf{k} = 1$

(d) $\mathbf{i} \cdot \mathbf{j} = \mathbf{j} \cdot \mathbf{k} = \mathbf{k} \cdot \mathbf{i} = 0$

(e) $\mathbf{a} \cdot (\mathbf{b} + \mathbf{c}) = \mathbf{a} \cdot \mathbf{b} + \mathbf{a} \cdot \mathbf{c}$

(f) If $\mathbf{a} = x_1\mathbf{i} + y_1\mathbf{j} + z_1\mathbf{k}$ and $\mathbf{b} = x_2\mathbf{i} + y_2\mathbf{j} + z_2\mathbf{k}$,
then $\mathbf{a} \cdot \mathbf{b} = x_1 x_2 + y_1 y_2 + z_1 z_2$

Exercise 5c (Miscellaneous)

1 Given that $\mathbf{p} = \begin{pmatrix} 3 \\ 4 \end{pmatrix}$ and $\mathbf{q} = \begin{pmatrix} -7 \\ 24 \end{pmatrix}$, find p, q and $\mathbf{p} \cdot \mathbf{q}$. Hence find the cosine of the angle between $\mathbf{p}$ and $\mathbf{q}$.

2 Given that $\mathbf{u} = 3\mathbf{i} + 4\mathbf{j}$ and $\mathbf{v} = 4\mathbf{i} - 3\mathbf{j}$, and that $\mathbf{c} = \mathbf{u} + \mathbf{v}$ and $\mathbf{d} = \mathbf{u} - \mathbf{v}$, find u, v, c and d. Find also the scalar products $\mathbf{u} \cdot \mathbf{v}$ and $\mathbf{c} \cdot \mathbf{d}$. The points U, V and C are such that $\overrightarrow{OU} = \mathbf{u}$, $\overrightarrow{OV} = \mathbf{v}$ and $\overrightarrow{OC} = \mathbf{c}$. Draw a diagram showing the points U, V and C. Describe the quadrilateral OUCV.

3 The vectors $\mathbf{p}$ and $\mathbf{q}$ have magnitudes 10 and 4 respectively, and they are inclined to the x-axis at angles of 30° and 60°, respectively. Express $\mathbf{p} + \mathbf{q}$ as a column vector and find its magnitude, correct to 3 significant figures.

4 Draw a diagram illustrating the unit vector $\mathbf{u} = (\sin \alpha)\mathbf{i} + (\cos \alpha)\mathbf{j}$, indicating clearly the (acute) angle α. Write down, in terms of α, two unit vectors which are perpendicular to $\mathbf{u}$.

5 Find the cosines of the angles which the vector $20\mathbf{i} + 12\mathbf{j} + 9\mathbf{k}$ makes with the x-, y- and z-axes.

6 The position of a point P is given by $\overrightarrow{OP} = \mathbf{r}$, where $\mathbf{r} = 20t\mathbf{i} + (80t - 5t^2)\mathbf{j}$.

Find the (two) values of t for which P is on the x axis. Find also the vector $\dfrac{d\mathbf{r}}{dt}$ for these values of t.

7 Given that $\dfrac{d^2\mathbf{r}}{dt^2} = -10\mathbf{j}$, and that when $t = 0$, $\mathbf{r} = 0$ and $\dfrac{d\mathbf{r}}{dt} = 10\mathbf{i}$, find $\mathbf{r}$ in terms of t.

8 The point P, where $\overrightarrow{OP} = \mathbf{r}$, and $\mathbf{r} = 3(\cos 2t)\mathbf{i} + 3(\sin 2t)\mathbf{j}$, moves in a two-dimensional plane. Show that, for all values of t

(a) $r = 3$ (b) $\mathbf{r} \cdot \dfrac{d\mathbf{r}}{dt} = 0$ (c) $\dfrac{d^2\mathbf{r}}{dt^2} = -4\mathbf{r}$

Describe the path of the point P.

9 Given that $\mathbf{r} = (\sqrt{2}t)\mathbf{i} + (5t - 5t^2)\mathbf{j}$, find two values of t for which $\mathbf{r}$ is perpendicular to $\dfrac{d\mathbf{r}}{dt}$.

10 Given that $\mathbf{v} - \mathbf{u} = \mathbf{a}t$ and $\mathbf{v} + \mathbf{u} = \dfrac{2\mathbf{s}}{t}$, prove that, for all values of t,

$$v^2 - u^2 = 2\mathbf{a} \cdot \mathbf{s}.$$

11 Given the points A (1,2,4), B (5,3,7) and C (3,7,4), write down as column vectors $\overrightarrow{AB}$ and $\overrightarrow{AC}$. Calculate
(a) the scalar product $\overrightarrow{AB} \cdot \overrightarrow{AC}$
(b) the length of $\overrightarrow{AB}$
(c) the length of $\overrightarrow{AC}$
Hence show by calculation that the angle BAC is 61.7° approximately. Calculate the area of triangle ABC.

*(SMP)

12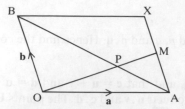

Figure 5.13

OAXB is a parallelogram. M is the midpoint of AX, and OM intersects AB at P.
$\overrightarrow{OA} = \mathbf{a}$, $\overrightarrow{OB} = \mathbf{b}$.
(a) Write down in terms of $\mathbf{a}$ and $\mathbf{b}$:
(i) $\overrightarrow{AX}$ (ii) $\overrightarrow{AM}$ (iii) $\overrightarrow{OM}$ (iv) $\overrightarrow{BA}$
(b) Given that $\overrightarrow{BP} = x\overrightarrow{BA}$, write down an expression for $\overrightarrow{OP}$ in terms of $\mathbf{a}$, $\mathbf{b}$ and x.
(c) Given that $\overrightarrow{OP} = y\overrightarrow{OM}$, write down an expression for $\overrightarrow{OP}$ in terms of $\mathbf{a}$, $\mathbf{b}$ and y.
(d) Show that your answers to (b) and (c) are identical when $x = y = \frac{2}{3}$.
*(M.E.I)

13 Given that $\mathbf{a} = \begin{pmatrix} 4 \\ -3 \end{pmatrix}$, $\mathbf{b} = \begin{pmatrix} 2 \\ 4 \end{pmatrix}$ and $\mathbf{c} = \begin{pmatrix} 22 \\ -11 \end{pmatrix}$, find
(a) a unit vector perpendicular to $\mathbf{a}$;
(b) the value of the constants m and n, for which $m\mathbf{a} + n\mathbf{b} = \mathbf{c}$. Evaluate the scalar product of $\mathbf{a}$ and $\mathbf{b}$ and hence find the cosine of the angle between the direction of $\mathbf{a}$ and the direction of $\mathbf{b}$.   *(C)

14 Given that the vectors $\mathbf{u} = 3\mathbf{i} + 2\mathbf{j}$ and $\mathbf{v} = 2\mathbf{i} + \lambda\mathbf{j}$, determine the values of λ so that
(a) $\mathbf{u}$ and $\mathbf{v}$ are at right angles;
(b) $\mathbf{u}$ and $\mathbf{v}$ are parallel;
(c) the acute angle between $\mathbf{u}$ and $\mathbf{v}$ is $\pi/4$.   (L: part)

15 The geometrical position vector $R_1 = \begin{pmatrix} 3 \\ -5 \end{pmatrix} + t\begin{pmatrix} 1 \\ 4 \end{pmatrix}$ gives, in two dimensions, a point on a line N_1 through $(3, -5)$ in the direction of the vector $\begin{pmatrix} 1 \\ 4 \end{pmatrix}$.
As t varies (including negative values) any point on the line can be reached.
State in a similar way for the line N_2, given by $R_2 = \begin{pmatrix} 1 \\ -1 \end{pmatrix} + s\begin{pmatrix} 2 \\ 4 \end{pmatrix}$, the coordinates of a point on it and a vector giving its direction. Obtain values of s and t which make $R_1 = R_2$ and deduce the coordinates of the point where N_1 and N_2 intersect.

Consider the lines N_3 and N_4 given in three dimensions by

$$R_3 = \begin{pmatrix} 1 \\ 2 \\ 4 \end{pmatrix} + u\begin{pmatrix} -1 \\ 0 \\ 1 \end{pmatrix} \quad \text{and} \quad R_4 = \begin{pmatrix} 1 \\ 0 \\ 1 \end{pmatrix} + v\begin{pmatrix} 2 \\ 2 \\ 3 \end{pmatrix}$$

For each line state the coordinates of a point on it and a vector giving its direction. Show algebraically that it is impossible to find values of u and v that make $\mathbf{R}_3 = \mathbf{R}_4$ and explain what this fact means geometrically.

*(MEI)

Revision Exercise 1

1 A train starts from rest and accelerates at a uniform rate of $0.8\,\mathrm{m\,s}^{-2}$. It is observed to pass two points A and B, a distance of 400 m apart, in five seconds. Find the velocities as it passes A and B.

2 The velocity of an object, $v\,\mathrm{m\,s}^{-1}$, after it has been falling for t s is shown in the following table:

t	0	0.5	1	2	3	4	5	6
x	0	3.9	6.3	8.6	9.5	9.8	9.9	10.0

Draw the velocity–time graph. Describe the motion in words and give an example of a situation in which motion of this kind might occur. From your graph estimate the acceleration when (a) $t = 1$, (b) $t = 2$, (c) $t = 3$.

3 In question 2, estimate the distance travelled in the first 6 seconds.

4 Two cars A and B are travelling in the same direction along a straight track; as they pass a fixed point P, their speeds are $u\,\mathrm{m\,s}^{-1}$ and $v\,\mathrm{m\,s}^{-1}$, respectively, where $v > u$. Car A is moving with constant acceleration $a\,\mathrm{m\,s}^{-2}$ and B with constant retardation $a\,\mathrm{m\,s}^{-2}$. Show that A will overtake B after $\dfrac{(v-u)}{a}$ seconds, and find their distance from P at this moment.

5 A motorcyclist accelerates uniformly from rest up to a speed V; she then travels at constant speed and finally slows to a halt with constant retardation. Given that the average speed over the whole journey is $5V/6$, show that the constant speed was maintained for two-thirds of the total time. (Do *not* assume that the acceleration and retardation are equal in magnitude.)

6 The acceleration of a car depends on the gear selected. A particular driver obtained the following performance, with a measured in $\mathrm{m\,s}^{-2}$, using each gear in turn for t seconds:

gear	first	second	third	top
acceleration	a	$3a$	$2a$	a

After the first $4t$ seconds, the car continued at constant speed. Neglecting the time taken changing gear, find in terms of a and t
(a) the constant speed;
(b) the average acceleration over the first $4t$ seconds;
(c) the distance travelled in the first $10t$ seconds.

7 A particle P, moving with constant acceleration, passes three fixed points, A, B and C, where $AB = x$ and $BC = x$. Given that P takes t seconds to travel from A to B, and $\frac{1}{2}t$ seconds from B to C, find, in terms of x and t, its acceleration, and the speed as it passes point A.

8 At time $t = 0$ a particle is projected vertically upwards with speed U, and at time $t = \dfrac{U}{2g}$ a second particle is projected vertically upwards from the same point with speed $2U$. Prove that the particles meet at time $t = \dfrac{3U}{4g}$.

Find the height above the point of projection at which the particles collide.

(L)

9 A bus starts from rest and moves along a straight road with constant acceleration f until its speed is V; it then continues at constant speed V. When the bus starts, a car is at a distance b behind the bus and is moving in the same direction with constant speed u. Find the distance of the car behind the bus at time t after the bus has started

(a) for $0 < t < \dfrac{V}{f}$ (b) for $t > \dfrac{V}{f}$

Show that the car cannot overtake the bus during the period $0 < t < \dfrac{V}{f}$ unless $u^2 > 2fb$.

Find the least distance between the car and the bus in the case when $u^2 < 2fb$ and $u < V$. State briefly what will happen if $u^2 < 2fb$ and $u > V$.

(JMB)

10 A particle is moving in a straight line with uniform retardation, and it passes a series of equally spaced marks along the line. The speeds of the particle at the nth and $(n + 1)$th marks are ku and u respectively, where $k > 1$. Find the speed of the particle at the $(n - 1)$th mark, and prove that its speed at the first mark is $u(k^2n - n + 1)^{\frac{1}{2}}$.

If the speed of the particle at the nth mark is the mean of its speeds at the first mark and the $(n + 1)$th mark, find a quadratic equation to determine k in terms of n, and deduce that $k = \dfrac{n}{(4 - n)}$.

(C)

11 Two particles A and B are moving in the same direction along adjacent parallel lines on horizontal ground, A with constant acceleration $4\,\mathrm{m\,s^{-2}}$ and B with constant acceleration $f\,\mathrm{m\,s^{-2}}$. At a given instant A is $2\,\mathrm{m}$ behind B; A is moving with speed $6\,\mathrm{m\,s^{-1}}$, whilst B is moving with speed $4\,\mathrm{m\,s^{-1}}$. Write down an expression for the distance between the two particles t s later.

If $f = \frac{19}{4}$, show that A passes B and then subsequently B passes A, and find the distance travelled by the particles between these two events.

If $f > 5$, show that A never catches up with B.

Describe briefly the motion if (a) $f = 4$, (b) $f = 5$.

(C)

12 A lorry starts from rest and moves with constant acceleration $\frac{1}{2}\,\mathrm{m\,s^{-2}}$. When it has travelled $100\,\mathrm{m}$, it passes a stationary car. Find the lorry's speed at that instant (*Continues over page.*)

As the lorry is passing the car, the car starts to move with constant acceleration $3 \, \mathrm{m \, s^{-2}}$ in the same direction as the lorry. Show that it passes the lorry when the lorry has travelled a total distance of 196 m.

After being in motion for 10 s, the car starts to brake with constant retardation $\frac{9}{8} \mathrm{m \, s^{-2}}$, whilst the lorry continues to accelerate as before. Write down an expression for the distance between the two vehicles after a *further* time T s, and find the value of T when the lorry passes the car again.

(C)

13 A particle moves along the positive x-axis. At time $t = 0$ it is observed to start with velocity $1 \, \mathrm{m \, s^{-1}}$ from the origin O. At time $t = 10$, it is observed to be 20 m from the origin and to have velocity $3 \, \mathrm{m \, s^{-1}}$ away from the origin. Show that these observations agree with each of the following theories:
(a) the acceleration of the particle is uniform;
(b) the acceleration at any time t is given by

$$\frac{3}{125}(t - 5)^2 \, \mathrm{m \, s^{-2}}$$

Show that if the acceleration of the particle at a distance x metres from the origin is given by $\dfrac{1}{100}(x + 10) \, \mathrm{m \, s^{-2}}$, then the particle can start with velocity of $1 \, \mathrm{m \, s^{-1}}$ at O and can have velocity $3 \, \mathrm{m \, s^{-1}}$ at a distance of 20 m from O. Show however, that it would then take almost 11 s to cover this distance.

(Hint: in the last part, use $\displaystyle\int \frac{1}{x} \, \mathrm{d}x = \ln x + c$)

(C)

14 A particle P moves along the horizontal x-axis and, at time t, its displacement is x, its velocity is v and its acceleration is $-(4v + 2)$. When $t = 0$, $x = 0$ and $v = \frac{1}{2}$.
(a) Find an expression for v in terms of t.
(b) Show that, as t increases, v approaches a limiting value, and state this value.
(c) Show that

$$\frac{\mathrm{d}x}{\mathrm{d}v} = -\frac{1}{4}\left(1 - \frac{1}{2v + 1}\right)$$

and hence, or otherwise, find an expression for x in terms of v.

(C)

15 A cyclist C moves in a straight line and has an acceleration of $0.1(10v - v^2) \, \mathrm{m \, s^{-2}}$, where $v \, \mathrm{m \, s^{-1}}$ is the speed of C. Given that, at time $t = 0$, C passes through a point O with speed $2 \, \mathrm{m \, s^{-1}}$, show that at time t seconds the speed of C is given by

$$v = \frac{10 e^t}{4 + e^t}$$

Hence, or otherwise, find the distance of C from O at time $t = 2$, giving your answer in metres, correct to 1 decimal place.

(L)

16 Two trains T_1 and T_2 are moving on perpendicular tracks which cross at the point O. Relative to O, the position vectors of T_1 and T_2 at time t are $\mathbf{r}_1$ and $\mathbf{r}_2$ respectively, where

$$\mathbf{r}_1 = Vt\mathbf{i} \quad \text{and} \quad \mathbf{r}_2 = 2V(t - t_0)\mathbf{j}$$

Here $\mathbf{i}$ and $\mathbf{j}$ are unit vectors, and V and t_0 are positive constants. Which train goes through O first, and how much later does the other train go through O?

Show that the trains are closest together when $t = 4t_0/5$, and calculate their distance apart at this time. Draw a diagram to show the positions of the trains at this time; show also the directions in which they are moving.

(MEI)

17 The vertices A, B and C of a triangle have position vectors $\mathbf{a}$, $\mathbf{b}$ and $\mathbf{c}$ respectively relative to an origin O. H is a point coplanar with A, B and C and with the position vector $\mathbf{h}$ such that $(\mathbf{a} - \mathbf{h}).(\mathbf{b} - \mathbf{c}) = 0$ and $(\mathbf{b} - \mathbf{h}).(\mathbf{c} - \mathbf{a}) = 0$. Deduce that $(\mathbf{c} - \mathbf{h}).(\mathbf{a} - \mathbf{b}) = 0$. Interpret this result geometrically.

In the case when A, B, C have coordinates (1.1), (5.5) and (3,4) respectively when referred to two-dimensional Cartesian axes, determine the coordinates of H. (JMB: part)

18 The position vectors of the points A, B, C with respect to the origin O are given by

$$\mathbf{a} = 3\mathbf{i} \quad \mathbf{b} = 2\mathbf{j} \quad \mathbf{c} = 9\mathbf{k} \quad \text{respectively.}$$

(a) Find the exact value of the cosine of angle ACB.

(b) Show that the area of triangle ACB is $16\frac{1}{2}$ units.

(c) Write down the volume of the tetrahedron OABC, and hence, or otherwise, find the perpendicular distance from O to the plane ABC. (The volume of a tetrahedron is $\frac{1}{3} \times$ base area $\times$ perpendicular height.)

(C)

19 O is the origin, A is the point $(1,3)$, B is the point $(-2,1)$ and P is the variable point given by $\overrightarrow{OP} = \overrightarrow{OA} + t\overrightarrow{AB}$. Calculate

(a) the scalar product $\overrightarrow{OA}.\overrightarrow{OB}$;

(b) the size, to the nearest degree, of the angle AOB;

(c) the value of t for which $\overrightarrow{OP}$ is perpendicular to $\overrightarrow{AB}$;

(d) the value of t for which $OP = AP$.

(C)

20 The vector x satisfies

$$\mathbf{x} + \frac{\lambda}{a^2}(\mathbf{a}.\mathbf{x})\mathbf{a} = \mathbf{b}$$

where $a \neq 0$. Given that $\lambda \neq -1$, show that

$$\mathbf{a}.\mathbf{x} = \frac{\mathbf{a}.\mathbf{b}}{1 + \lambda}$$

Hence, or otherwise, find x in terms of a, $\mathbf{a}$, $\mathbf{b}$ and λ.

In the case when $\lambda = -1$, show that either $\mathbf{b} = \mathbf{0}$, or $\mathbf{a}$ and $\mathbf{b}$ are perpendicular. (JMB)

Relative velocity

Chapter 6

Motion in two dimensions

6.1 Introduction

In this chapter we shall develop the ideas introduced in the preceding chapters. In particular, we shall use the idea that, if we can express the displacement of a point in terms of t, the time, then we can find the velocity by differentiating with respect to t, and from this we can find the acceleration by differentiating again. Conversely, if we know the acceleration in terms of time, we can work back to the velocity and then to the displacement, although in this case we shall need some further information about the system.

We shall also exploit the techniques, introduced in Chapter 5, for using vectors, and this will enable us to study displacements, velocities and accelerations in two or three dimensions.

The first task is to demonstrate what is meant by 'adding' velocities.

6.2 The addition of velocities

Imagine a man standing on the deck of an aircraft carrier which is sailing due North at $4 \, \text{m s}^{-1}$. If he stands still he will be carried 4 m northwards by the ship in one second and we could represent his displacement by the vector $A\vec{B}$ in Figure 6.1.

Figure 6.1

However if the man runs across the deck, at right angles to the direction of motion of the carrier at a speed of $3\,\mathrm{m\,s^{-1}}$, then, at the end of one second he would be at point C, 3 m to the right of B. A stationary observer would note that, in the course of the one second interval, the man had moved from point A to point C. (See Figure 6.2.)

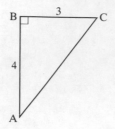

Figure 6.2

Since the same argument could be applied to any interval of time (provided the man and the ship maintain constant velocities, and the man does not fall over the side of the ship!), the observer would note that the man is moving in a straight line, with constant speed, in the direction of the vector $\overrightarrow{AC}$. The lengths of the sides of the triangle ABC represent distances travelled over a period of one second, and so we may regard the vectors $\overrightarrow{AB}$, $\overrightarrow{BC}$ and $\overrightarrow{AC}$ as representing the velocities of the ship relative to a stationary observer, the man relative to the ship and the velocity of the man relative to the stationary observer, respectively.

Using Pythagoras' theorem, in Figure 6.2, we can deduce that $AC = 5$, and that $\tan A = \frac{3}{4}$ (and hence that $A = 36.9°$). Consequently the resultant velocity of the man is $5\,\mathrm{m\,s^{-1}}$ on a bearing 037°, correct to the nearest degree.

More generally, if $\mathbf{V}$ represents the velocity of the carrier (as seen by a stationary observer) and if $\mathbf{v}$ represents the velocity of the man relative to the carrier, then the velocity of the man as seen by the stationary observer is given by the sum of the vectors $\mathbf{V}$ and $\mathbf{v}$, i.e. $\mathbf{V} + \mathbf{v}$ (see Figure 6.3).

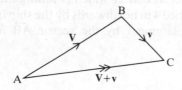

Figure 6.3

This argument can be applied to many situations involving a combination of two velocities, for example an aircraft being blown off course by the wind, or a boat being carried along by a current of water.

The lengths and angles in the vector triangle can be estimated from a scale drawing or calculated using (where appropriate) elementary trigonometry, including the sine and cosine rules.

Example 1

The pilot of a light aircraft sets a course which, in still air, would enable the aircraft to fly at $300 \, km \, h^{-1}$ on a bearing of 060°. It is blown off course by a wind whose velocity is $50 \, km \, h^{-1}$ due South. Calculate the resultant velocity.

The velocity of the aircraft relative to the ground is the vector sum of the velocity relative to the air and the wind velocity; these are represented by the vectors $\overrightarrow{AB}$ and $\overrightarrow{BC}$ respectively in Figure 6.4. The resultant velocity is represented by $\overrightarrow{AC}$.

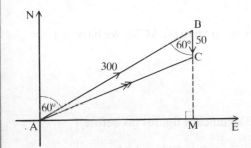

Figure 6.4

Applying the cosine rule to triangle ABC, we have

$$AC^2 = AB^2 + BC^2 - 2 \times AB \times BC \times \cos B$$
$$= 300^2 + 50^2 - 2 \times 300 \times 50 \times \cos 60°$$
$$= 90\,000 + 2500 - 15\,000$$
$$= 77\,500$$

$\therefore \quad AC = 278$ (correct to 3 significant figures)

Hence the resultant speed of the aircraft (usually called the ground speed) is $278 \, km \, h^{-1}$.

Using the sine rule in triangle ABC

$$\frac{\sin A}{BC} = \frac{\sin B}{AC}$$

$\therefore \quad \sin A = \dfrac{\sin 60°}{AC} \times 50$

$\therefore \qquad A = 9°$ (correct to the nearest degree)

Hence the resultant velocity is on a bearing of 069°.

Alternative method (not using the cosine rule)

In Figure 6.4, note that BC is produced to meet the horizontal axis at M, and that $\angle AMB = 90°$.

In triangle ABM,

$$MB = 300 \cos 60°$$
$$= 300 \times \tfrac{1}{2}$$
$$= 150$$

Hence

$$MC = 100$$

Also

$$AM = 300 \sin 60°$$
$$= 300 \times \tfrac{1}{2}\sqrt{3}$$
$$= 150\sqrt{3}$$

By applying Pythagoras' theorem to triangle ACM, we have

$$AC^2 = AM^2 + MC^2$$
$$= (150\sqrt{3})^2 + 100^2$$
$$= 67\,500 + 10\,000$$
$$= 77\,500$$
$$\therefore \quad AC = 278 \quad \text{[correct to 3 significant figures (as before)]}$$

Hence the ground speed of the aircraft is $278 \, \text{km h}^{-1}$.

Also

$$\tan \angle \, \text{CAM} = \frac{MC}{AM}$$

$$= \frac{100}{150\sqrt{3}}$$

$$\therefore \quad \angle \, \text{CAM} \quad 21.0° \quad \text{(correct to 3 significant figures)}$$

Hence the resultant velocity is on a bearing of 069° (as before).

Qu. 1 A man rows a boat at a speed of $2 \, \text{m s}^{-1}$ and he steers due East. He is carried off his intended course by a current flowing due South at $1.5 \, \text{m s}^{-1}$. Calculate the resultant velocity of the boat.

6.3 Relative velocity

In Figure 6.5, A and B are two (moving) points, whose position vectors are **a** and **b** respectively; **a** and **b** are functions of t, the time. The velocities of A and B, in the notation of Chapter 5, are $\dfrac{d\mathbf{a}}{dt}$ and $\dfrac{d\mathbf{b}}{dt}$ respectively; for convenience, these velocities will be written as $\mathbf{v}_a$ and $\mathbf{v}_b$.

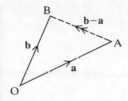

Figure 6.5

The position of B, relative to A, is given by the displacement vector $\overrightarrow{AB}$. Now

$$\overrightarrow{OA} + \overrightarrow{AB} = \overrightarrow{OB}$$
$$\therefore \qquad \overrightarrow{AB} = \overrightarrow{OB} - \overrightarrow{OA}$$
$$= \mathbf{b} - \mathbf{a}$$

The velocity of B *relative to* A (written $_b\mathbf{v}_a$) is found by differentiating the vector $\overrightarrow{AB}$ with respect to t, i.e.

$$_b\mathbf{v}_a = \frac{\mathrm{d}(\overrightarrow{AB})}{\mathrm{d}t}$$

$$= \frac{\mathrm{d}(\mathbf{b} - \mathbf{a})}{\mathrm{d}t}$$

$$= \frac{\mathrm{d}\mathbf{b}}{\mathrm{d}t} - \frac{\mathrm{d}\mathbf{a}}{\mathrm{d}t}$$

$$\therefore \qquad _b\mathbf{v}_a = \mathbf{v}_b - \mathbf{v}_a$$

Some readers may find it helpful to regard this as $\mathbf{v}_b + (-\mathbf{v}_a)$, i.e.

> the velocity of B relative to A
> = the velocity of B plus the velocity of A *reversed*

If one imagines the view from the window of a train travelling at, say 80 mph, then the surrounding countryside appears to be rushing past the observer at 80 mph in the opposite direction, so a moving object seen from this viewpoint appears to have its actual velocity, plus the velocity of the train, *reversed*.

Notice that, in Figure 6.3, **V** and **V** + **v** were the actual velocities of the aircraft carrier and the man respectively, and $(\mathbf{V} + \mathbf{v}) + (-\mathbf{V}) = \mathbf{v}$ is the velocity of the man relative to the carrier.

Example 2

A passenger on a train which is travelling due East at 100 mph, observes a car which is travelling at 60 mph along a motorway running North East. Calculate the velocity of the car relative to the observer on the train.

The data can be summarised as follows:

	Speed	Bearing	Velocity
Passenger	100	090°	$\mathbf{v}_p$
Car	60	045°	$\mathbf{v}_c$

The velocity of the car, relative to the passenger, $_c\mathbf{v}_p$, is given by

$$_c\mathbf{v}_p = \mathbf{v}_c - \mathbf{v}_p$$

and the velocity triangle is shown in Figure 6.6.

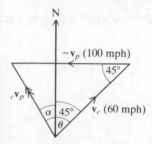

Figure 6.6

By the cosine rule

$$_cv_p{}^2 = 60^2 + 100^2 - 2 \times 60 \times 100 \times \cos 45°$$

$$\therefore \quad _cv_p = 71.5 \quad \text{(correct to 3 significant figures)}$$

(The reader is reminded that $\mathbf{v}$, in bold type, represents the velocity but v, in italic type, represents the speed.)

Using the sine rule

$$\frac{\sin \theta}{100} = \frac{\sin 45°}{_cv_p}$$

$$\therefore \quad \sin \theta = \frac{\sin 45°}{_cv_p} \times 100$$

$$\therefore \quad \theta = 81° \quad \text{(correct to the nearest degree)}$$

Hence, in the notation shown in Figure 6.6

$$\alpha = 36°$$

and so the bearing is 324°.

Hence the velocity of the car relative to the observer on the train is 71.5 mph on a bearing of 324°.

Example 3

Two points P and Q move in straight lines across the VDU screen of a microcomputer. Their position vectors at time t seconds are

$$\mathbf{p} = 5t\mathbf{i} + 30\mathbf{j}$$
$$\mathbf{q} = 3.2t\mathbf{i} + 2.4t\mathbf{j}$$

respectively, where $\mathbf{i}$ and $\mathbf{j}$ are unit vectors parallel to fixed axes on the screen. Find:

(a) vectors which represent the velocities of P and Q;
(b) the vector $_p\mathbf{V}_q$ representing the velocity of P relative to Q;
(c) the position vector of P relative to Q;
(d) the distance QP, in terms of t; hence find
   (i) the time when P and Q are nearest to each other, and
   (ii) this shortest distance.

Show also that when P is nearest to Q, the vector $\overrightarrow{QP}$ is perpendicular to the vector $_p\mathbf{V}_q$.

(a) The velocity vectors of P and Q are:

$$\mathbf{v}_p = \frac{d\mathbf{p}}{dt} = 5\mathbf{i} \quad \text{and} \quad \mathbf{v}_q = \frac{d\mathbf{q}}{dt} = 3.2\mathbf{i} + 2.4\mathbf{j}$$

(b) The velocity of P relative to Q is

$$\begin{aligned} _p\mathbf{V}_q &= \mathbf{v}_p - \mathbf{v}_q \\ &= 5\mathbf{i} - (3.2\mathbf{i} + 2.4\mathbf{j}) \\ &= 1.8\mathbf{i} - 2.4\mathbf{j} \end{aligned}$$

[Note that the answer to part (b) could be obtained by differentiating $\overrightarrow{QP}$ with respect to t.]

(c) The position vector of P relative to Q is $\overrightarrow{QP}$, where

$$\begin{aligned} \overrightarrow{QP} &= \mathbf{p} - \mathbf{q} \\ &= (5t\mathbf{i} + 30\mathbf{j}) - (3.2t\mathbf{i} + 2.4t\mathbf{j}) \\ &= 1.8t\mathbf{i} + (30 - 2.4t)\mathbf{j} \end{aligned}$$

(d) The distance between P and Q is QP. This can be obtained from $\overrightarrow{QP}$ by Pythagoras' theorem as follows:

$$\begin{aligned} QP^2 &= (1.8t)^2 + (30 - 2.4t)^2 \\ &= 3.24t^2 + 900 - 144t + 5.76t^2 \\ &= 9t^2 - 144t + 900 \\ &= 9(t^2 - 16t + 100) \end{aligned}$$

and, completing the square,†

$$QP^2 = 9[(t - 8)^2 + 36]$$

---

† Readers who are unfamiliar with the technique of 'completing the square' can obtain this result by calculus.

Since $(t - 8)$ is squared it cannot be negative, so its least value is zero, and this occurs when $t = 8$. Hence the least value of QP^2 is $9 \times 36 = 324$, and so the least value of the distance QP is 18 units. Hence

(i)   Q and P are nearest when $t = 8$, and
(ii) this shortest distance is 18 units.

From (c), $\overrightarrow{QP} = 1.8t\mathbf{i} + (30 - 2.4t)\mathbf{j}$, so when $t = 8$

$$\overrightarrow{QP} = 14.4\mathbf{i} + 10.8\mathbf{j}$$

and the scalar product of $\overrightarrow{QP}$ and $_p\mathbf{V}_q$ is

$$(14.4\mathbf{i} + 10.8\mathbf{j}) \cdot (1.8\mathbf{i} - 2.4\mathbf{j}) = 14.4 \times 1.8 - 10.8 \times 2.4$$
$$= 25.92 - 25.92$$
$$= 0$$

Since their scalar product is zero, the vectors $\overrightarrow{QP}$ and $_p\mathbf{V}_q$ are perpendicular, when $t = 8$.

(The reader should try to explain the answer to the last part of Example 3, by referring to the path of P relative to Q.)

Exercise 6a

1 A motor boat travels at $5\,\mathrm{m\,s^{-1}}$ relative to the water and the driver steers due North. Find the resultant velocity in the following currents:
(a) due North at $2\,\mathrm{m\,s^{-1}}$ (b) due South at $2\,\mathrm{m\,s^{-1}}$
(c) due East at $4\,\mathrm{m\,s^{-1}}$ (d) North East at $4\,\mathrm{m\,s^{-1}}$

2 A car is travelling at 50 mph when a passenger throws an object horizontally from it, at right angles to the direction in which the car is travelling, with an initial speed of 20 mph. Find the actual velocity of the object at the moment when it is projected.

3 An aircraft takes off with sufficient fuel for a 2 hour flight. Its air speed is 200 mph and it is flying against a constant headwind of 40 mph. Calculate the distance to its 'point of no return', that is the furthest distance the aircraft can fly and be able to return to its starting point.

4 Repeat question 3, using
flying time $= T$ hours
air speed of aircraft $= V$ mph
wind speed $= w$ mph

5 Two cars which are travelling at 40 mph and 30 mph, respectively, collide; find the relative speed when they collide if they were travelling
(a) in the same direction (with the slower car in front);
(b) in opposite directions;
(c) at right angles.

6 (In this question all distances are in km, times are in hours and the speeds in $\mathrm{km\,h^{-1}}$.)
A ship P sets out from a port whose position relative to a set of axes is (5,0), and after t hours its position vector is $\mathbf{p} = 5\mathbf{i} + 3t\mathbf{j}$. A second ship Q sets out,

at the same time as P, from a port whose coordinates are (0,3) and its position vector after t hours is $\mathbf{q} = 3\mathbf{j} + t(\mathbf{i} + \mathbf{j})$.

Find, in terms of $\mathbf{i}$ and $\mathbf{j}$, the vectors which represent

(a) the velocity of P;

(b) the velocity of Q;

(c) the displacement of Q from P after t hours;

(d) the velocity of Q relative to P.

7 The pilot of an airliner flying at 300 knots on a bearing of 050°, spots a light aircraft whose ground speed is 200 knots on a bearing of 120°. Find the velocity of the light aircraft relative to the airliner.

8 A passenger on a train which is travelling due South at 50 mph, observes a car which appears to be travelling at 40 mph due East. Find the actual velocity of the car.

9 A parachutist, falling vertically at $10 \, \mathrm{m \, s^{-1}}$, observes another who appears to be moving at $2 \, \mathrm{m \, s^{-1}}$ along a path inclined at 20° above the horizontal (relative to the first parachutist). Find the velocity of the second parachutist.

10 A cat and a mouse are in a room which is 20 m long and 10 m wide (see Figure 6.7).

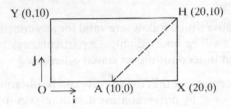

Figure 6.7

Initially the mouse is at the point A whose coordinates are (10,0) and the cat is at Y (0,10). The mouse runs at $2 \, \mathrm{m \, s^{-1}}$, in a straight line, heading for its hole H, which is at the point (20,10). When the mouse starts to run, the cat also sets off with a speed of $V \, \mathrm{m \, s^{-1}}$ along the wall YH. Show that the position vectors of the cat $\mathbf{c}$ and the mouse $\mathbf{m}$ after t seconds are

$$\mathbf{c} = Vt\mathbf{i} + 10\mathbf{j}$$
$$\mathbf{m} = 10\mathbf{i} + (\sqrt{2}\mathbf{i} + \sqrt{2}\mathbf{j})t$$

Given that the cat and the mouse arrive at the hole simultaneously, prove that $t = 5\sqrt{2}$, and find the value of V. Write down, in terms of $\mathbf{i}$ and $\mathbf{j}$, an expression for the velocity of the cat relative to the mouse.

6.4 Motion in two dimensions with constant acceleration

A particle moves in two dimensions, and its velocity, $\mathbf{v} \, \mathrm{m \, s^{-1}}$, after t seconds is shown in the following table:

t	0	1	2	3	4	5
$\mathbf{v}$	$50\mathbf{i}$	$50\mathbf{i} + 10\mathbf{j}$	$50\mathbf{i} + 20\mathbf{j}$	$50\mathbf{i} + 30\mathbf{j}$	$50\mathbf{i} + 40\mathbf{j}$	$50\mathbf{i} + 50\mathbf{j}$

From the table we see that the initial velocity was $50\,\mathrm{m\,s^{-1}}$, parallel to the x-axis and that, in each interval of 1 second, a velocity of $10\,\mathrm{m\,s^{-1}}$, parallel to the y-axis is added to this. In other words, the particle is moving with a constant acceleration of $10\mathbf{j}\,\mathrm{m\,s^{-2}}$; as long as this is maintained the velocity after t seconds will be $50\mathbf{i} + 10t\mathbf{j}$.

More generally, if the initial velocity is $\mathbf{u}\,\mathrm{m\,s^{-1}}$, and if the particle moves with a constant acceleration of $\mathbf{a}\,\mathrm{m\,s^{-2}}$ (i.e. in each second a velocity of $\mathbf{a}\,\mathrm{m\,s^{-1}}$ is added), then the velocity $\mathbf{v}\,\mathrm{m\,s^{-1}}$ after t seconds is given by

$$\mathbf{v} = \mathbf{u} + \mathbf{a}t$$

This result is illustrated by the vector triangle in Figure 6.8.

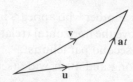

Figure 6.8

[As before (see §3.1), this result, and those which follow, are valid for any consistent set of units, consequently the units will be omitted unless a particular system is required; time is normally measured in seconds unless stated otherwise.]

From earlier chapters, the reader will remember that the velocity can be obtained from the function giving the displacement, by differentiating it with respect to t, the time; the acceleration, in turn, can be obtained by differentiating the velocity. Conversely, if the acceleration is given, the velocity and displacement can be found by integration. Using the notation

t = time (usually in seconds)
$\mathbf{u}$ = velocity when $t = 0$
$\mathbf{v}$ = velocity at time t
$\mathbf{a}$ = the (constant) acceleration
$\mathbf{r}$ = displacement

we can write

$$\frac{d\mathbf{v}}{dt} = \mathbf{a}$$

and integrating with respect to t

$$\mathbf{v} = \mathbf{a}t + \mathbf{c}$$

where $\mathbf{c}$ is a (vector) constant of integration.

When $t = 0$, the velocity is $\mathbf{u}$, so $\mathbf{c} = \mathbf{u}$.

Therefore

$$\mathbf{v} = \mathbf{u} + \mathbf{a}t$$

. . . . (1)

(This is the formula derived intuitively earlier.) Writing $\mathbf{v}$ as $\dfrac{d\mathbf{r}}{dt}$ we have

$$\frac{d\mathbf{r}}{dt} = \mathbf{u} + \mathbf{a}t$$

and integrating this with respect to t gives

$$\mathbf{r} = \mathbf{u}t + \tfrac{1}{2}\mathbf{a}t^2 \qquad\qquad \text{. . . (2)}$$

where $\mathbf{r}$ is the displacement from the origin at time t (and hence $\mathbf{r} = \mathbf{0}$ when $t = 0$, i.e. the constant of integration is zero).

As in Chapter 3, we can eliminate $\mathbf{u}$ from equations (1) and (2), giving

$$\begin{aligned}
\mathbf{r} &= (\mathbf{v} - \mathbf{a}t)t + \tfrac{1}{2}\mathbf{a}t^2 \\
&= \mathbf{v}t - \tfrac{1}{2}\mathbf{a}t^2
\end{aligned} \qquad\qquad \text{. . . (3)}$$

Also, eliminating $\mathbf{a}$ from equations (2) and (3), by adding, we obtain

$$\begin{aligned}
2\mathbf{r} &= \mathbf{u}t + \mathbf{v}t \\
&= (\mathbf{u} + \mathbf{v})t \\
\therefore \qquad \mathbf{r} &= \tfrac{1}{2}(\mathbf{u} + \mathbf{v})t
\end{aligned} \qquad\qquad \text{. . . (4)}$$

Qu. 2 (Before attempting this question, the reader is advised to revise the proof of equation (5) in §3.2 and the scalar product §5.7.)

Make $\mathbf{a}$ the subject of the formula $\mathbf{v} = \mathbf{u} + \mathbf{a}t$, and, by considering the scalar product $\mathbf{a} \cdot \mathbf{r}$, prove that

$$v^2 = u^2 + 2(\mathbf{a} \cdot \mathbf{r})$$

Summary

The equations of motion, in vector notation, for a particle moving with constant acceleration are as follows:

$$\mathbf{v} = \mathbf{u} + \mathbf{a}t$$
$$\mathbf{r} = \mathbf{u}t + \tfrac{1}{2}\mathbf{a}t^2$$
$$\mathbf{r} = \mathbf{v}t - \tfrac{1}{2}\mathbf{a}t^2$$
$$\mathbf{r} = \tfrac{1}{2}(\mathbf{u} + \mathbf{v})t$$
$$v^2 = u^2 + 2(\mathbf{a} \cdot \mathbf{r})$$

Note that these equations may be used for motion in *three* dimensions.

Example 4

A particle, moving in two dimensions, has initial velocity $(6\mathbf{i} + 8\mathbf{j})$ and constant acceleration $-2\mathbf{j}$. Write down the vectors which represent its velocity and acceleration after t seconds. Find also
(a) the time when it is moving parallel to the x-axis;
(b) the time(s) when its path intersects the x-axis.

Using $\mathbf{v} = \mathbf{u} + \mathbf{a}t$, the velocity $\mathbf{v}$ after t seconds is given by

$$\mathbf{v} = (6\mathbf{i} + 8\mathbf{j}) + (-2\mathbf{j})t$$
$$= 6\mathbf{i} + (8 - 2t)\mathbf{j} \qquad \qquad \ldots \quad (1)$$

The displacement, at time t, can be found from $\mathbf{r} = \mathbf{u}t + \frac{1}{2}\mathbf{a}t^2$:

$$\mathbf{r} = (6\mathbf{i} + 8\mathbf{j})t + \frac{1}{2}(-2\mathbf{j})t^2$$
$$\therefore \quad \mathbf{r} = 6t\mathbf{i} + (8t - t^2)\mathbf{j} \qquad \qquad \ldots \quad (2)$$

(a) The particle is moving parallel to the x-axis when the velocity is a multiple of $\mathbf{i}$ (and and coefficient of $\mathbf{j}$ is zero). Hence from (1)

$$8 - 2t = 0$$
$$2t = 8$$
$$\therefore \qquad t = 4$$

The particle is moving parallel to the x-axis when $t = 4$.

(b) The path of the particle intersects the x-axis when the displacement, $\mathbf{r}$, is a multiple of $\mathbf{i}$. Hence from (2)

$$8t - t^2 = 0$$
$$t(8 - t) = 0$$
$$\therefore \qquad t = 0 \text{ or } 8$$

The path of the particle intersects the x-axis when $t = 0$ and again when $t = 8$.

Qu. 3 A particle is moving in two dimensions with constant acceleration. It has initial velocity $(3\mathbf{i} + 4\mathbf{j})$, and when $t = 5$ its velocity is $(3\mathbf{i} - \mathbf{j})$. Find
(a) the acceleration;
(b) the change in the displacement, over the five seconds from $t = 0$ to $t = 5$;
(c) the speed when $t = 10$.

6.5 General motion in two dimensions

In the last section we developed the equations which can be used for motion in two (or three) dimensions with *constant* acceleration. If the acceleration is not constant these equations are not valid; however the techniques of integration and differentiation may still be used to relate displacement, velocity and acceleration, as in the next example.

Example 5

The displacement, $\mathbf{r}$, from the origing, of a particle moving in two dimensions, is given by

$$\mathbf{r} = 10t\mathbf{i} + \left(\frac{16}{t + 1}\right)\mathbf{j}$$

Find the vectors which represent velocity and acceleration when $t = 3$.

At time t,

$$\mathbf{r} = 10t\mathbf{i} + \left(\frac{16}{t+1}\right)\mathbf{j}$$

To find the velocity, we differentiate with respect to t, i.e.

$$\mathbf{v} = \frac{d\mathbf{r}}{dt} = 10\mathbf{i} - \frac{16}{(t+1)^2}\mathbf{j} \qquad \qquad \ldots \ (1)$$

and to find the acceleration, we differentiate the velocity, $\mathbf{v}$, with respect to t, i.e.

$$\mathbf{a} = \frac{d\mathbf{v}}{dt} = \frac{32}{(t+1)^3}\mathbf{j} \qquad \qquad \ldots \ (2)$$

To find the velocity and acceleration when $t = 3$, we substitute $t = 3$ in equations (1) and (2), giving

$$\mathbf{v} = 10\mathbf{i} - \mathbf{j}$$

and

$$\mathbf{a} = \tfrac{1}{2}\mathbf{j}$$

Hence when $t = 3$, the velocity is $(10\mathbf{i} - \mathbf{j})$ and the acceleration is $\tfrac{1}{2}\mathbf{j}$.

Suppose a particle moves along a curve in a two-dimensional plane and that at time t it is at the point P with coordinates (x,y). Then its position vector $\mathbf{r}$ is

$$\mathbf{r} = x\mathbf{i} + y\mathbf{j}$$

Its velocity, $\mathbf{v}$, at this instant, is given by

$$\mathbf{v} = \frac{d\mathbf{r}}{dt} = \frac{dx}{dt}\mathbf{i} + \frac{dy}{dt}\mathbf{j}$$

In other words, its velocity is the vector sum of velocities $\dfrac{dx}{dt}$ and $\dfrac{dy}{dt}$ parallel to the x-axis and y-axis, respectively; see Figure 6.9.

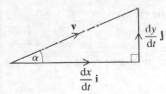

Figure 6.9

In Figure 6.9

$$\tan \alpha = \frac{dy}{dt} \div \frac{dx}{dt}$$

$$= \frac{dy}{dx}$$

Now $\dfrac{dy}{dx}$ is the gradient of the tangent, at the point P, to the curve along which the particle is moving; see Figure 6.10.

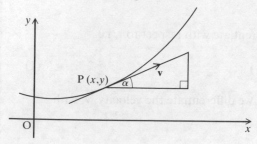

Figure 6.10

Hence, not very surprisingly, the velocity is directed along the path of the particle. The same cannot, however, be said about the acceleration, as the next two examples illustrate.

Example 6

The acceleration of a particle, moving in a two-dimensional plane, is $-10\mathbf{j}$. Given that the initial velocity is $20\mathbf{i}$ and that the initial displacement is $40\mathbf{j}$, find the velocity and displacement in the subsequent motion, in terms of t. Sketch the path of the particle, for $t = 0$ to 3, indicating the directions of the velocity and acceleration when $t = 2$.

Let the velocity at time t be $\mathbf{v}$, then

$$\frac{d\mathbf{v}}{dt} = -10\mathbf{j}$$

Integrating with respect to t,

$$\mathbf{v} = -10t\mathbf{j} + \mathbf{c}$$

We know that when $t = 0$, $\mathbf{v} = 20\mathbf{i}$, so $\mathbf{c} = 20\mathbf{i}$. Hence

$$\mathbf{v} = 20\mathbf{i} - 10t\mathbf{j}$$

To find the displacement from the origin, $\mathbf{r}$, in terms of t, we integrate again.

$$\mathbf{r} = 20t\mathbf{i} - 5t^2\mathbf{j} + \mathbf{c'}$$

When $t = 0$, $\mathbf{r} = 40\mathbf{j}$, so $\mathbf{c'} = 40\mathbf{j}$. Hence

$$\begin{aligned}\mathbf{r} &= 20t\mathbf{i} - 5t^2\mathbf{j} + 40\mathbf{j}\\ &= 20t\mathbf{i} + (40 - 5t^2)\mathbf{j}\end{aligned}$$

So, if the particle is at P (x,y) at time t, then

$$x = 20t \quad \text{and} \quad y = 40 - 5t^2$$

The coordinates of P, for $t = 0, 1, 2,$ and 3, are shown in the table below:

t	0	1	2	3
x	0	20	40	60
y	40	35	20	-5

When $t = 2$, the acceleration $\mathbf{a} = -10\mathbf{j}$ and the velocity $\mathbf{v} = 20\mathbf{i} - 20\mathbf{j}$.

A sketch of the path is shown in Figure 6.11, showing the directions of the velocity and the acceleration when $t = 2$.

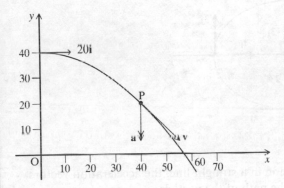

Figure 6.11

Example 7

The position vector, $\mathbf{r}$, of a particle moving in a plane, is

$$\mathbf{r} = (4 \cos t)\mathbf{i} + (3 \sin t)\mathbf{j}$$

Find its velocity and acceleration in terms of t.

Sketch the path of the particle and indicate the directions of the velocity and acceleration when $t = 0, \pi/2, \pi$ and $3\pi/2$.

In the usual notation

$$\mathbf{r} = (4 \cos t)\mathbf{i} + (3 \sin t)\mathbf{j}$$

Differentiating with respect to t,

$$\mathbf{v} = -(4 \sin t)\mathbf{i} + (3 \cos t)\mathbf{j}$$

and differentiating again,

$$\mathbf{a} = -(4 \cos t)\mathbf{i} - (3 \sin t)\mathbf{j}$$

(Notice that $\mathbf{a} = -\mathbf{r}$ in this particular example.)

When $t = 0$, $\mathbf{r} = 4\mathbf{i}$, $\mathbf{v} = 3\mathbf{j}$ and $\mathbf{a} = -4\mathbf{i}$.

When $t = \pi/2$, $\mathbf{r} = 3\mathbf{j}$, $\mathbf{v} = -4\mathbf{i}$ and $\mathbf{a} = -3\mathbf{j}$.

When $t = \pi$, $\mathbf{r} = -4\mathbf{i}$, $\mathbf{v} = -3\mathbf{j}$ and $\mathbf{a} = 4\mathbf{i}$.

When $t = 3\pi/2$, $\mathbf{r} = -3\mathbf{j}$, $\mathbf{v} = 4\mathbf{i}$ and $\mathbf{a} = 3\mathbf{j}$.

The path of the particle is shown below (Figure 6.12); the single-headed arrows indicate the directions of the velocities and the double-headed ones indicate the directions of the accelerations (the lengths of the arrows are not drawn to scale, and so they have no significance).

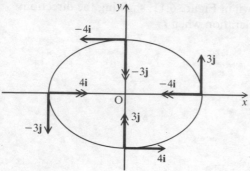

Figure 6.12

Notice that, unlike motion in a straight line, the acceleration vectors are not directed along the path of the particle.

Exercise 6b

Throughout this exercise, it should be assumed that the units of length, speed and acceleration are m, m s^{-1} *and* m s^{-2}, *respectively.*

1 In this question the abbreviations listed in §6.4 are used. Given that:
   (a) $\mathbf{u} = 5\mathbf{i} + 3\mathbf{j}$, $\mathbf{a} = 4\mathbf{j}$, $t = 2$, find $\mathbf{v}$;
   (b) $\mathbf{v} = 7\mathbf{i} + \mathbf{j}$, $\mathbf{a} = -10\mathbf{j}$, $t = 2$, find $\mathbf{r}$;
   (c) $\mathbf{u} = 2\mathbf{i} + 3\mathbf{j}$, $\mathbf{v} = 10\mathbf{i} + 8\mathbf{j}$, $t = 4$, find $\mathbf{r}$.
2 A projectile is fired in a vertical plane from a point O; its initial velocity is 100 m s^{-1} at an angle of 30° to the horizontal. Express this velocity in the form $u\mathbf{i} + v\mathbf{j}$, where $\mathbf{i}$ and $\mathbf{j}$ are unit vectors in the horizontal and vertical directions respectively.

The projectile is subject to a constant acceleration of 10 m s^{-2} vertically downwards; express this in the form $\alpha\mathbf{j}$. Hence write down the position vector, $\mathbf{r}$, relative to the initial point O, of the projectile after t seconds. Find the time which elapses before it strikes the horizontal plane through O. Find also the distance of the projectile from O at this instant.

3 A particle moves in a two-dimensional plane and its position vector, $\mathbf{r}$, after t seconds, is given by

$$\mathbf{r} = (t + \sin t)\mathbf{i} + (1 - \cos t)\mathbf{j}$$

Find its speed in terms of t, and find the values of t between 0 and 4π for which the particle is momentarily at rest.

4 A particle moves in a two-dimensional plane and its position vector, $\mathbf{r}$, at time t, is

$$\mathbf{r} = (5 \cos 2t)\mathbf{i} + (5 \sin 2t)\mathbf{j}$$

Show that

(a) the path of the particle is a circle, centre O, of radius 5 units;
(b) its speed is constant;
(c) its acceleration is $-4\mathbf{r}$.

What can be deduced from (c) about the direction of the acceleration?

5 A particle moves in a two-dimensional plane and its position vector, $\mathbf{r}$, after t seconds is

$$\mathbf{r} = (1 + \cos t)\mathbf{i} + (\sin t)\mathbf{j}$$

Show that

(a) its speed is constant;
(b) its acceleration, $\mathbf{a}$, satisfies the equation $\mathbf{r} + \mathbf{a} = \mathbf{i}$.

What can be deduced from (b) about the direction of the acceleration?

Exercise 6c (Miscellaneous)

1 A boat heads due North, with its engine set so that it would travel at 6 knots in still water. It is carried off course by a current, and after 2 hours it has travelled 10 nautical miles along a course with bearing 010°. (1 knot = 1 nautical mile per hour.) Find the velocity of the current.

2 An airliner is flying horizontally at 480 mph on a bearing of 090°, when its pilot spots a military jet fighter, flying at the same altitude. The fighter is headed on a course with bearing 045°, but to the pilot of the airliner, it appears to be on 330°. Calculate the speed of the fighter.

3 At noon, two boats P and Q are at points whose position vectors are $4\mathbf{i} + 8\mathbf{j}$ and $4\mathbf{i} + 3\mathbf{j}$ respectively. Both boats are moving with constant velocity; the velocity of P is $4\mathbf{i} + \mathbf{j}$ and that of Q is $2\mathbf{i} + 5\mathbf{j}$. (All distances are measured in kilometres, and the time is in hours.)

What is the velocity of Q relative to P? Draw a diagram showing the initial positions of P and Q, and the path of Q relative to P; on it indicate the point where Q is nearest to P and hence calculate the shortest distance between the boats.

4 A particle is moving in a plane and its position vector, $\mathbf{r}$, at time t is

$$\mathbf{r} = (25t - 4t^2)\mathbf{i} + (50 - 2t^2)\mathbf{j}$$

Sketch its path for $t = 0$ to 5, and find its velocity and acceleration in terms of t. Indicate on the sketch the directions of the velocity and acceleration when $t = 4$.

5 A particle is moving in three dimensions and its position vector at time t is

$$\mathbf{r} = (\cos t)\mathbf{i} + (\sin t)\mathbf{j} + t\mathbf{k}$$

(where the unit of length is one metre).
Show that
(a) its distance from the z-axis is constant, and find this distance;
(b) its speed is constant and equal to $\sqrt{2}\,\mathrm{m\,s^{-1}}$;
(c) the acceleration is always horizontal and constant.
Describe the motion in words.

6 The coordinates of a particle P, moving in two dimensions are $(2 \sin t, \sin 2t)$. Find its velocity at time t. Sketch its path for $t = 0$ to 2π. (Note that if a calculator is used to find the coordinates, it must be in radian mode.) When is the velocity parallel to the x-axis? When is the acceleration parallel to the x-axis? Indicate on the sketch the directions of the velocity and acceleration when $t = \pi/4$.

7 A stone is projected horizontally from a point A, with an initial velocity of $u\mathbf{i}\,\mathrm{m\,s^{-1}}$. While it is in the air its acceleration is $-10\mathbf{j}\,\mathrm{m\,s^{-2}}$. Write down its position vector after t seconds, in terms of u, t, $\mathbf{i}$ and $\mathbf{j}$. After 4 seconds it is observed to strike a point B, whose angle of depression from A is 40°. Calculate the initial speed u.

8 A canoeist who paddles at 5 knots in still water wishes to cross an estuary straight from A to B and then straight back again from B to A. There is a tide flowing at 4 knots from West to East (assume that all the water is moving with this velocity) and the bearing of A from B is 038°.

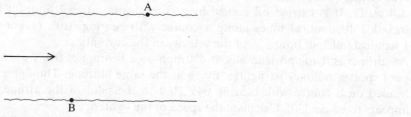

Figure 6.13

With a scale of 2 cm to 1 knot, draw a velocity diagram for each part of the journey, and find by measurement the resultant velocity of the canoeist in each case.

*(SMP)

9 An UFO is observed to leave a tower from a height of 80 metres and to move so that its displacement t seconds later is given by $3t\mathbf{i} + 4t\mathbf{j} + 5t^2\mathbf{k}$, where $\mathbf{i}, \mathbf{j}$, $\mathbf{k}$ are unit vectors West, North and vertically downwards.
(a) State its velocity after t seconds.
(b) Show that initially it is travelling horizontally and calculate its bearing.

(c) Show that its speed after 2 seconds is $5\sqrt{17}\,\mathrm{m\,s^{-1}}$.

(d) Calculate its distance from its starting point after 2 seconds.

(e) Find when it hits the ground.

*(SMP)

10 A particle initially at rest at the point (2,2) has acceleration $\begin{pmatrix} t \\ 3 \end{pmatrix}$ in $\mathrm{m\,s^{-2}}$ after t seconds.

Find vector expressions for its velocity and position after t seconds. Find its change in position in the first $\frac{1}{100}$ second.

After how many seconds is it moving in a direction inclined at 45° to the x-axis?

*(O & C)

11 A ship P is travelling due East at $30\,\mathrm{km\,h^{-1}}$ and a ship Q is travelling due South at $40\,\mathrm{km\,h^{-1}}$. Both ships keep constant speed and course. At noon they are each 10 km from the point of intersection, O, of their courses, and moving towards O.

Find the coordinates, with respect to axes Ox Eastwards and Oy Northwards, of P and Q at time t hours past noon, and find the distance PQ at this time.

Find the time at which P and Q are closest to one another.

Find the magnitude and direction of the velocity of Q relative to P, indicating the direction on a diagram.

Show that, at the position of closest approach, the bearing of Q from P is $\theta°$ South of East, where $\tan\theta = \frac{3}{4}$.

(JMB)

12 An aircraft takes off from the end of a runway in a Southerly direction and climbs at an angle of $\tan^{-1}(\frac{1}{2})$ to the horizontal at a speed of $225\sqrt{5}\,\mathrm{km\,h^{-1}}$. Show that t seconds after take-off the position vector $\mathbf{r}$ of the aircraft with

respect to the end of the runway is given by $\mathbf{r} = \dfrac{t}{16}\,(2\mathbf{i} + \mathbf{k})$ where $\mathbf{i}$, $\mathbf{j}$, $\mathbf{k}$

represent vectors of length 1 km in directions South, East and vertically upwards.

At time $t = 0$ a second aircraft flying horizontally South-West at a speed of $720\sqrt{2}\,\mathrm{km\,h^{-1}}$ has position vector $-1.2\mathbf{i} + 3.2\mathbf{j} + \mathbf{k}$. Find its position vector at time t in terms of $\mathbf{i}$, $\mathbf{j}$, $\mathbf{k}$ and t. Show that there will be a collision unless courses are changed and state at what time it will occur.

(SMP)

13 A ship leaves a port at noon (12.00 hours) and steams at $12\,\mathrm{km\,h^{-1}}$ due North. At 12.30 hours a second ship leaves the same port and steams at $18\,\mathrm{km\,h^{-1}}$ on a bearing N 30° E. Using a graphical method, or otherwise, find the least distance, to the nearest $\frac{1}{10}$ km, between the ships (after both have left the port) and the time to the nearest minute at which this occurs.

(O & C)

14 A ship A is travelling on a course of 060° at a speed of $30\sqrt{3}$ km h^{-1} and a ship B is travelling on a course of 030° at 20 km h^{-1}. At noon B is 260 km due East of A. Using unit base vectors **i** and **j** pointing East and North respectively find $\overrightarrow{AB}$ in terms of **i**, **j** and t at time t hours in the afternoon. Hence, or otherwise, calculate the least distance between A and B, to the nearest kilometre, and the time at which they are nearest to one another.

(L)

15 A wide river flows from North to South with a speed of 3 m s^{-1}. A boat P crossing the river moves from West to East with a speed of 4 m s^{-1}. Using the West–East direction and the South–North direction as your coordinate axes Ox and Oy respectively, write down in vector form the velocities of the river and the boat.

Find the velocity and speed of the boat relative to the river.

A second boat Q has a velocity $(3\mathbf{i} + 7\mathbf{j})$ relative to the river. Calculate the velocity of P relative to Q and find the cosine of the angle between the velocity of Q and the velocity of P.

Initially both boats are on the West bank of the river with Q being 40 m to the South of P. They wish to meet on the river and both boats must sail with the velocities given above. Show that P must sail 2.5 s after Q.

(L)

Chapter 7

Projectiles

7.1 Introduction

In this chapter we shall study the motion of a projectile moving in two dimensions, under the influence of gravity. The effect of gravity is to produce an acceleration, vertically downwards, which is usually denoted by g (as in §3.3). The numerical value of g is approximately $9.81 \, \mathrm{m \, s^{-2}}$, although it is frequently convenient to use $10 \, \mathrm{m \, s^{-2}}$, since this simplifies the arithmetic. We may regard g as constant, provided we are dealing with short range trajectories near the Earth's surface.

The effect of air-resistance will be ignored, since this requires mathematical techniques which are beyond the scope of this book. This is a reasonable approximation for aerodynamic projectiles moving with fairly low speeds.

7.2 Projectiles

The reader is reminded that if a vector $\mathbf{F}$, of magnitude F, is inclined at an angle α to the horizontal, then it can be expressed in terms of the unit vectors $\mathbf{i}$ and $\mathbf{j}$, as follows (see Chapter 5):

$$\mathbf{F} = (F\cos\alpha)\mathbf{i} + (F\sin\alpha)\mathbf{j}$$

Also, note that if two vectors $\mathbf{p} = x\mathbf{i} + y\mathbf{j}$ and $\mathbf{q} = x'\mathbf{i} + y'\mathbf{j}$ are equal, then $x = x'$ and $y = y'$.

Suppose that a particle is projected with an initial velocity $u \, \mathrm{m \, s^{-1}}$, at an angle α with the horizontal. Then, neglecting air-resistance, it moves in a vertical plane with a vertical acceleration g (downwards). Using the notation of Chapter 5, we have

$$\mathbf{u} = (u\cos\alpha)\mathbf{i} + (u\sin\alpha)\mathbf{j}$$

and $\mathbf{a} = -g\mathbf{j}$

Hence, using $\mathbf{r} = \mathbf{u}t + \frac{1}{2}\mathbf{a}t^2$, its displacement, t seconds after projection is given by

$$\mathbf{r} = [(u\cos\alpha)\mathbf{i} + (u\sin\alpha)\mathbf{j}]t + \frac{1}{2}(-g\mathbf{j})t^2$$

99

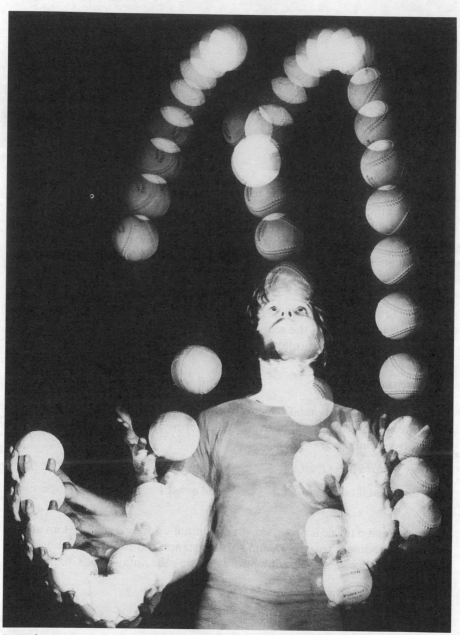

Projectiles

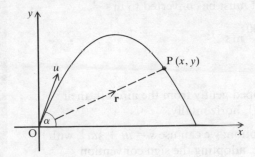

Figure 7.1

Writing $\mathbf{r} = x\mathbf{i} + y\mathbf{j}$, (see Figure 7.1) we have

$$x\mathbf{i} + y\mathbf{j} = (u\cos\alpha)t\mathbf{i} + (u\sin\alpha)t\mathbf{j} - \tfrac{1}{2}gt^2\mathbf{j}$$

Equating the horizontal components gives

$$x = (u\cos\alpha)t \qquad \qquad \cdots \quad (1)$$

and equating the vertical components gives

$$y = (u\sin\alpha)t - \tfrac{1}{2}gt^2 \qquad \qquad \cdots \quad (2)$$

The horizontal and vertical components of the velocity at time t can be found by differentiating, i.e.

$$\frac{\mathrm{d}x}{\mathrm{d}t} = u\cos\alpha \qquad \qquad \cdots \quad (3)$$

and

$$\frac{\mathrm{d}y}{\mathrm{d}t} = u\sin\alpha - gt \qquad \qquad \cdots \quad (4)$$

(Alternatively, these may be obtained from the formula $\mathbf{v} = \mathbf{u} + \mathbf{a}t$.)

In most questions on projectiles, the reader will find it convenient to start by writing any of the equations (1) to (4) which may be required, by applying the appropriate constant acceleration formulae to the horizontal and vertical motion. In the examples which follow, T is used to denote the 'time of flight', corresponding to the range on a horizontal plane.

Example 1

An aircraft delivering food parcels releases them from a height of 125 m, while it is flying horizontally at $180\,\mathrm{km\,h^{-1}}$. Calculate the horizontal distance from the target area, (i.e. the point where they first strike the ground) at which the parcels should be dropped. (Take g to be $10\,\mathrm{m\,s^{-2}}$.) Find also the velocity of the parcels when they land.

First, the speed of the aircraft must be converted to $m\,s^{-1}$:

$$180\,km\,h^{-1} = \frac{180 \times 1000}{3600}\,m\,s^{-1}$$

$$= 50\,m\,s^{-1}$$

Provided the parcels are dropped gently from the aircraft, their initial velocity will be $50\,m\,s^{-1}$, horizontally.

Next, consider the vertical motion: we can use $s = ut + \frac{1}{2}at^2$, with $u = 0$, $a = 10$ and $s = 125$ (i.e. adopting the sign convention 'downwards is positive').

$$125 = \tfrac{1}{2} \times 10 \times t^2$$
$$= 5t^2$$
$$\therefore \quad t^2 = 25$$
$$t = 5$$

Hence, the parcels take 5 s to reach the ground.

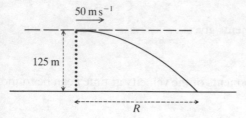

Figure 7.2

Now consider the horizontal motion: if we ignore air-resistance the horizontal component of the velocity is constant; consequently the horizontal displacement, R m, is given by

$$R = 50 \times 5$$
$$= 250$$

Hence the parcels travel 250 m horizontally while they are falling.

The vertical component of the velocity can be obtained from $v = u + at$, with $u = 0$, $a = 10$ and $t = 5$, i.e.

$$v = 10 \times 5$$
$$= 50$$

Hence the vertical component of the velocity is $50\,m\,s^{-1}$, and we know that the horizontal component is also $50\,m\,s^{-1}$, so the resultant velocity is

$$\sqrt{(50^2 + 50^2)} = 70.7\,m\,s^{-1} \quad \text{(correct to 3 significant figures)}$$

and the direction of motion at this instant will be at 45° to the horizontal.

In Example 1 it was convenient to use the convention 'downwards is positive', but in the next three examples *upwards is positive* will be more suitable.

Example 2

A projectile is fired from a point O on a horizontal plane with an initial velocity of $20\,\text{m s}^{-1}$ at an angle of 40° to the horizontal, and it lands at a point P on the plane. Calculate the length OP (i.e. the range).

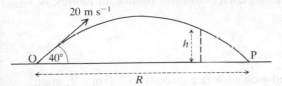

Figure 7.3

The initial velocity has components $20\cos 40°$ horizontally and $20\sin 40°$ vertically.

Consider the vertical motion:
the height, h, metres, at time t seconds after the projectile is fired, can be found from $s = ut + \frac{1}{2}at^2$, i.e.

$$h = (20\sin 40°)t - \tfrac{1}{2}gt^2 \qquad\qquad \cdot\;\cdot\;\cdot \;\;(1)$$

(Note that the acceleration is $-g$ because we are using the 'upwards is positive' convention; the reader is advised to leave g in the equation and only substitute $g = 9.81\,\text{m s}^{-2}$ when a calculation is essential.)

In this question we need to know the range, OP; now the height of P above O is zero, so we put $h = 0$ in equation (1). Therefore

$$0 = (20\sin 40°)T - \tfrac{1}{2}gT^2$$

where T seconds is the time of flight.

Hence

$$(20\sin 40°)T = \tfrac{1}{2}gT^2$$

$$\therefore \qquad\qquad gT = 40\sin 40°$$

$$\therefore \qquad\qquad T = \frac{40\sin 40°}{g}$$

$$= 2.62$$

Hence the time of flight is 2.62 s.

[Readers who are using calculators will find it convenient to store the value of T in the calculator's memory for later use.]

Now consider the horizontal motion:

the horizontal component of the velocity is $20 \cos 40°$, and the acceleration is zero.

Hence, writing $OP = R$.

$$R = (20 \cos 40°)T$$
$$= 40.2$$

Hence the horizontal range, correct to three significant figures, is 40.2 m.

Example 3

An artillery shell is fired with a muzzle velocity of $200 \, \text{m s}^{-1}$, at an angle of $60°$ to the horizontal. It is observed that the angle of elevation of the point of impact, from the gun, is $20°$. Calculate the horizontal and vertical components of the displacement of the point of impact from the gun. (Take g to be $9.81 \, \text{m s}^{-2}$.)

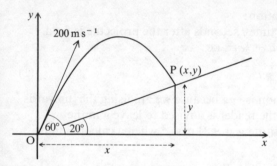

Figure 7.4

Let the coordinates of the point of impact, P, relative to axes through the gun be (x,y), then if the time of flight is T s, we have:

horizontally: $x = (200 \cos 60°)T$ . . . (1)

vertically: $y = (200 \sin 60°)T - \tfrac{1}{2}gT^2$ . . . (2)

Since OP is inclined at $20°$ to the horizontal, $y = x \tan 20°$

$$(200 \sin 60°)T - \tfrac{1}{2}gT^2 = (200 \cos 60°)T \times \tan 20°$$

∴ $\tfrac{1}{2}gT = 200 \sin 60° - 200 \cos 60° \tan 20°$

$$T = \frac{2[200 \sin 600 - 200 \cos 60° \tan 20°]}{g}$$

$$T = 27.9$$

The values of x and y can now be calculated by substituting this value of T into equations (1) and (2), above, giving

$$x = 2790 \text{ and } y = 1020 \quad \text{(correct to three significant figures)}$$

Hence the horizontal and vertical components of the displacement of the point of impact from the gun are 2790 m and 1020 m, respectively.

Example 4

A projectile is given an initial speed of 40 m s^{-1} and it is required to hit a spot which is 50 m away, on a horizontal line through the point of projection. Calculate the two possible angles of projection. (Take g to be 9.81 m s^{-2}.)

Let the angle of projection be θ and the time of flight T s.

Using the usual notation, and proceeding as in the previous examples, we have

horizontally: $50 = (40 \cos \theta)T$ . . . (1)

vertically: $0 = (40 \sin \theta)T - \frac{1}{2}gT^2$ . . . (2)

From equation (2)

$$gT = 80 \sin \theta$$

$$\therefore \quad T = \frac{80 \sin \theta}{g}$$

and substituting this into equation (1) gives

$$50 = 40 \cos \theta \times \left(\frac{80 \sin \theta}{g} \right)$$

$$= \frac{3200 \cos \theta \sin \theta}{g}$$

$$\therefore \quad \cos \theta \sin \theta = \frac{50g}{3200}$$

But, for any value of θ,

$2 \sin \theta \cos \theta = \sin 2\theta$[†], so $\cos \theta \sin \theta = \frac{1}{2} \sin 2\theta$,

$$\therefore \quad \frac{1}{2} \sin 2\theta = \frac{50g}{3200}$$

[†] See *Pure Mathematics Book 1* by J. K. Backhouse and S. P. T. Houldsworth, revised by P. J. F. Horril, published by Longman, Chapter 17.

$$\therefore \quad \sin 2\theta = \frac{100g}{3200}$$

$$= \frac{g}{32}$$

$$\therefore \qquad 2\theta = 17.85° \quad \text{or} \quad (180° - 17.85°)$$

Hence $\quad \theta = 8.9°$ or $81.1°$, correct to the nearest $0.1°$.

Exercise 7a

(*Take g to be* $9.81 \, \text{m s}^{-2}$.)

1 A ball is thrown with an initial velocity of $20 \, \text{m s}^{-1}$, at an angle of $30°$ to the horizontal. Write down its coordinates (x, y), relative to horizontal and vertical axes through the point of projection, t seconds after it is projected. Draw a graph of its trajectory, plotting the points corresponding to $t = 0, 0.5, 1.0, 1.5, 2.0, 2.5, 3.0, 3.5, 4$. From your graph estimate (a) the greatest height reached, and (b) the horizontal range.

2 Repeat question 1, but with an angle of projection of $60°$. What do you notice about the ranges in these two questions? If the thrower wished to increase the range, using the same initial speed, what would you suggest?

3 Calculate the greatest height and the horizontal range in question 1.

4 Calculate the greatest height and the horizontal range in question 2.

5 Two tennis balls are thrown with the same initial speed, $u \, \text{m s}^{-1}$, at angles of $30°$ and $60°$, respectively. Find expressions, in terms of u, for the greatest heights they reach. Prove that these heights are in the ratio $1:3$.

6 A javelin thrower always uses the same angle of projection. Find the percentage increase in the length of the throw, on a horizontal plane, which could be obtained by increasing the speed of release by 10 per cent. (Ignore the height of the javelin at the moment it is released.)

7 A hammer thrower, who has been trained to release the hammer when its path makes an angle of $40°$ with the horizontal, wishes to achieve a throw of $85 \, \text{m}$. What 'speed of release' is required? (Ignore the height of the hammer at the moment it is released.)

8 Find the percentage increase in performance which the thrower in question 7 could achieve by increasing the angle of projection to $45°$ (but without changing the initial speed).

9 A stone is thrown horizontally from the top of a cliff, and it is seen to fall into the sea 4 s later. From the top of the cliff the angle of depression of the splash made by the stone is $40°$. Calculate its initial speed.

10 Two students, Andrew and Barbara, calculate the range of a projectile whose initial velocity is $30 \, \text{m s}^{-1}$ at $35°$ to the horizontal. Andrew takes g to be $10 \, \text{m s}^{-2}$ and Barbara uses $9.81 \, \text{m s}^{-2}$. Express the 'error' in Andrew's answer as a percentage of Barbara's, giving your answer correct to 2 significant figures.

11 A cricket ball is thrown with an initial speed of $20 \, \text{m s}^{-1}$. Find the two possible angles of projection which would give it a horizontal range of $25 \, \text{m}$. (Hint: see Example 4.)

12 A shotputter releases the shot at a height of 2 m above the ground, and it strikes the ground at a horizontal distance of 18 m. If the shot was projected at an angle of 40° with the horizontal, calculate the speed with which it was released.

13 The initial velocity of a projectile is $u\mathbf{i} + v\mathbf{j}$. Find in terms of u, v and g
    (a) the time of flight,   (b) the range.

14 A missile is dropped from an aircraft when it is flying horizontally at 600 km h^{-1}, at a height of 500 m. The initial velocity of the missile relative to the aircraft is zero, but the rocket motor of the missile gives it a horizontal component of acceleration $\frac{1}{2}g$. Calculate the horizontal distance travelled by the missile when it strikes the ground.

15 A basketball player throws the ball at a basket which is at a horizontal distance of 5 m and 1 m above the point from which the ball was thrown. If the speed of projection is 10 m s^{-1}, calculate the two possible angles of projection.

7.3 Projectiles (using algebra)

In this section we shall derive formulae for the more common results associated with projectiles. As in the previous section, we shall use the constant acceleration formulae (see §3.2). In addition to the standard abbreviations, we shall use the following:

R = the horizontal range
T = the time of flight
α = the angle of projection (above the horizontal)
U = the initial speed of the projectile
H = the greatest height (above a horizontal plane through the point of projection)

The point P (x,y) will be used for the position of the projectile, t seconds after it is released, relative to horizontal and vertical axes through the point of projection, (see Figure 7.5).

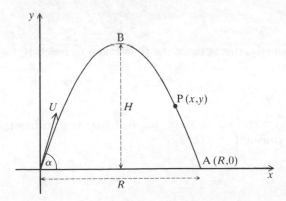

Figure 7.5

THE GREATEST HEIGHT, H

Let the time taken to reach B, the highest point on the trajectory, be t_1; the value of t_1 can be found by considering the vertical motion, and putting $u = U \sin \alpha$, $a = -g$, $t = t_1$ and $v = 0$, in the equation $v = u + at$, as follows:

$$0 = U \sin \alpha - gt_1$$

Hence

$$gt_1 = U \sin \alpha$$

$$t_1 = \frac{U \sin \alpha}{g}$$

The greatest height, H, can now be found by using this value for t, in the formula $s = \frac{1}{2}(u + v)t$, with $u = U \sin \alpha$ and $v = 0$, i.e.

$$H = \frac{1}{2}(U \sin \alpha) \times \left(\frac{U \sin \alpha}{g} \right)$$

$$\therefore \quad H = \frac{U^2 \sin^2 \alpha}{2g}$$

The horizontal displacement at this instant is $(U \cos \alpha)t_1 = \dfrac{U^2 \sin \alpha \cos \alpha}{g}$

Qu. 1 Calculate the greatest height of the projectiles in Exercise 7a, questions 1 and 2, using the formula for H above.

THE TIME OF FLIGHT, T

We find the time of flight (i.e. the time taken to reach point A in Figure 7.5) by applying the formula $s = ut + \frac{1}{2}at^2$ to the vertical motion, with $s = 0$, $u = U \sin \alpha$, $a = -g$ and $t = T$, i.e.

$$0 = (U \sin \alpha)T - \frac{1}{2}gT^2$$

Hence $\frac{1}{2}gT^2 = UT \sin \alpha$

$$\therefore \quad T = \frac{2U \sin \alpha}{g}$$

(Notice that, not very surprisingly, this is twice the time taken to reach B, the highest point on the trajectory.)

THE RANGE, R

With this value of T, we can now find R, using the fact that the horizontal component of the velocity is constant ($= U \cos \alpha$):

$$R = (U \cos \alpha)T$$

$$= \frac{2U^2 \sin \alpha \cos \alpha}{g}$$

(Notice that this is twice the horizontal displacement at point B.)

But, for any angle α, $2 \sin \alpha \cos \alpha = \sin 2\alpha$, so the formula for R can be written as follows:

$$R = \frac{U^2 \sin 2\alpha}{g}$$

It is not necessary for the reader to memorise these formulae, but one should be able to derive them from the constant acceleration formulae, as shown above.

THE MAXIMUM RANGE

One advantage of having these results expressed as formulae is that we can examine their properties more closely; for instance, in the formula

$$R = \frac{U^2 \sin 2\alpha}{g}$$

we know that $0 \leqslant \sin 2\alpha \leqslant 1$, ($\sin 2\alpha = 0$ when $\alpha = 0$, and 1 when $\alpha = 45°$), so the maximum value of R, for a given speed of projection is

$$R_{max} = \frac{U^2}{g}$$

and this is attained by having an angle of projection of $45°$.

Qu. 2 Find the minimum muzzle velocity required for a gun if it is to have a range of 1 km.

THE ANGLE(S) OF PROJECTION

For given values of U and R, the angle(s) of projection required can be calculated, as follows:

$$R = \frac{U^2 \sin 2\alpha}{g}$$

$$\therefore \quad \sin 2\alpha = \frac{Rg}{U^2}$$

Provided $Rg \leqslant U^2$, this equation can be solved for the angle 2α; indeed it has *two* solutions, one acute and the other obtuse (unless $Rg = U^2$, in which case $\alpha = 45°$); if the acute angled solution is

$$2\alpha = A$$

then the obtuse one is

$$2\alpha = 180° - A$$

So the possible angles of projection are $\frac{1}{2}A°$ and $(90 - \frac{1}{2}A)°$.

The following example on the next page illustrates this point.

Example 5

A shell is fired from a gun with a muzzle velocity of 120 m s^{-1}. Find the angle(s) of projection which will enable the shell to strike a target at a distance of 1 km, on a horizontal plane through the point of projection.

Using the formula

$$R = \frac{U^2 \sin 2\alpha}{g}$$

and putting $R = 1000$, $U = 120$ and $g = 9.81$, we have

$$1000 = \frac{120^2 \sin 2\alpha}{9.81}$$

Hence

$$\sin 2\alpha = \frac{1000 \times 9.81}{120^2}$$

$$= 0.681\,25$$

and so

$$2\alpha = 42.94° \quad \text{or} \quad 180° - 42.94° \qquad (= 137.06°)$$

$$\therefore \qquad \alpha = 21.5° \quad \text{or} \quad 68.5° \quad \text{(correct to the nearest 0.1°)}$$

Hence the two possible angles of projection are 21.5° and 68.5°.

It will not come as a surprise to anyone who has ever thrown a ball that, provided the target is within range, there are always two possible trajectories, a high one and a low one; from the work we have just done, we can see that the two possible angles of projection are always complementary, that is their sum is 90°.

Qu. 3 Find the two angles of projection which, for a given initial speed, will enable a gun to hit a target at half the maximum range.

7.4 The equation of the trajectory

Let $P(x,y)$ be the position of a projectile t seconds after being released, then, as in §7.2, we have

$$x = Ut \cos \alpha \qquad \qquad \cdots \quad (1)$$

$$y = Ut \sin \alpha - \tfrac{1}{2}gt^2 \qquad \qquad \cdots \quad (2)$$

From equation (1),

$$t = \frac{x}{U \cos \alpha}$$

and substituting this into equation (2) gives

$$y = U \sin \alpha \left(\frac{x}{U \cos \alpha} \right) - \tfrac{1}{2} g \left(\frac{x}{U \cos \alpha} \right)^2$$

$$= \frac{x \sin \alpha}{\cos \alpha} - \frac{g x^2}{2 U^2 \cos^2 \alpha} \qquad \qquad \cdots \quad (3)$$

$$= x \left(\frac{\sin \alpha}{\cos \alpha} - \frac{g x}{2 U^2 \cos^2 \alpha} \right)$$

$$= x \left(\frac{2 U^2 \sin \alpha \cos \alpha - g x}{2 U^2 \cos^2 \alpha} \right)$$

$$- \frac{g x}{2 U^2 \cos^2 \alpha} \left(\frac{2 U^2 \sin \alpha \cos \alpha}{g} - x \right)$$

Now the first term inside the brackets is R, the range of the projectile, so the equation becomes

$$y = \left(\frac{g}{2 U^2 \cos^2 \alpha} \right) x (R - x)$$

This equation, which expresses the y-coordinate of P in terms of the x-coordinate, is the equation of the trajectory. The right-hand side is a quadratic function of x, and readers who have studied this topic in Pure Mathematics, will realise that it is the equation of a **parabola**. Notice that $y = 0$ when $x = 0$ or $x = R$, in other words the graph passes through the origin and through the point $(R, 0)$ as we would expect.

Equation (3) can also be written in the form

$$y = x \tan \alpha - \left(\frac{g x^2}{2 U^2} \right) \sec^2 \alpha \qquad \qquad \cdots \quad (4)$$

This form of the equation can be very useful, especially if the $\sec^2 \alpha$ is replaced by $(1 + \tan^2 \alpha)$; the advantage of this is that the only trigonometrical ratio used is $\tan \alpha$.

Example 6

A squash player strikes a ball, giving it a speed of $20 \, \text{m s}^{-1}$, towards a vertical wall, which is at a horizontal distance of $10 \, \text{m}$. The ball hits the wall at a point which is $5 \, \text{m}$ above the point of projection. Find the two possible angles of projection.

Using Equation (4) above and putting $\sec^2 \alpha = 1 + \tan^2 \alpha$, we have

$$y = x \tan \alpha - \frac{g x^2}{2 U^2} (1 + \tan^2 \alpha)$$

Inserting the data gives

$$5 = 10 \tan \alpha - \frac{100g}{2 \times 400}(1 + \tan^2 \alpha)$$

Hence

$$5 = 10X - \frac{g}{8}(1 + X^2) \quad \text{where } X = \tan \alpha$$

$$\therefore \quad 40 = 80X - g(1 + X^2)$$

which gives us the quadratic equation

$$gX^2 - 80X + (g + 40) = 0$$

Putting $g = 9.81$, this becomes

$$9.81X^2 - 80X + 49.81 = 0$$

This can be solved using the quadratic formula[†]

$$X = \frac{-b \pm \sqrt{(b^2 - 4ac)}}{2a}$$

with $a = 9.81$, $b = -80$ and $c = 49.81$, which gives

$$X(= \tan \alpha) = \frac{80 \pm \sqrt{(6400 - 4 \times 9.81 \times 49.81)}}{2 \times 9.81}$$

$$\therefore \qquad \alpha = 82.4° \quad \text{or} \quad 34.2°$$

The possible angles of projection are 82.4° and 34.2°, correct to the nearest 0.1°.

Although it is usually convenient to take axes through the point of projection this is not always the case: in the next example it is advantageous to use axes whose origin is the highest point of the trajectory.

Example 7

A ball is thrown so that it just clears three vertical walls standing on horizontal ground whose heights are h m, $2h$ m and h m, respectively. The walls are separated by equal distances of a m. Prove that the horizontal distance from the middle wall to the point where the ball lands is $a\sqrt{2}$ m.

Let the highest point of the trajectory be the origin, O, and let the x-axis be the horizontal line through O in the plane of the trajectory, and the y-axis be the vertical line through O, directed *downwards*, (see Figure 7.6).

[†]See *Pure Mathematics Book 1* by J. K. Backhouse and S. P. T. Houldsworth, revised by P. J. F. Horril, published by Longman, Chapter 10.

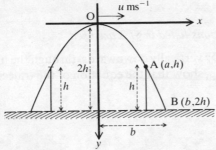

Figure 7.6

With respect to these axes the coordinates of the top of the third wall, point A in the diagram, are (a, h) and let the coordinates of point B, where the ball lands, be $(b, 2h)$. (We have to find b.)

At O the ball is moving horizontally; let its speed be $u\,\text{m s}^{-1}$. By considering the horizontal and vertical motions, we can write down the coordinates of the point (x, y), through which the ball passes, t seconds after passing O.

$$x = ut$$
$$y = \tfrac{1}{2}gt^2$$

and, eliminating t, we have

$$y = \tfrac{1}{2}g\left(\frac{x}{u}\right)^2$$

or $2u^2y = gx^2$

Substituting the coordinates of A and B respectively, we have

$$2u^2h = ga^2$$

and $4u^2h = gb^2$

Dividing, to eliminate h, gives

$$\frac{4u^2h}{2u^2h} = \frac{gb^2}{ga^2}$$

and hence

$$2 = \frac{b^2}{a^2}$$

Therefore

$$b^2 = 2a^2$$

and so

$$b = a\sqrt{2}$$

Hence the horizontal distance from the middle wall to the point where the ball lands is $a\sqrt{2}\,\text{m}$.

Exercise 7b

Throughout this exercise the abbreviations listed in §7.3 are used.

1 Taking axes Ox horizontally and Oy vertically downwards, through the highest point on the path of a projectile, show that the equation of the trajectory can be expressed in the form

$$y = \frac{gx^2}{2U^2 \cos^2 \alpha}$$

2 Using the axes in question 1, show that the equation of the trajectory of a projectile whose range is R, and whose greatest height is H, can be written

$$R^2 y = 4Hx^2$$

3 Using axes Ox horizontally and Oy vertically upwards through the point of projection, show that the equation of the trajectory of a projectile whose range is R and whose greatest height is H can be expressed in the form

$$R^2 y = 4Hx(R - x)$$

Hence find the horizontal displacements of the two points on the trajectory whose height is $\frac{3}{4}H$.

4 A projectile falls a m short of its target when the angle of projection is α, and a m beyond its target when it is fired with the same muzzle velocity at an angle of projection β. If θ is the angle required for a direct hit, show that

$$\sin 2\theta = \tfrac{1}{2}(\sin 2\alpha + \sin 2\beta)$$

5 Show that, provided the range is less than U^2/g, there are two possible angles of projection for a given range. If the times of flight for the corresponding trajectories are t_1 and t_2, show that

$$R = \tfrac{1}{2}g t_1 t_2$$

6 A projectile is fired at an angle of 60° to the horizontal, from a point A on a horizontal plane; it lands at point B on the plane. At points P and Q on its trajectory, its direction of motion is inclined to the horizontal at angles of $+30°$ and $-30°$, respectively. Prove that $AB = 3PQ$.

7 A gun is situated on a hillside, whose line of greatest slope is inclined at an angle β to the horizontal. The plane of the trajectory contains the line of greatest slope. By writing the coordinates of the point where the projectile lands in the form $(r \cos \beta, r \sin \beta)$, or otherwise, show that r, the range on the inclined plane is given by

$$r = \left(\frac{2U^2}{g \cos^2 \beta} \right) \sin (\alpha - \beta) \cos \alpha$$

[Hint: $\sin \alpha \cos \beta - \cos \alpha \sin \beta = \sin (\alpha - \beta)$]

8 Show that the maximum range of a projectile on a horizontal plane is U^2/g. A target vehicle, which is initially out of range of a gun, drives straight towards the gun along the horizontal line through the gun-position, with a

velocity of $V\,\mathrm{m\,s^{-1}}$. The instant when it is within range, the gun is fired with a muzzle velocity of $10V\,\mathrm{m\,s^{-1}}$. Show that θ, the angle of projection required to hit the target, is given by the equation

$$5 \sin 2\theta = 5 - \sin \theta$$

By considering the graphs of $y = 5 \sin 2\theta$ and $y = 5 - \sin \theta$, or otherwise, show that the possible angles of projection required are approximately $32°$ and $62°$.

(Readers who are familiar with the Newton–Raphson method for solving equations should use it.)

9 A projectile is required to hit a target whose coordinates, relative to horizontal and vertical axes through the point of projection, are (a,b). If the muzzle velocity is $\sqrt{(2ga)}$, show that the angles of projection are given by the following quadratic equation in $\tan \alpha$:

$$a \tan^2 \alpha - 4a \tan \alpha + (4b + a) = 0$$

Show that it is impossible to hit the target if $b > \frac{3}{4}a$.

10 If the muzzle velocity in question 9 is $U\,\mathrm{m\,s^{-1}}$, show that the angles of projection are given by

$$(ga^2) \tan^2 \alpha - (2aU^2) \tan \alpha + (2U^2b + ga^2) = 0$$

By considering the condition for this quadratic equation to have identical roots, or otherwise, show that the points which are just within range lie on the curve whose equation is

$$(2gU^2)y = U^4 - g^2x^2$$

Write down the coordinates of the points where this curve intersects the axes and comment on the significance of these points. Sketch the curve and indicate how the paths of individual trajectories are related to it.

An athlete can 'put the shot' with a release speed of $14\,\mathrm{m\,s^{-1}}$, the shot leaving his hand $2\,\mathrm{m}$ above the ground, which is horizontal. Taking g to be $9.81\,\mathrm{m\,s^{-2}}$, estimate the greatest horizontal distance he should be able to achieve.

Exercise 7c (Miscellaneous)

1 A stone was thrown up at $60°$ from the horizontal at a building 36 metres away and broke a window 3 seconds later.
(a) State the initial horizontal speed and show that the initial vertical speed was approximately $20.8\,\mathrm{m\,s^{-1}}$.
(b) Calculate the height of the window above the point from which the stone was thrown. [Take g to be $10\,\mathrm{m\,s^{-2}}$.]
(c) State with reason whether the stone was rising or falling when it hit the window.

*(SMP)

2 A particle is projected from a point O on horizontal ground with velocity $V \,\mathrm{m\,s^{-1}}$ at an angle of elevation α and passes through a point P which is at a distance x m horizontally from O and at a height y m above the ground. Prove that

$$y = x \tan \alpha - \left(\frac{gx^2}{2V^2}\right) \sec^2\alpha$$

If $x = 30$, $y = 10$ and $V = \frac{1}{2}(21\sqrt{10})$, find the possible values of $\tan \alpha$. [Take g to be $9.8 \,\mathrm{m\,s^{-2}}$.]

(O & C)

3 During some blasting operations men shelter at a distance of 100 m from an explosion. The charge throws debris in all directions with speeds of up to $50 \,\mathrm{m\,s^{-1}}$.

How soon after detonation should the men be under cover and how soon after this may they emerge to avoid being struck by debris? Neglect air-resistance and assume that debris does not bounce.

(MEI)

4 O is a point on horizontal ground. H is the point h m vertically above O. A particle A is projected horizontally from H with speed $u \,\mathrm{m\,s^{-1}}$. Simultaneously a second particle B is projected from O with speed $v \,\mathrm{m\,s^{-1}}$ at an elevation α so that both particles move to the right of OH and in the same vertical plane. Show that, if the particles collide, then $u = v \cos \alpha$.

Given that $h = 78.4$, $u = 39.2$, $v = 49.0$ and $\tan \alpha = \frac{3}{4}$, verify that a collision occurs and find the time to the collision.

Show also that if, instead, particle A is projected 2 s after particle B, then both particles strike the ground simultaneously. [Take g to be $9.8 \,\mathrm{m\,s^{-2}}$.]

(O & C)

5 A particle is thrown directly upwards from the foot of a tower 25 m high with a speed of $8 \,\mathrm{m\,s^{-1}}$. At the same instant a second particle is projected horizontally from the top of the tower with a speed $6 \,\mathrm{m\,s^{-1}}$. Find their distance of closest approach and the time from the start of the motion at which it occurs.

Show that, when the particles are closest together, their relative velocity is perpendicular to the line joining them.

(O & C)

6 From a point O, a particle P is projected under gravity with speed V, in a direction making an angle α with the horizontal. The particle strikes the horizontal plane through O at the point A. Find the time of flight and the distance OA.

At the instant when P is at the highest point of its path a second particle Q is projected from O with speed U in a direction making an angle β with the horizontal. The particle Q strikes the plane at A at the same instant as P. Given that $\tan \alpha = \frac{4}{3}$, find U, in terms of V, and show that $\tan \beta = \frac{1}{3}$.

(C)

7 A and B are two points 900 m apart on a horizontal plane, and O is the point 100 m vertically above A. A particle is projected from O with speed V m s^{-1} at an angle of elevation of 45°, and it moves freely under gravity until it hits the plane at B. Find the value of V. Find also the angle between the direction of motion of the particle and the horizontal just before the impact at B.

A second particle is projected from O with the same initial speed V m s^{-1} as the first particle, but at an angle of elevation α such that $\alpha \neq 45°$. This particle also hits the horizontal plane at B. Find the value of $\tan \alpha$.
[Take g to be 10 m s^{-2}.]

(C)

8 A particle is projected under gravity from a point O with initial speed V at an angle of inclination α to the horizontal. The axes Ox and Oy are respectively horizontal and vertically upwards through O. At time t the particle is at the point with position vector $\begin{pmatrix} x \\ y \end{pmatrix}$ and has velocity $\begin{pmatrix} u \\ v \end{pmatrix}$.

Write down equations giving $\begin{pmatrix} u \\ v \end{pmatrix}, \begin{pmatrix} x \\ y \end{pmatrix}$ in terms of V, α, g and t.

Hence determine for the particle
(a) the maximum height it attains;
(b) the range OB, where B is the point where it strikes the horizontal plane through O.

For a given value of V the angle α is chosen so that the maximum height of the particle during its flight is h.

Show that the range OB is then $\sqrt{\left[\dfrac{8h}{g}(V^2 - 2gh) \right]}$.

(JMB)

9 A particle is projected with initial velocity $u\mathbf{i} + v\mathbf{j}$, $\mathbf{i}$ and $\mathbf{j}$ being unit vectors in the horizontal and upward vertical directions respectively, and moves with constant gravitational acceleration of magnitude g. Establish the formulae

$$\dot{\mathbf{r}} = u\mathbf{i} + (v - gt)\mathbf{j}$$
and $\quad \mathbf{r} = ut\mathbf{i} + (vt - \tfrac{1}{2}gt^2)\mathbf{j}$

where the position vector of the particle at time t is relative to the point of projection.

$$\left[\text{Note} \quad \dot{\mathbf{r}} = \frac{\mathrm{d}}{\mathrm{d}t}\mathbf{r} \right]$$

The particle is projected from the origin with initial velocity $28\mathbf{i} + 100\mathbf{j}$ towards an inclined plane whose line of greatest slope has the equation $\mathbf{r} = 480\mathbf{i} + \lambda(2\mathbf{i} + \mathbf{j})$ where λ is a parameter.

Show that the particle strikes the plane after 20 seconds and determine the distance along the line of greatest slope from the point where $\lambda = 0$ to the point of impact
[Take g to be 9.8 m s^{-2}.]

(JMB)

In questions 10, 12, 14 and 15, take g to be $10\,\text{m}\,\text{s}^{-2}$.

10 A particle is projected with speed V and angle of elevation α from a point O. Show that the equation of the path of the particle, referred to horizontal and vertical axes Ox and Oy respectively in the plane of the path, is

$$y = x\tan\alpha - (gx^2\sec^2\alpha)/(2V^2)$$

A particle is projected at an elevation α, where $\tan\alpha = 3$, from a point A on a horizontal plane distant $100\,\text{m}$ from the foot of a vertical tower of height $50\,\text{m}$. The particle just clears the tower and lands at a point B on the horizontal plane. Determine the initial speed of the particle and the distance AB. Find also the greatest height reached by the particle above the plane.

(L)

11 The coordinate axes in a vertical plane are arranged with the x-axis horizontal and the positive y-axis vertically upwards. A particle is projected from the origin O with initial speed V at an angle θ to the horizontal in the x–y plane. Find the position of the particle after a time t. Hence find the maximum height H of the particle above the horizontal plane through O and the range R on this plane.

For a given V find the value of θ for which R is greatest.

Given that the greatest permitted height that the particle can reach is $V^2/(8g)$, find the greatest range on the horizontal plane and the corresponding angle of elevation.

(L)

12 A particle is projected with speed u and at an angle of elevation α from a point O. Show that, at time t after projection, the position vector of the particle relative to O is $\mathbf{r}$, where

$$\mathbf{r} = (u\cos\alpha)t\mathbf{i} + [(u\sin\alpha)t - \tfrac{1}{2}gt^2]\mathbf{j}$$

and $\mathbf{i}$ and $\mathbf{j}$ are unit vectors directed horizontally and vertically upwards respectively.

Given that $u = 40\,\text{m}\,\text{s}^{-1}$ and that the particle strikes a target A on the same horizontal level as O, where $OA = 60\,\text{m}$, find the least possible time, to the nearest tenth of a second, that elapses before the particle hits the target A.

(L)

13 A particle is projected from a point O, with speed U at an angle of elevation α, and moves freely under gravity. Write down expressions for the horizontal and vertical components of
(a) the velocity of the particle at time t,
(b) the displacement of the particle from O at time t,
where time t is measured from the instant of projection.

The particle strikes the horizontal plane through O at time $3T/2$. Show that $T = (4U\sin\alpha)/(3g)$. Given also that, at time T, the particle is moving at right angles to its original direction of motion, show that

$$U^2 = \frac{3g^2T^2}{4}$$

(L)

14 A particle is projected at an angle of elevation α from a point A on horizontal ground. When travelling upwards at an angle β to the horizontal, the particle passes through a point B. The line AB makes an angle θ with the horizontal. Show that

$$2 \tan \theta = \tan \alpha + \tan \beta$$

The point B is at a horizontal distance of 30 m from point A and is at a height of 20 m above ground. At B, the particle is travelling upwards at an angle $\tan^{-1}(\frac{1}{3})$ to the horizontal. Find the angle of projection and the initial speed of the particle.

(L)

15 A particle is projected with speed v at an angle α to the horizontal from a point X on a horizontal plane. It hits the plane again at a point A. Find the range XA.

If the range XA is equal to the greatest height reached by the particle, show that $\tan \alpha = 4$.

Hence determine v given that the greatest height reached by the particle is $\frac{1000}{34}$ m.

(L)

Force = mass × acceleration

Chapter 8

Newton's Laws of Motion

8.1 Introduction

So far, we have studied in detail the meaning of velocity and acceleration, but we have not considered the forces which cause these phenomena. For our understanding of the effect of forces on massive objects, we are indebted to Sir Isaac Newton (1642 1727), who, with great insight, propounded the laws of motion which explain the relationship between a force and the motion it causes. These laws have proved extremely successful in explaining many of the physical phenomena which we observe on Earth, and indeed many interplanetary phenomena as well. In the 20th century it has been realised that they are not suitable for very small scale (i.e. subatomic) events, nor are they suitable for velocities approaching the speed of light. But provided we accept these limitations, Newton's model is an extremely good one.

As in other branches of mathematics and science, many of the words employed in Mechanics are also used in a rather imprecise way in everyday speech. When we use them in a mathematical context it is necessary to define them very carefully so that quantitative methods can be applied to the concepts they convey. Before we embark on Newton's Laws of Motion, we must discuss the meaning of force, mass and momentum.

8.2 Force

Most readers will have an intuitive idea about what is meant by a force: if a car breaks down, then it may have to be pushed or pulled to the nearest garage, that is, a horizontal force must be applied. If we pick up a heavy case and walk about with it, we must exert a vertical force on it: if we let go, it falls back to the ground. If we try to push a heavy desk along the floor our effort will be opposed by the **friction** between the desk and the floor: if we are not strong enough the desk does not move. If we push hard enough the desk starts to move, but the friction will still counter our effort. The task can be made easier by mounting the desk on small wheels, thereby reducing the effect of the friction.

Imagine a tug of war contest between two evenly matched teams: both teams may be trying very hard, but if the teams exert forces of equal magnitude nothing happens! If however, one team loses its footing, or if the rope snaps under the strain, the effect will be immediately obvious to all concerned! A similar situation exists if a crane supports a heavy load in mid-air. The upthrust provided by the crane balances the downward force (i.e. the weight) of the load. If now the load is placed on a platform, the force previously supplied by the crane must be provided by the platform (such a force is called a **reaction**). The platform must be strong enough to provide such a reaction otherwise it will collapse.

A very common type of force is that provided by a compressed (or stretched) spring. To keep the spring compressed, a force must be applied to it. It can be proved experimentally that the relationship between the amount of compression (or extension) and the magnitude of the force is very simple: the two quantities are directly proportional, and this enables us to compare forces by measuring the amount by which they compress (or extend) a spring.

Forces, then, may take many forms: pushes, pulls, the attraction of the Earth, the attraction of a magnet, tension in a spring, friction and reactions are some of the most common ones we shall encounter in this subject.

8.3 Mass

Mass is a difficult, but vital, concept in applied mathematics. It must not be confused with weight, although the two are closely related.

Imagine two identical trucks, standing on horizontal railway lines, one empty and the other fully loaded with coal. In each case the **weight** of the truck is supported by the railway track. The trucks could be moved by applying a suitable horizontal force, but it is a matter of common sense that the empty truck will be easier to move than the loaded one: for a given effort the empty truck will pick up speed more rapidly than the full one. We say that the loaded truck has greater **mass** than the empty one. Mass is the measure of an object's reluctance to move: its **inertia**.

The basic unit of mass in the Système International d'Unités (SI units) is the kilogram, and this has been fixed by international agreement. In theory, other standard kilogram masses could be made by comparing them with the block of platinum iridium which is preserved to define the basic unit. Masses are compared (at least in theory) by placing them on the scale pans of an old-fashioned balance; objects which balance are said to have equal mass.

It is important to understand that the mass of an object does not depend on the pull of gravity. The mass of an object would not change if it were transported to the Moon (where gravity is about $\frac{1}{6}$th of its value on Earth), or even if it were carried into deep space where gravity could be neglected.

8.4 Weight

The weight of an object is the force exerted on it by gravity; this can be measured by using a suitably calibrated spring balance. Unlike mass, then, weight *does* depend on gravity. Consequently, if an object were weighed on a spring balance on the Earth, and if it were then transported to the Moon and weighed again on the same spring balance, its weight would appear to be reduced to about ⅙th of its value on the Earth. Plainly weight is related to mass: a small mass will have less weight than a large mass when they are compared at the same location.

8.5 Momentum

In everyday language we often say that an object 'gathers momentum' as it moves faster and faster. We use the same word in applied mathematics, but we define it precisely, as follows:

momentum = mass × velocity

Momentum is an extremely important quantity in applied mathematics, and we shall study it in detail in Chapter 11. For our present purposes it will be sufficient to be familiar with its definition and to note that it is a vector quantity, i.e. it has both magnitude and direction.

Qu. 1 A ball whose mass is 200 grams falls vertically and strikes a horizontal plane with downward velocity of $30 \, \text{m s}^{-1}$. It rebounds, leaving the plane with a velocity of $20 \, \text{m s}^{-1}$, vertically upwards. Using **j** to represent the unit vector vertically upwards, find the change in the momentum.

8.6 Newton's first law of motion

Newton was a genius, perhaps the greatest genius in the field of mathematics and science the world has ever seen. Like geniuses in other spheres, his special characteristic was the ability to cut through the muddle of contradictory evidence which confuses ordinary mortals, and grasp the essential, underlying truth.

Newton's first law of motion exemplifies the clarity of his thinking. In present day language (Newton published it in Latin), it says:

Every body remains in a state of rest, or of uniform motion in a straight line, unless it is acted on by a force.

In the 17th century, this must have seemed a very strange idea, since everyday experience indicates that moving objects slow down unless they are pushed. Newton was able to see what everyone else had missed: this slowing down is due to another force, usually friction. For us, living in the 20th century, Newton's first law is easier to accept, because we have seen on television, or read about, objects moving through space without any driving force, or we have seen skaters gliding along on a modern ice rink, with very little friction to slow them down.

In effect, the first law defines, in qualitative terms, what we mean by a force: it is any agent which causes a stationary object to move, or a moving object to speed up, slow down, or change direction; i.e. a force is something which causes an object to accelerate.

8.7 Newton's second law of motion

The second law quantifies the relationship between a force and the acceleration it causes. It says (again in modern English):

> The rate of change of momentum of an object is proportional to the force applied (or the resultant force if there is more than one).

In mathematical terms this becomes

$$\mathbf{F} = k\frac{d}{dt}(m\mathbf{v}) \qquad \qquad \dots \ (1)$$

where $\mathbf{F}$ is the resultant force, m the mass of the object and $\mathbf{v}$ its velocity. The factor of proportionality, k, is a number which depends on the units used for $\mathbf{F}$, m and $\mathbf{v}$. In some forms of motion, especially that of rockets, very large amounts of fuel are burned, indeed rocket propulsion depends on ejecting this fuel at very high velocity. In this book, however, we shall only be concerned with objects whose total mass is constant. In this case equation (1) becomes

$$\mathbf{F} = km\frac{d\mathbf{v}}{dt}$$

and since $\dfrac{d\mathbf{v}}{dt} = \mathbf{a}$, the acceleration of the object, we can write this equation in the form

$$\mathbf{F} = km\mathbf{a} \qquad \qquad \dots \ (2)$$

8.8 Units

In this book, we shall use SI units when working with Newton's second law of motion. As already explained (see §8.3) mass is measured in kilograms. Acceleration has been discussed in some detail in earlier chapters, and the reader will know that it is measured in $m\,s^{-2}$. We now define the unit of force, called the **newton** (the abbreviation being N):

> A newton is that force which, when applied to a mass of one kilogram, gives it an acceleration of $1\,m\,s^{-2}$.

With the unit of force defined in this way, the constant k in equation (2) above, is 1, which enables us to write Newton's second law of motion in a particularly simple form, namely

$$\mathbf{F} = m\mathbf{a}$$

. . . (3)

Notice that $\mathbf{F}$ and $\mathbf{a}$ are both vector quantities, so they both have the same direction. (However, as we saw in the last chapter, the direction of the acceleration is not necessarily the same as the direction of motion, so Newton's second law does *not* imply that the direction of the force and the direction of motion are always the same.)

We have already noted (see §8.4) that the weight of an object is the force with which the Earth attracts it. Since the Earth gives it an acceleration of $g\,\mathrm{m\,s^{-2}}$, it follows from Newton's second law of motion that an object of mass m kg weighs mg newtons. The reader should note that the force required to support, say, a 1 kg bag of sugar, is just under 10 N.

Example 1

A block of mass 5 kg is initially at rest on a smooth horizontal plane and a horizontal force of 20 N is applied to it for 10 s. Find the speed of the block after this time.

[The use of the word smooth, which appears in many questions like this, indicates that the friction may be neglected.]

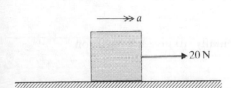

Figure 8.1

Using Newton's second law of motion, we have

$$20 = 5a$$

where $a\,\mathrm{m\,s^{-2}}$ is the acceleration of the block. Hence

$$a = 4$$

Since the acceleration is constant, we may use the formula $v = u + at$ to find the speed when $t = 10$ (we are given that $u = 0$).

$$\therefore \quad v = 4 \times 10$$
$$= 40$$

Hence, after 10 s, the block is moving at $40\,\mathrm{m\,s^{-1}}$.

8.9 Resolving forces

In many cases there are several forces which are applied simultaneously to the objects in question. If so, the force **F** in equation (3) on page 125 should be regarded as the *resultant* force (i.e. the vector sum of the individual forces). Strictly speaking, in Example 1 above there were two further forces, the weight *W*, acting vertically downwards and a reaction *R*, vertically upwards.

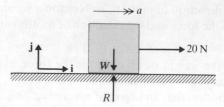

Figure 8.2

Using the unit vectors **i** and **j** in the usual way, we could write

$$(20\mathbf{i} + R\mathbf{j} - W\mathbf{j}) = 5a\mathbf{i}$$

If we then scalar multiply by **i** (and remembering that $\mathbf{i}.\mathbf{i} = 1$ and $\mathbf{i}.\mathbf{j} = 0$) we obtain

$$20 = 5a$$

as before. Now scalar multiplying both sides by **j** gives

$$R - W = 0$$

We say that we have 'resolved horizontally' to give the equation

$$20 = 5a$$

and we have 'resolved vertically' to give the equation

$$R - W = 0$$

Hence

$$R = W$$

i.e. the vertical reaction, *R*, is equal in magnitude to the weight *W* ($= mg$).

Qu. 2 A force of 50 N acts at an angle of 65° to the upward vertical. Resolve this force (a) horizontally, (b) vertically.

Note that, in the interest of clarity, the symbols for units are frequently omitted from diagrams; it is assumed that all the quantities shown are in the appropriate SI units. For example, in Figure 8.3, the weight is shown as 30*g* (newtons).

Example 2

A heavy block of mass 30 kg is pushed along a fixed horizontal plane by a force of magnitude 100 N acting along a line which is inclined at an angle of 30° to the horizontal (see Figure 8.3). The motion of the block is opposed by a horizontal friction force of 50 N. Find the vertical reaction of the plane on the block, and the acceleration.

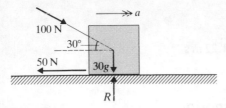

Figure 8.3

There are four forces acting on the block: the force pushing it along, its weight, the friction, and the vertical reaction R. (Note that the reaction R is certainly vertical, but its line of action does not necessarily pass through the middle of the block; this, however, does not affect our working.) The weight is $30g$ newtons. The vector sum of these forces is

$$\begin{pmatrix} 100\cos30° \\ -100\sin30° \end{pmatrix} + \begin{pmatrix} 0 \\ -30g \end{pmatrix} + \begin{pmatrix} -50 \\ 0 \end{pmatrix} + \begin{pmatrix} 0 \\ R \end{pmatrix}$$

$$= \begin{pmatrix} 100\cos30° - 50 \\ -100\sin30° - 30g + R \end{pmatrix}$$

Let the acceleration of the block be $a\,\mathrm{m\,s^{-2}}$, then the acceleration vector can be written $\begin{pmatrix} a \\ 0 \end{pmatrix}$. Using Newton's second law of motion, we have

$$\begin{pmatrix} 100\cos30° - 50 \\ -100\sin30° - 30g + R \end{pmatrix} = 30\begin{pmatrix} a \\ 0 \end{pmatrix}$$

Resolving horizontally (i.e. equating the top lines in this vector equation) gives

$$100\cos30° - 50 = 30a \qquad \qquad \text{. . . (1)}$$

Resolving vertically (i.e. equating the bottom lines)

$$-100\sin30° - 30g + R = 0 \qquad \qquad \text{. . . (2)}$$

[In most questions it will be convenient to go straight to this pair of equations.]

From equation (2)

$$R = 30g + 100\sin30°$$

and, taking g to be 9.81,

$$R = 344.3$$

Hence the vertical reaction is 344 N, correct to three significant figures.

From equation (1)

$$30a = 36.6$$

and hence $a = 1.22$

The acceleration of the block is $1.22 \, \text{m s}^{-2}$.

8.10 Newton's third law of motion

Newton's third law of motion states that:

> When two objects exert a force on each other, these forces are equal in magnitude, but opposite in direction.

In Example 2 above, the plane exerts an upward force of 344 N on the block; Newton's third law of motion says that the force on the plane, due to the block is also of magnitude 344 N, but it acts vertically downwards. Notice also that the plane exerts a horizontal friction force of 50 N on the block (acting to the left in Figure 8.3); so, by Newton's third law the block exerts a friction force of 50 N on the plane (but acting to the right in Figure 8.3). If the plane were not fixed, this friction force might cause the plane to move to the right, as the block is pushed along.

8.11 Motion on an inclined plane

Although we shall frequently resolve horizontally and vertically, there are some situations where it is more convenient to resolve in some other direction. If we have a vector $\mathbf{V}$ and we wish to resolve it in the direction of a line l, inclined to $\mathbf{V}$ at an angle α (see Figure 8.4), then we scalar multiply $\mathbf{V}$ by $\mathbf{u}$, the unit vector in the direction of line l, giving $\mathbf{V} \cdot \mathbf{u} \, (= V \cos \alpha)$.

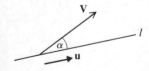

Figure 8.4

Qu. 3 A heavy rock is placed on a plane which is inclined to the horizontal at an angle of 20°. The weight of the rock is 1000 N. Resolve the weight (a) parallel, and (b) perpendicular to the line of greatest slope.

Example 3

A child on a sledge slides down a snow-covered hill which is inclined to the horizontal at an angle of 20°. The combined mass of the child and the sledge is 40 kg, and their motion is opposed by a friction force of 80 N. If the sledge's initial speed is zero, how fast will it be moving when it has travelled 100 m?

Figure 8.5

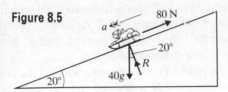

The diagram shows the forces acting on the sledge and child (it is convenient to regard them as a single object), and the acceleration down the slope, $a \, \text{m s}^{-2}$.

The resolved part of the weight which acts down the slope is $40g \sin 20°$. So, resolving parallel to the inclined plane gives a resultant force down the plane of $(40g \sin 20° - 80)$ N. Applying Newton's second law of motion gives

$$40g \sin 20° - 80 = 40a$$

and taking g to be 9.81, we have a = 1.355†.

To find the speed, v, when the sledge has travelled 100 m with this acceleration, we use the constant acceleration formula, $v^2 = u^2 + 2as$, giving

$$v^2 = 2 \times 1.355 \times 100$$

Hence $v = 16.5$ (correct to three significant figures)

Hence, when the sledge has travelled 100 m, its speed is 16.5 m s⁻¹.

Try to follow these guidelines when answering any question on this topic:

1 Always draw a clear diagram showing all the relevant forces, and the acceleration(s).
2 Define the algebraic symbols clearly.
3 Explain the method used, for example 'by Newton's second law of motion'.
4 If the problem involves several particles or other objects, it is usually advisable to consider each particle separately.
5 State the answer(s) clearly, giving the units and, in the case of vector quantities, quoting both the magnitude and the direction.

---

† Note that since the value of a is not explicitly required as an answer to the question, it is better *not* to correct it to three significant figures: indeed, if a calculator is being used, the value of a, with all the significant figures available, should be stored for use in the calculation of v.

Exercise 8 (Miscellaneous)

Take g to be $9.81 \, \text{m s}^{-2}$ *unless the question gives some other value.*

1 A car whose mass is 1 tonne is travelling along a horizontal road at $108 \, \text{km h}^{-1}$, when the driver applies the brakes, causing the car to skid to a halt under the action of a friction force of 2000 N. Calculate the distance it travels while it is skidding.

2 A lorry of mass 3 tonnes, travelling at $25 \, \text{m s}^{-1}$, starts to climb a hill which is inclined at an angle α to the horizontal, where $\sin \alpha = 0.2$. Assuming that the tractive force, T N, between its tyres and the road is constant, find T, given that the speed is reduced to $15 \, \text{m s}^{-1}$ in a distance of 500 m.

3 A crate of mass 500 kg is placed in the back of a van; it is known that the crate will slip if a horizontal force exceeding 2000 N is applied to it. If the van is travelling on a horizontal road at $54 \, \text{km h}^{-1}$, find the minimum stopping distance which would allow the van to stop at a uniform rate without the crate slipping.

4 A lift of mass 1000 kg starts from rest and ascends vertically. It accelerates uniformly for 5 s and then travels at constant speed for 13 s. After that it slows down at a uniform rate, coming to rest in a further 4 s. Given that the total distance travelled is 105 m, find the tension in the lift cable during each of the three stages.

5 A car of mass M kg is travelling at constant speed up a hill which is inclined to the horizontal at an angle α. Find the tractive force exerted by the car in terms of M, g and α. (Other forces, for example wind resistance, may be neglected.) Assuming that the tractive force is unchanged when the angle of the slope changes to β, where $0 < \beta < \alpha$, find the acceleration of the car in terms of g, α and β.

6 A car of mass M kg, which can exert a maximum tractive force of P N, is used to tow a caravan of mass m kg. Show that the maximum acceleration of the outfit on a horizontal road is $\dfrac{P}{(M + m)} \, \text{m s}^{-2}$. Find also the tension in the coupling under these conditions.

7 A bead of mass 100 grams is free to slide on a smooth straight wire, which is inclined to the horizontal at an angle of 60°. A horizontal force of magnitude 2 N is applied to the bead, causing it to move up the wire, (the line of action of the force being in the vertical plane containing the wire). Find
(a) the acceleration of the bead;
(b) the force exerted on the wire by the bead.

8 A jet plane whose mass is 50 tonnes touches down at $216 \, \text{km h}^{-1}$ and uses 'vectored thrust' to decelerate; this consists of directing the thrust of each of its two engines forward horizontally, but at an angle of 25° either side of its line of motion. If the thrust of each engine is 75 000 N, and no other forces are used to slow the plane down, calculate the length of runway used by the plane while coming to a halt.

9 An aircraft has a total mass of 50 tonnes and when taking off it accelerates uniformly from 0 to $200 \, \text{km h}^{-1}$ in a distance of 2000 m. Given that the total thrust from its engines is 60 000 N, calculate the magnitude of the resistance

to its motion. Discuss *briefly* whether it is reasonable to assume that the engine thrust, the acceleration and the resistance are constant while the plane is taking off.

10 Three forces are given in newtons by the following vectors:

$$\begin{pmatrix} 10 \\ 20 \end{pmatrix} \qquad \begin{pmatrix} 5 \\ -5 \end{pmatrix} \qquad \begin{pmatrix} -55 \\ 15 \end{pmatrix}$$

If they act simultaneously on a particle of mass 10 kg, find the acceleration, expressing this as a column vector.

11 Forces P, Q and R have magnitudes 100 N, 50 N and 40 N respectively, and they are inclined to the positive x-axis at angles of 30°, 120° and $-45°$, respectively. If they act simultaneously on a particle of mass 20 kg, find the magnitude and direction of the acceleration.

12 A tug-boat towing two barges in line along a straight canal accelerates at a constant rate of 0.1 m s^{-2}. The mass of the tug-boat is 50 tonnes and the mass of each barge is 100 tonnes. The tug's engines provide a forward thrust of 100 kN, but the tug-boat and each of the barges is opposed by water-resistance of R N. Find R, and the tension T_1 between the tug and the first barge, and the tension T_2 between the two barges.

13 A particle of mass 100 grams moves along a smooth curve in a two-dimensional plane so that its position (x,y) t seconds after passing the origin is given by

$$x = 20t + 5t^2 \qquad y = 40t - 5t^2$$

Show that the force which causes this motion is constant and find its magnitude and direction.

14 A particle of mass m kg moves in a horizontal plane and its position vector, $\mathbf{r}$, relative to a fixed point O in the plane is given by

$$\mathbf{r} = (a\cos nt)\mathbf{i} + (a\sin nt)\mathbf{j}$$

where a and n are positive constants and t is the time in seconds which has elapsed since the particle passed through the point $(a,0)$. Show that the force acting on the particle is (man^2) N and that it is always directed towards the point O.

15 A particle of unit mass moves in a horizontal plane, and its position relative to a fixed point O in the plane is given by the vector $\mathbf{r}$, where

$$\mathbf{r} = t\mathbf{i} + (\sin t)\mathbf{j}$$

Find the force, in vector form, when $t = \pi/2$ and $t = 3\pi/2$. Sketch the path of the particle, and on the sketch indicate the forces at these two instants.

Chapter 9

Connected particles – friction

9.1 Connected particles

In this section we shall be considering the motion of heavy particles connected by light, inextensible, strings; in this context the word 'light' means that we can neglect the mass of the string. To consider the effect of this, let us first consider a 'heavy' string: in Figure 9.1 a segment of the string, mass m, is acted on by tensions T_1 and T_2.

Figure 9.1

If these tensions are unequal they will cause the segment to accelerate, and, by Newton's second law of motion the acceleration, a will be given by the equation

$$T_1 - T_2 = ma$$

However, if the string is light, m is taken to be zero, and hence

$$T_1 - T_2 = 0$$
$$\therefore \quad T_1 = T_2$$

Hence in a light string the tension is constant throughout its length.

In many cases these strings will pass over pulleys. At this stage we shall assume that these pulleys are light (i.e. their mass can be neglected) and frictionless. As a consequence no effort is required to make them turn, so the tension either side of such a pulley will be the same. (In Chapter 18, we shall consider heavy pulleys.)

In all problems involving connected particles, it is important to apply Newton's laws of motion to each individual particle: the following example illustrates the method.

Example 1

Two particles of mass 2 kg and 6 kg are connected by a light, inextensible string which passes over a smooth, light pulley. The string on either side of the pulley hangs vertically. Initially the system is held at rest, and then it is released. Find the acceleration of the particles, the tension in the string and the vertical force which must be applied to the pulley to support the system.

[Take g to be 9.81 m s^{-2}.]

Let the tension in the string be T N (as explained earlier, this is constant throughout the length of the string) and let the acceleration of the particles be a m s^{-2}. Figure 9.2 shows the forces acting on the particles, and a force R N supporting the pulley.

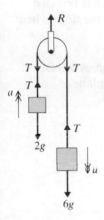

Figure 9.2

Applying Newton's second law of motion to each particle in turn, gives

$$T - 2g = 2a$$

and $6g - T = 6a$

Adding these equations (to eliminate T), we have

$$4g = 8a$$
$$\therefore \quad a = \tfrac{1}{2}g$$

Substituting this into the first of the original equations, gives

$$T - 2g = 2 \times \tfrac{1}{2}g$$
$$\therefore \quad T = 3g$$

Hence the acceleration of the particles is $\tfrac{1}{2}g$ m s^{-2} (≈ 4.91 m s^{-2}), and the tension in the string is $3g$ N (≈ 29.4 N).

Since the pulley is stationary, when Newton's second law of motion is applied to it we obtain

$$R - 2T = 0$$
$$R = 2T$$
$$= 6g$$

Hence the force supporting the system is $6g$ N (≈ 58.9 N).

Example 2

A heavy load, mass M kg, is pulled up a rough plane which is inclined at an angle of $30°$ to the horizontal, by a light inextensible cable, which passes over a smooth light pulley at the top of the plane (see Figure 9.3). At the other end of the cable there is a heavy block of mass $2M$ kg, hanging vertically. The friction force between the load and the inclined plane is $\frac{1}{4}Mg$ N. When the system is released from rest, the block descends vertically, and it strikes the floor when it has descended h m. Find
(a) the speed with which the block strikes the floor;
(b) the total distance moved by the block before it comes to rest.
(Assume that the block does not reach the top of the plane.)

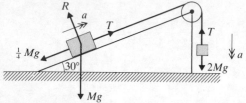

Figure 9.3

The following forces act on the load:
    its weight Mg N, vertically downwards
    the tension in the cable, T N
    the friction $\frac{1}{4}Mg$ N acting down the plane
    the reaction R N of the plane on the load (perpendicular to the plane)

Resolving up the plane, and using Newton's second law of motion, we have

$$T - \tfrac{1}{4}Mg - Mg\sin 30° = Ma$$

where a m s^{-2} is the acceleration of the load up the plane.
Hence

$$T - \tfrac{3}{4}Mg = Ma \qquad \qquad . \quad . \quad . \quad (1)$$

(Resolving perpendicular to the plane would give
$R - Mg\cos 30° = 0$, and hence $R = Mg\cos 30°$, but this is not required in this question.)

Now consider the forces acting on the block, as it descends; these are

its weight, $2Mg$, acting downwards
the tension T in the cable, acting upwards

Applying Newton's second law of motion to the downward motion of the block, we have

$$2Mg - T = 2Ma \qquad \qquad \qquad \ldots \ldots \quad (2)$$

('The acceleration is a because the cable is inextensible.)

From equation (2), we have

$$T = 2Mg - 2Ma$$

and substituting this for T in equation (1) gives

$$2Mg - 2Ma - \tfrac{3}{4}Mg = Ma$$
$$\therefore \qquad \qquad \tfrac{5}{4}Mg = 3Ma$$
$$\therefore \qquad \qquad \qquad a = \tfrac{5}{12}g$$

(a) Since the acceleration is constant, we may use the formula

$$v^2 = u^2 + 2as,$$

with $a = \tfrac{5}{12}g$, $u = 0$ and $s = h$, to find the speed with which the block strikes the floor, i.e.

$$v^2 = 2 \times \tfrac{5}{12}gh$$
$$= \tfrac{5}{6}gh$$
$$\therefore \qquad v = \sqrt{(\tfrac{5}{6}gh)}$$

The block strikes the floor at $\sqrt{(\tfrac{5}{6}gh)}\,\mathrm{m\,s}^{-1}$.

(b) At this moment, the load has moved a distance h m up the plane and it is moving at $\sqrt{(\tfrac{5}{6}gh)}\,\mathrm{m\,s}^{-1}$. The tension in the cable now becomes zero, and the equation of motion of the load, i.e. equation (1) with $T = 0$, becomes

$$-\tfrac{3}{4}Mg = Ma$$
$$a = -\tfrac{3}{4}g$$

so the load has a retardation of $\tfrac{3}{4}g\,\mathrm{m\,s}^{-2}$.

Suppose the load moves a further distance x m, before it comes to rest, then, using the formula $v^2 = u^2 + 2as$, we have

$$0 = (\tfrac{5}{6}gh) + 2 \times (-\tfrac{3}{4}g)x$$
$$\therefore \qquad \tfrac{3}{2}gx = \tfrac{5}{6}gh$$
$$\therefore \qquad \quad x = \tfrac{5}{9}h$$

Adding h to this, the total distance moved up the plane, by the load, is $\tfrac{14}{9}h$ m.

Exercise 9a

In this exercise, the 'strings' should be regarded as light and inextensible, and the 'pulleys' should be taken to be light and smooth, unless stated otherwise. In questions 1–5, take g to be $10 \, \text{m s}^{-2}$.

1 Two particles A and B have masses 5 kg and 3 kg, respectively. They are connected by a string which passes over a fixed pulley. The parts of the string which are not in contact with the pulley hang vertically. Find the magnitude of the vertical force which must be applied to B to keep the system at rest. This force is now removed, and the system is allowed to move freely under gravity. Find the acceleration of the particles, and the tension in the string.

2 A particle A, of mass 6 kg, lying on a rough horizontal table, is connected by a string, which passes over a pulley at the edge of the table, to a particle B, of mass 4 kg, which hangs vertically. When the system is released from rest, the motion of A is resisted by a friction force of 5 N. Find the acceleration of the particles, and the tension in the string. (The particles and the pulley all lie in the same vertical plane.)

3 A particle A, of weight 30 N, is connected to a particle B, of weight 10 N, by a light inextensible cable which passes over a cogwheel. The parts of the cable which are not in contact with the cogwheel hang vertically. The cogwheel is not light, and as a result, when the system is released from rest the tension in the part of the cable supporting A is twice that in the part supporting B. Find the acceleration of the particles, and the tensions in the cable.

4 A particle A, of mass 2 kg, is pulled up a plane, which is inclined to the horizontal at an angle of 30°, by a string which passes over a pulley at the top of the plane. A particle B, of mass 4 kg, is attached to the other end of the string, and the part of the string between the pulley and this particle hangs vertically.

The system is released from rest and, as it moves, the motion of A is opposed by a friction force of 10 N. Find the acceleration and the tension.

When particle B has moved 1 m it strikes the floor and does not rebound. How much further does particle A move, before coming instantaneously to rest?

5 Figure 9.4 shows a system which consists of two pulleys and a string passing over them. One end of the string is fixed to a point P in the ceiling, and one of

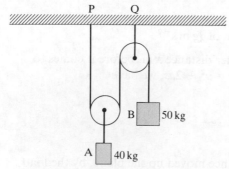

Figure 9.4

the pulleys is attached to a point Q in the ceiling; the other pulley is free to move vertically. The parts of the string not in contact with the pulleys hang vertically.

The movable pulley carries an object A, of mass 40 kg, and the free end of the string carries an object B, of mass 50 kg. Show that if A moves x cm upwards, then B moves $2x$ cm downwards; hence show that the magnitudes of their accelerations are in the ratio $1:2$.

The system is released from rest; find the acceleration of each object and the tension in the string.

In questions 6–10, leave g in your answers. It should be assumed that all the quantities are in SI units.

6 Two particles A and B have masses M and m, respectively (where $M > m$). They are connected by a string which passes over a fixed pulley. The parts of the string which are not in contact with the pulley hang vertically. Initially the system is held at rest; it is then released and allowed to move freely under gravity. Find the acceleration of the particles, and the tension in the string, in terms of M, m and g.

7 A particle of mass M, lying on a rough horizontal plane, is connected by a string, which passes over a pulley at the edge of the table, to a particle of mass m; the part of the string between the pulley and this particle hangs vertically. When the system is released from rest, the motion of A is resisted by a friction force of F (where $mg > F$). Find, in terms of M, m, F and g, expressions for the acceleration of the particles, and the tension in the string. (The particles and the pulley all lie in the same vertical plane.)

8 A particle A, of mass m, is pulled up a plane, which is inclined to the horizontal at an angle of α, by a string which passes over a pulley at the top of the plane. A particle B, of mass M, is attached to the other end of the string, and this hangs vertically.

The system is released from rest and, as it moves, the motion of A is opposed by a friction force of F. Find expressions for the acceleration and the tension, in terms of m, M, g, F and α.

When particle B has moved a distance h it strikes the floor and does not rebound. How much further does particle A move?

9 Figure 9.5 shows a block of mass M resting on a smooth horizontal table. It is connected, by strings which pass over pulleys at the edges of the table, to two loads, of mass m and $2m$ which hang vertically (as shown).

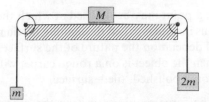

Figure 9.5

Show that, when the system is released, the magnitude of the acceleration, a, of each particle is given by

$$a = \frac{mg}{3m + M}$$

and find the tension in each string.

10 Figure 9.6 shows a double wedge, whose smooth faces are inclined to the horizontal at angles α and β; a smooth pulley is fixed to the vertex. The wedge is not free to move.

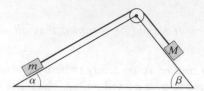

Figure 9.6

Particles of mass m and M, connected by a string, are placed on the wedge, with the string passing over the pulley (as shown). When the system is released the particle of mass M begins to move down the slope. Show that its acceleration, a, is given by

$$a = \frac{(M \sin \beta - m \sin \alpha)g}{(M + m)}$$

and find an expression for the tension in the string.

9.2 Friction

It is tempting to think of friction as a nuisance which should be eliminated; in a working machine this is usually the case. However, without friction, many forms of motion would be impossible; it is, for example, much more difficult to walk along an icy road than a dry one.

Imagine trying to push a heavy object, say a large chest of drawers, across a rough horizontal surface. At first it may be difficult to make the object move: the force applied is matched by the friction force between the object and the floor. While no movement takes place, this friction force is equal in magnitude to the force applied, but in the opposite direction.

If the force applied is gradually increased, a point may be reached at which the object begins to slip. The friction force at this instant is called the **limiting friction**. The magnitude of the limiting friction will depend on the nature of the surfaces that are in contact; the limiting friction when the object is on a rough carpet will clearly be greater than when the object is on a polished, tiled, surface.

Once the equilibrium has been broken, the force required to keep the object moving is slightly less than that required to break the equilibrium. For this reason a distinction is sometimes made between **static friction**, when there is no motion, and **dynamic friction**, when there *is* relative motion. However the difference between the two is not very great, and so we shall not bother with this distinction in this book.

The ratio of the limiting friction force, F, to the normal reaction, R, between the object and the surface, is called the **coefficient or friction**. ('Normal' means perpendicular to the surface.) This ratio is always denoted by the Greek letter μ, thus

$$F/R = \mu$$

i.e. $F = \mu R$

9.3 The laws of friction

Friction between dry surfaces was investigated experimentally by the distinguished French physicist Charles Augustin de Coulomb (1736–1806). The laws of friction which he discovered are still very useful, although they should not be used for lubricated surfaces, or for situations where there is a high relative velocity between the surfaces in contact. Coulomb's laws of friction are:

(1) The friction force between two surfaces opposes the relative motion.
(2) The magnitude of the friction force is independent of the area of the surfaces in contact.
(3) The magnitude of the friction force does not depend on the relative velocity of the surfaces.
(4) The magnitude of the limiting friction is proportional to the normal reaction, i.e. $F = \mu R$.

The significance of using the *normal* reaction may not, at first, be appreciated, but it becomes very important when the forces are not all horizontal, as we shall see.

These laws can be verified, and the coefficient of friction calculated experimentally by setting up a plane whose angle of elevation, α, can be varied. Place a heavy block of mass M on the plane and connect it via a pulley to a load of mass m, hanging vertically (see Figure 9.7).

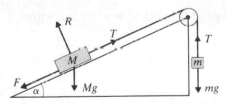

Figure 9.7

Now increase the load, m, until the friction force, F, reaches its limiting value (i.e. $F = \mu R$, where R is the normal reaction) and the block begins to slip up the plane. The coefficient of friction, μ, can now be calculated, as follows.

Let the tension in the string be T, then since the acceleration is zero, Newton's second law of motion gives

for the load,

$$mg - T = 0 \qquad \qquad \text{. . . (1)}$$

for the block, resolving parallel to the plane

$$T - \mu R - Mg \sin \alpha = 0 \qquad \qquad \text{. . . (2)}$$

for the block, resolving perpendicular to the plane

$$R - Mg \cos \alpha = 0 \qquad \qquad \text{. . . (3)}$$

From equation (1), we have $T = mg$; substituting this into equation (2), and rearranging, we have

$$mg = \mu R + Mg \sin \alpha$$

Also, from equation (3), $R = Mg \cos \alpha$, hence

$$mg = \mu Mg \cos \alpha + Mg \sin \alpha$$

Dividing by g and making μ the subject of this equation gives

$$\mu = \frac{m - M \sin \alpha}{M \cos \alpha}$$

Example 3

A car whose mass is 1.5 tonnes is initially moving at $30 \, \text{m s}^{-1}$. While travelling along a horizontal road, it skids to a halt from this speed in a distance of $100 \, \text{m}$. Taking g to be $10 \, \text{m s}^{-2}$, calculate the value of μ, the coefficient of friction between the tyres and the road surface.

Using the same value for μ, calculate the distance required to skid to a halt from the same initial speed, while descending a hill inclined to the horizontal at an angle α, where $\sin \alpha = 0.2$.

First, we find the retardation, using the formula $v^2 = u^2 + 2as$, with $v = 0$, $u = 30$ and $s = 100$, i.e.

$$0 = 900 + 200a$$
$$\therefore \quad a = -4.5$$

The retardation of the car is $4.5 \, \text{m s}^{-2}$.

This retardation is caused by the (limiting) friction force and this is equal to μR, where R is the normal reaction. However, in the vertical direction there are only two forces, namely R (upwards) and the weight $1500g$ newtons (downwards), and there is no vertical acceleration, so $R = 1500g \; (= 15\,000)$. Hence the retarding force is $15\,000\mu$ newtons. Using this in Newton's second law of motion gives

$$15\,000\mu = 1500 \times 4.5$$
$$\therefore \qquad 10\mu = 4.5$$
$$\therefore \qquad \mu = 0.45$$

The coefficient of friction is 0.45.

We now consider the motion of the car while it is skidding down the hill. The forces acting on the car are shown in Figure 9.8.

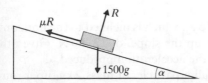

Figure 9.8

Resolving perpendicular to the plane gives

$$R - 1500g \cos \alpha = 0$$
$$\therefore \quad R = 15\,000 \cos \alpha$$

The force acting *down* the plane is $(15\,000 \sin \alpha - \mu R)$. So applying Newton's second law of motion to the forces down the plane, we have

$$15\,000 \sin \alpha - 0.45 \times 15\,000 \cos \alpha = 1500\,a$$

where a is the acceleration down the plane. Hence

$$a = 10 \sin \alpha - 4.5 \cos \alpha$$
$$= -2.41 \quad \text{(correct to the three significant figures)}$$

(The minus sign indicates, as we might have expected, that this is a retardation.)

We can now find the distance s, travelled while skidding down the slope, by using $v^2 = u^2 + 2as$, (with $v = 0$, $u = 30$ and $a = -2.41$), which gives

$$900 = 2 \times 2.41s$$
$$\therefore \quad s = 187 \quad \text{(correct to three significant figures)}$$

Hence, while travelling down the hill, the car requires 187 m to skid to a halt from a speed of $30 \, \text{m s}^{-1}$.

Exercise 9b

In questions 1–8, take g to be $9.81\,\mathrm{m\,s}^{-2}$.

1 A car travelling along a horizontal road, skids to a halt. Given that the skid marks are 100 m long, and that the coefficient of friction is 0.8, calculate the initial speed of the car.

2 A heavy chest, of mass 100 kg, is pushed along a rough horizontal floor by a horizontal force of 1000 N. Given that the coefficient of friction is 0.6, calculate the acceleration.

3 If, in question 2, the chest is pushed along the same surface by a force of 1000 N, acting downwards at an angle of 20° to the horizontal, find the magnitude of the acceleration produced.

4 Repeat question 2 given that the force acts upwards at an angle of 20° and *pulls* the chest along the floor.

5 A log of mass 1000 kg is pulled up a slope, which is inclined to the horizontal at an angle of 20°, by a force, acting up the slope, of 12 000 N. Given that $\mu = 0.75$, calculate the magnitude of the acceleration produced.

6 A force of 15 N is just sufficient to pull a particle, of mass 2 kg, up a rough plane, inclined at an angle of 40° to the horizontal. Given that the force acts up the plane, calculate the value of the coefficient of friction.

7 A heavy block slides down a rough plane at constant speed. Given that the plane is inclined at an angle of 25° to the horizontal, find the coefficient of friction.

8 A child pulls a toboggan, whose mass is 25 kg, up a snow-covered hill by exerting a force of 100 N directed up the slope. Given that the hill is inclined at an angle of 20° to the horizontal, find the coefficient of friction.
The child, whose mass is 35 kg, then slides down the hill on the toboggan. If the toboggan is initially at rest, find the speed attained when it has travelled 100 m.

9 A heavy block, of mass M, slides down a rough plane at constant speed. If the plane is inclined at an angle α to the horizontal, and the coefficient of friction is μ, show that

$$\mu = \tan \alpha$$

10 A heavy block, of mass M, is pulled along a rough horizontal surface, whose coefficient of friction is μ, by a constant horizontal force P. Show that the acceleration a produced is given by

$$a = (P/M) - \mu g$$

9.4 Further examples on connected particles and friction

No new concepts are introduced in this section, but two examples of solutions to harder questions are provided.

Example 4

A heavy block of mass $5m$ lies on a rough horizontal table; the coefficient of friction between the block and the table is $\frac{1}{2}$. The particle is connected, by a light inextensible string which passes over a light pulley fixed to the edge of the table, to a particle of mass $2m$, and this in turn is connected by another light inextensible string to a particle of mass m. The two particles hang vertically and the string is taut (see Figure 9.9). The system is released from rest; find
(a) the acceleration,
(b) the tension in each string.

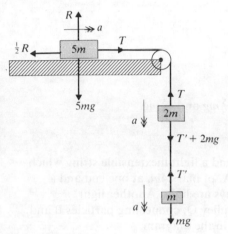

Figure 9.9

The strings are light and inextensible, so the acceleration is the same (in magnitude) for all three objects; let this be a. The pulley is light, so the tension in the string will not change as it passes over the pulley; let this tension be T, and let the tension in the string connecting the two particles be T', as shown in the diagram.

We now apply Newton's second law of motion to each of the three objects:

the block, mass $5m$

resolve vertically $\qquad\qquad R - 5mg = 0$
resolve horizontally $\qquad\qquad T - \frac{1}{2}R = 5ma$
hence $\qquad\qquad T - \frac{1}{2} \times 5mg = 5ma \qquad \ldots \ (1)$

the particle, mass $2m$

resolve vertically $\qquad 2mg + T' - T = 2ma \qquad \ldots \ (2)$

the particle, mass m

resolve vertically $\qquad\qquad mg - T' = ma \qquad \ldots \ (3)$

(a) Both T and T' can be eliminated by adding equations (1), (2) and (3):

$$\tfrac{1}{2}mg = 8ma$$
$$\therefore \quad a = \tfrac{1}{16}g$$

The acceleration is $\tfrac{1}{16}g$.

(b) Substituting this for a in equation (1) gives

$$T - \tfrac{1}{2} \times 5mg = 5m \times \tfrac{1}{16}g$$

Hence

$$T = (\tfrac{5}{16} + \tfrac{5}{2})mg$$
$$= \tfrac{45}{16}mg$$

and from equation (3)

$$T' = mg - ma$$
$$= mg - \tfrac{1}{16}mg$$
$$= \tfrac{15}{16}mg$$

The tensions in the strings are $\tfrac{45}{16}mg$ and $\tfrac{15}{16}mg$.

Example 5

Pulley P in Figure 9.10 is fixed and a light inextensible string which passes over it carries a particle A, of mass $4m$, at one end and a pulley Q at the other; both pulleys are light. Another light inextensible string passes over pulley Q, connecting particles B and C, of mass $2m$ and m, as shown in the diagram.

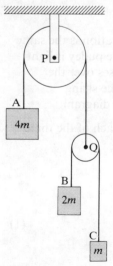

Figure 9.10

Find the acceleration of particle A when the system is released from rest.

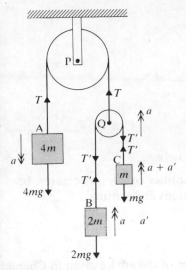

Figure 9.11

Let the tensions be T and T', as shown in Figure 9.11. Since pulley Q is light (i.e. its mass is zero), Newton's second law of motion gives

$$T - 2T' = 0 \times a$$

i.e. $T = 2T'$ . . . (1)

Let the acceleration of B, *relative to pulley Q*, be a', i.e. the *actual accelerations* of B and C are $a - a'$ and $a + a'$, as shown in Figure 9.11. Applying Newton's second law of motion to each particle in turn gives:

particle A	$4mg - T = 4ma$	. . . . (2)
particle B	$T' - 2mg = 2m(a - a')$	. . . . (3)
particle C	$T' - mg = m(a + a')$	. . . . (4)

Multiply equation (4) by 2 and add it to equation (3) to eliminate a'

$$3T' - 4mg = 4ma$$

Hence

$$T' = \tfrac{1}{3}(4mg + 4ma)$$
$$= \tfrac{4}{3}m(g + a)$$

and hence, from equation (1)

$$T = \tfrac{8}{3}m(g + a)$$

Substituting this in equation (2), we have

$$4mg - \tfrac{8}{3}m(g + a) = 4ma$$

Dividing by $4m$ gives

$$g - \tfrac{2}{3}(g + a) = a$$

Multiply both sides by 3:

$$3g - 2(g + a) = 3a$$
$$\therefore \qquad g - 2a = 3a$$
$$\therefore \qquad g = 5a$$
$$\therefore \qquad a = \tfrac{1}{5}g$$

Therefore the acceleration of particle A is $\tfrac{1}{5}g$.

[Note that the remaining unknown quantities could, if necessary, be found by returning to the original equations and putting $a = \tfrac{1}{5}g$.]

9.5 A reminder

We end this chapter by reminding the reader of the advice given in Chapter 8, which is applicable to most exercises in applied mathematics.

1 Always draw a clear diagram showing all the relevant forces, and the acceleration(s).
2 Define the algebraic symbols clearly.
3 Explain the method used, for example 'by Newton's second law of motion'.
4 If the problem involves several particles, it is usually advisable to consider each particle separately.
5 State the answer(s) clearly, giving the units and, in the case of vector quantities, quoting both the magnitude and the direction.

Exercise 9c (Miscellaneous)

In questions 1–6 take g to be 9.81 m s^{-2}.

1 An aircraft, with its undercarriage retracted, crash lands at 80 m s^{-1} on a horizontal field which provides a coefficient of friction of 0.5. How far from the point of touchdown will the aircraft slide before coming to rest?
2 A car is travelling at 30 m s^{-1} on a straight greasy road; the coefficient of friction between the road and the car's tyres, when they are locked, is 0.4. The driver spots a hazard 100 m ahead and makes an emergency stop with the wheels locked. Find the speed with which the car hits the hazard.
3 A children's slide in an amusement park is inclined to the horizontal at an angle $\tan^{-1} 0.75$; its coefficient of friction for a child wearing ordinary clothing is 0.3. A small girl slides down it, starting from rest at a vertical height of 6 m. Calculate the time she takes to reach ground level. (Neglect the height of the final 'run-off' above ground level.)
4 A particle A, of mass 2 kg, is pulled along a rough horizontal table by a light inextensible string, which passes over a light pulley attached to the table's edge; a particle B, of mass 3 kg, is attached to the other end of the string, and the part of the string between the pulley and this particle hangs vertically. The coefficient of friction between particle A and the table is 0.9.

The system is released from rest; find the tension in the string and the acceleration.

5 Repeat question 4, given that the weight of particle A is 10 N, and the weight of particle B is 20 N.

6 A particle A, of mass 2 kg, is pulled up a rough plane, inclined to the horizontal at an angle of 30°, by a light inextensible string, which passes over a light pulley attached to the top of the incline; a particle B, of mass 3 kg, is attached to the other end of the string, and the part of the string between the pulley and this particle hangs vertically. The coefficient of friction between particle A and the plane is 0.75.
The system is released from rest; find the time it takes for particle B to reach the floor, 1 m below its initial position. (Assume that particle A does not reach the pulley.)

7 A heavy block of mass M is pulled up a rough plane, which is inclined to the horizontal at an angle α, by a constant force P, acting up the plane; the force is sufficiently large to make the block slide up the plane. Given that the coefficient of friction is μ, show that the acceleration, a, of the block is given by

$$a = \frac{[P - Mg(\sin \alpha + \mu \cos \alpha)]}{M}$$

8 A particle A, of mass M, is pulled along a rough horizontal table by a light inextensible string, which passes over a light pulley attached to the table's edge; a particle B, of mass m, is attached to the other end of the string, and this hangs vertically. The coefficient of friction between particle A and the table is μ, where $m > \mu M$. Show that, when the system is released from rest, its acceleration, a, is given by

$$a = \frac{(m - \mu M)g}{(m + M)}$$

9 A wedge, whose cross-section is an equilateral triangle, is fixed to a horizontal plane. A smooth light pulley is attached to the highest point. Two particles whose weights are 10 N and 20 N are connected by a light inextensible string which passes over the pulley and one of the particles is placed on each of the inclined faces of the wedge. The coefficient of friction between each of the particles and the surface of the wedge is $\frac{1}{6}\sqrt{3}$. The system is released from rest; show that the tension in the string is $\frac{20}{3}\sqrt{3}$ N, and that the acceleration is $\frac{1}{12}g\sqrt{3}$ m s^{-2}.

10 A light inextensible string passes over a light smooth pulley and, at each end of it, there is a scale pan of negligible weight. A particle of mass $2m$ is placed in one of the pans, and two particles of mass m and $5m$ are placed in the other; the system is then released from rest. Find the acceleration, and the tension in the string.
When the scale pans have moved a distance h the particle of mass $5m$ falls off. Show that the system comes momentarily to rest when it has moved a further distance $\frac{3}{2}h$.

In questions 11–14 take g to be 10 m s^{-2}.

11 The diagram shows three masses of 5 kg, 4 kg and 11 kg respectively, connected by light inextensible strings, one of which passes over a smooth fixed pulley. The system is released from rest.

Calculate
  (a) the acceleration of the masses;
  (b) the tension in the string joining the 4 kg mass to the 11 kg mass;
  (c) the tension in the string joining the 4 kg mass to the 5 kg mass.

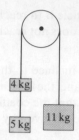

Figure 9.12   *(C)

12 The diagram shows a smooth fixed pulley and three masses connected by light strings. Find the acceleration of the masses when the system is released from rest.

Given that the string joining A and B breaks 2 seconds after the start of motion, calculate the further distance that B will descend before coming to instantaneous rest.

(It may be assumed that C does not reach the pulley.)

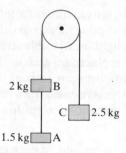

Figure 9.13   *(C)

13 A particle of mass 10 kg is pulled along a rough horizontal plane by a force of magnitude P N and inclined at 30° to the horizontal. If the coefficient of friction between the particle and the plane is $1/\sqrt{3}$, find the least value of P.

(L)

14

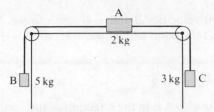

Figure 9.14

The diagram shows a particle A of mass 2 kg resting on a horizontal table. It is attached to particles B of mass 5 kg and C of mass 3 kg by light inextensible strings hanging over light smooth pulleys. The plane ABC is vertical. Initially, the strings are taut and pass over the pulleys at right angles to the edges of the table. In the ensuing motion from rest find the common acceleration of the particles and the tension in each string before A reaches an edge

(a) when the table is smooth;

(b) when the table is rough and the coefficient of friction between the particle A and the table is $\frac{1}{2}$.

(L)

15 A particle of mass $2m$ is on a plane inclined at an angle $\tan^{-1}\left(\frac{3}{4}\right)$ to the horizontal. The particle is attached to one end of a light inextensible string. This string runs parallel to a line of greatest slope of the plane, passes over a small smooth pulley at the top of the plane and then hangs vertically carrying a particle of mass $3m$ at its other end. The system is released from rest with the string taut. Find the acceleration of each particle and the tension in the string when the particles are moving freely, given that

(a) the plane is smooth;

(b) the plane is rough and the coefficient of friction between the particle and the plane is $\frac{1}{4}$.

(L)

Revision Exercise 2

1 Particles A and B move with constant velocities $2\mathbf{i}$ and $-4\mathbf{j}$ respectively. At time $t = 0$, A is at the origin and B is at the point with position vector $5\mathbf{i} + 3\mathbf{j}$. Find
   (a) the velocity of B relative to A;
   (b) the distance between A and B at time $t = 2$.

The velocity of a third particle C relative to A is in the direction of the vector $3\mathbf{i} + 2\mathbf{j}$ and relative to B is in the direction of the vector $\mathbf{i} + \mathbf{j}$. If the velocity of C is $a\mathbf{i} + b\mathbf{j}$, find the values of a and b.

(L)

2 A man cycling at a constant speed u on level ground finds that when his velocity is $u\mathbf{j}$ the velocity of the wind appears to be $v(3\mathbf{i} - 4\mathbf{j})$ where $\mathbf{i}$ and $\mathbf{j}$ are unit vectors in the East and North direction respectively. When the velocity of the man is $\frac{1}{5}u(-3\mathbf{i} + 4\mathbf{j})$ the velocity of the wind appears to be $w\mathbf{i}$. Find v and w in terms of u and prove that the true velocity of the wind is

$$\tfrac{1}{20}u(3\mathbf{i} + 16\mathbf{j})$$

Find the velocity of the wind relative to the man when his velocity is $u\mathbf{i}$.

(L)

3 An aircraft, which can fly at $300\,\text{km}\,\text{h}^{-1}$ in still air, has to fly directly from a point A to a point B, $600\,\text{km}$ North West of A. A steady wind is blowing from due South at a speed of $100\,\text{km}\,\text{h}^{-1}$. Using a graphical method, or any other method, find the course which the pilot should steer and the time taken for the journey. (Give your answers to an accuracy appropriate to your method.)

When the aircraft gets to B the pilot alters course and steers in a North Easterly direction. Find the distance and bearing of the plane, from B, two hours after leaving B.

(C)

4 The water in a wide river flows due South at a speed of $3\,\text{km}\,\text{h}^{-1}$. The wind is blowing from the direction N 60° E at a speed of $10\,\text{km}\,\text{h}^{-1}$. Find, graphically or otherwise, the magnitude (in $\text{km}\,\text{h}^{-1}$) and direction of the velocity of the wind relative to the water, giving your answers to an accuracy appropriate to your method.

The velocity of a boat, relative to the water, is of magnitude $5\,\text{km}\,\text{h}^{-1}$ and is in a South Easterly direction. Find the magnitude (in $\text{km}\,\text{h}^{-1}$) and direction of the velocity of the boat relative to the wind.

(C)

5 At noon a boat is at a point A and a helicopter is at a point B, $20\,\text{km}$ due West of A. The boat moves at constant velocity due North with speed $60\,\text{km}\,\text{h}^{-1}$. In order to intercept the boat, the helicopter flies at constant velocity *relative to the air* of magnitude $100\,\text{km}\,\text{h}^{-1}$ and direction θ East of North. Calculate the required value of $\tan\theta$ and the time at which the helicopter reaches the boat when
   (a) there is no wind;
   (b) the wind is $20\,\text{km}\,\text{h}^{-1}$ from the North.

In the case when the wind is $50 \, \text{km h}^{-1}$ from the North interception is impossible. Find the value of $\cos \theta$ which must be chosen if the helicopter is to approach the boat as closely as possible and show that the distance of closest approach is $\frac{20}{11}\sqrt{21} \, \text{km}$.

(JMB)

6 A particle is projected, from a point O, at an angle of $60°$ above the horizontal. The particle passes through the point A whose coordinates are $(a\sqrt{3}, a)$ relative to horizontal and upward vertical axes Ox, Oy in the plane of projection. Find the speed of projection and show that the greatest height reached is $\frac{9}{8}a$.

Find the time of flight from O to A and the magnitude and direction of the velocity at A.   (C)

7 (*Take g to be* $10 \, \text{m s}^{-2}$ *in this question.*)

A particle P projected with speed V and angle of elevation α from a point O on horizontal ground moves freely under gravity. When P is vertically above a point A on the ground its height is $900 \, \text{m}$ and its velocity components are $50 \, \text{m s}^{-1}$ horizontally and $40 \, \text{m s}^{-1}$ vertically upwards. Calculate V, $\tan \alpha$ and the distance OA.

At the instant when P is vertically above A, a second particle Q is projected from O in such a way that P and Q hit the ground at the same place at the same time. Calculate the initial speed of Q and the tangent of the angle of elevation at which Q is projected.

(C)

8 A particle P is projected under gravity from a point O. The velocity of projection is inclined at an angle α above the horizontal, where $\tan \alpha = \frac{4}{3}$. After a time T, the line OP is inclined at an angle β above the horizontal, where $\tan \beta = \frac{3}{4}$. Show that the speed of projection is $\frac{10}{9}gT$.

Find, in terms of g and T,
(a) the height of P above the level of O at time T and the time when P is next at this height;
(b) the speed of P at time T.

(C)

9 (*Take g to be* $10 \, \text{m s}^{-2}$ *in this question.*)

A particle is projected from a point O which is $50 \, \text{m}$ above horizontal ground. The speed of projection is $40 \, \text{m s}^{-1}$ and the particle hits the ground at a point R which is at a *horizontal* distance of $200 \, \text{m}$ from O. The angle of elevation at which the particle is projected is α. Show that one of the possible values of $\tan \alpha$ is $\frac{3}{2}$ and find the other possible value.

Given that $\tan \alpha = \frac{3}{5}$, find
(a) the time of flight;
(b) the direction of motion of the particle just before impact with the ground at R;
(c) the maximum height of the particle above ground level during the flight.

(C)

10 A shell is projected at an elevation $\pi/6$ from a point A on a horizontal plane. At the same instant of projection a target moves from B, another point on the plane, where $AB = l$, with uniform speed V, where $V^2 = 2gl/\sqrt{3}$, in the direction AB. Show that, if the shell is to hit the target, then the speed of projection must be $\sqrt{3}V$.

If the target had risen vertically from B with uniform speed V, find the speed of projection, at the same elevation, for the shell to hit the target. Find the height above the plane of the point of impact.

(L)

11 A particle of mass 3 kg is acted on by forces measured in newtons represented by the vectors $2\mathbf{i} + 5\mathbf{j}$, $9\mathbf{i} - 2\mathbf{j}$, $p\mathbf{i} + q\mathbf{j}$. If its acceleration in m s^{-2} is $5\mathbf{i} - 2\mathbf{j}$, what are the values of p and q?

The particle starts at the point $(1,1)$ with velocity $4\mathbf{i} + 10\mathbf{j}$, where distance is measured in metres and velocity in m s^{-1}. What is the greatest positive displacement parallel to the y-axis in the subsequent motion, and after how many seconds does it occur?

*(O & C)

12 A gas balloon, with its load and occupants, has a total mass of 400 kg. If it is descending with a constant speed of 5 m s^{-1}, how large is the combined force of buoyancy and air-resistance? (Take g to be 9.81 m s^{-2}.)

In order to slow down the rate of descent, a mass of x kg of ballast is thrown out, sufficient to produce a retardation of 0.2 m s^{-2}. If the buoyancy and air-resistance remain unaltered, what is the value of x?

(SMP)

13 (*Take g to be* 10 m s^{-2} *in this question.*)

A lift starts from rest, ascends vertically and then stops. The lift carries weighing apparatus, and readings are taken of the apparent weight of an object of mass 20 kg during the motion of the lift.

During the first 2 seconds the reading was 218 N. Calculate the acceleration of the lift during this period.

During the next 15 seconds the reading was 200 N, and during the final t seconds the reading was 173 N. Calculate the value of t and the total distance moved by the lift.

*(C)

14 A particle of mass m moves along the x-axis under the action of a force F which is directed along the positive x-axis and at time t, is given by $F = mk \sin (nt)$, where n and k are positive constants. At time $t = 0$ the particle is projected from the origin with speed u along the positive x-axis. Find the acceleration, velocity and displacement from O at time t.

(L)

15 A particle P, of mass 10 kg, has position vector

$[(4 \cos 5t)\mathbf{i} + (4 \sin 5t)\mathbf{j}]$ metres

at time t seconds. Describe the path followed by P and find the magnitude of the force acting on P at time t seconds and the unit vector in the direction of this force. Find also the angle between this force and the positive direction of the x-axis.

Obtain the distance of the closest approach of P to the point with position vector $(5\mathbf{i} + 12\mathbf{j})$ metres.

(L)

16 Two particles A and B, of mass m and $3m$ respectively, are connected by a light inextensible string of length l. Particle A is placed on a horizontal table at a perpendicular distance l from its edge, the coefficient of friction between particle A and the table being $\frac{1}{3}$. Particle B, which is at the edge of the table, is just pushed over the edge and subsequently falls under the action of gravity and the tension in the string. Assuming that there is no friction between the string and the table, show that the speed of A when it reaches the table edge is $2\sqrt{(\frac{1}{3}gl)}$.

Find also the time taken by particle A to reach the edge of the table and the tension in the string while A is moving on the table.

(L)

17 A triangular wedge is fixed with one face horizontal. The vertical cross-section ABC is an isosceles triangle with vertex A, and with BC horizontal and below the level of A. The angles at B and C are each equal to 30° and $AB = AC = 2l$. Two particles P and Q, of mass $4m$ and m respectively, are joined by a light inextensible string of length $2l$ passing over a smooth pulley at A and lying in the vertical plane ABC. The coefficient of friction between each particle and the wedge is $\frac{1}{6}\sqrt{3}$. Initially P is at A on the face through AB and Q is at C; the system is released from rest. When the particles are at the same horizontal level the string breaks.

Show that at this instant the speed of each particle is $\sqrt{(\frac{1}{10}gl)}$.

Find also the speed of P when it reaches B.

(C)

18 A rough plane is fixed at an angle α to the horizontal where $\tan \alpha = \frac{3}{4}$. A particle P of mass $5m$ is in contact with the plane and is attached to a light inextensible string which passes over a small light frictionless pulley at the highest point O of the plane. The other end of the string is attached to a particle Q of mass M hanging freely. The coefficient of friction between the particle and the plane is $\frac{1}{4}$. The system is initially held at rest with the string taut and in the vertical plane containing the line of greatest slope of the plane through O. Show that on being released the system stays at rest if $2m \leqslant M \leqslant 4m$.

Find the acceleration of the particles when $M = 7m$ and show that the tension in the string is then $\frac{21}{4}mg$.

(O & C)

19 A mass of 5 kg is moved along a rough horizontal table by means of a light inextensible string which passes over a smooth light pulley at the edge of the table and is attached to a mass of 1 kg hanging over the edge of the table. The system is released from rest with the string taut, each portion being at right

angles to the edge of the table. If the masses take twice the time to acquire the same velocity from rest that they would have done had the table been smooth, prove that the coefficient of friction is $\frac{1}{10}$.

After falling a distance $\frac{1}{5}$ m, the 1 kg mass reaches the floor and comes to rest. Prove that the 5 kg mass then moves a further $\frac{1}{6}$ m before coming to rest.

(Assume throughout that the 5 kg mass does not leave the table.)

(O & C)

20 A light inextensible string passes over a smooth light fixed pulley and masses of 3 kg and 7 kg are attached to its ends. The system is held at rest with the string taut, those parts not in contact with the pulley being vertical, and then released. Find the acceleration of each mass and the tension in the string.

After the 7 kg mass has descended a distance of one metre it strikes an inelastic horizontal table. Show that the time taken for this to happen is $\frac{5}{7}$ s from the start of the motion, assuming that the acceleration due to gravity is 9.8 m s^{-2}.

Find the time during which the 7 kg mass is at rest on the table.

(Assume throughout that the 3 kg mass does not reach the pulley.)

(O & C)

Chapter 10

Work, energy, power

10.1 Introduction

In this chapter we shall meet three new concepts: work, energy and power. These terms are used in everyday speech, but when they are used in applied mathematics they have precisely defined meanings.

10.2 The work done by a constant force

Imagine a climber using ropes to ascend a vertical cliff face. If he is simply suspended on the end of the rope, then he must exert a force equal to his own weight; let us take this to be 800 N. Suppose now he proceeds to climb up 20 m. No doubt the climber will feel this is hard work, and mathematically we say that he has done 800×20 units of work, i.e. multiply the force exerted by the displacement.

Units

The work done when the force exerted is 1 N and the displacement, in the direction of the force, is 1 m, is called a joule[†] (abbreviation J).

> The work done by a constant force F newtons when it moves through a displacement of x metres *in the direction of the force* is Fx joules.

If however the force does not act in the same direction as the displacement, say when a heavy chest is pulled along a rough horizontal floor by a force inclined at a constant angle to the horizontal, then the work done is the product of the displacement and the *component of the force in the direction of the displacement*. In other words, if the force is of magnitude F newtons and it acts at an angle α to the direction of the displacement x m, then the work done by the force is $Fx \cos \alpha$ (see Figure 10.1 on page 157).

[†] The unit of work is named after the distinguished physicist James Joule (1818–1889).

Conservation of energy

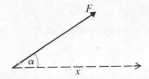

Figure 10.1

It is frequently convenient to express this in vector notation, using the idea of the scalar product of two vectors (see §5.7). If the force is represented by vector **F** and the displacement by vector **x**, then the work done is given by the scalar product **F . x** ($= Fx \cos \alpha$).

Note these two points.

(a) Work done is a scalar, not a vector, quantity, i.e. it has magnitude, but it does not have a direction.
(b) The definition of work done makes no reference to the rate at which the task is completed (this will come later when we consider 'power').

Qu. 1 Find the work done when a load of bricks, whose mass is 50 kg, is raised from ground level to a height of 15 m. (Take g to be $9.81 \, \text{m s}^{-2}$.)

Qu. 2 A force of 100 N acting at 60° to the horizontal is used to pull a chest across a horizontal floor, for a distance of 20 m. Calculate the work done by the force.

Qu. 3 A force, in newtons, represented by the vector $5\mathbf{i} + 10\mathbf{j}$ undergoes a displacement in metres, represented by the vector $3\mathbf{i} - \mathbf{j}$. Calculate the work done by the force.

10.3 The work done by a force of variable magnitude

Let us consider a variable force F acting along the x-axis and a displacement from $x = a$ to $x = b$.

The magnitude of the force F varies in the course of the displacement, i.e. F is a function of x, so we must first consider a *small displacement* δx. The force may be regarded as constant over a small displacement, consequently the work done is $F\delta x$. Now we sum this over all possible values of x from $x = a$ to $x = b$, and let δx tend to zero, i.e. we find the definite integral

$$\int_a^b F \, \mathrm{d}x \quad \dagger$$

---

†For a more detailed discussion of definite integrals see *Pure Mathematics Book 1* by J. K. Backhouse and S. P. T. Houldsworth, revised by P. J. F. Horril, published by Longman, Chapter 8.

Example 1

A particle is moved along the x-axis by a force P acting along the x-axis, given by

$$P = \frac{100}{x^2}$$

from $x = 1$ to $x = 5$. Calculate the work done, assuming that the force is in newtons and the displacement in metres.

$$
\begin{aligned}
\text{Work done} &= \int_1^5 P \, dx \\
&= \int_1^5 \frac{100}{x^2} \, dx \\
&= \left[-\frac{100}{x} \right]_1^5 \\
&= -\frac{100}{5} - \left(-\frac{100}{1} \right) \\
&= -20 + 100 \\
&= 80
\end{aligned}
$$

The work done is 80 joules.

Note that work done can be negative: if for instance a load of 800 N is *lowered* 20 m, the work done is $-16\,000$ J. Or again, if the particle in Example 1 moved from $x = 5$ to $x = 1$, the work done would be -80 J. Returning to the climber in §10.2, we say that the climber has done $+16\,000$ J of work *against* gravity; we can also say that, since the climber's weight is in the opposite direction to the displacement, the work done by gravity is $-16\,000$ J.

When working in more than one dimension, the definition of work done, used above, must be refined. If the (variable) force is represented by the vector $\mathbf{F}$, and if $\mathbf{r}$ represents the position vector of a point on the path along which the point of application of $\mathbf{F}$ moves, then

$$\text{work done} = \int_A^B \mathbf{F} \cdot d\mathbf{r}$$

where the limits of the definite integral are determined by the coordinates of the initial point, A, and the final point, B.

Definite integrals of this type are, in general, beyond the scope of this book. One case, however, is of special importance, namely work done against the attraction of gravity.

Suppose that a man carrying a heavy case weighing 500 N, walks up a slope, whose cross-section is shown in Figure 10.2, from point A to point B, where the vertical height, h, of B above A is 20 m.

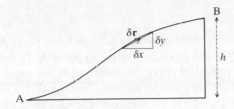

Figure 10.2

The force exerted by the man on the case is 500**j** (i.e. 500 N vertically upwards).
Let

$$\delta\mathbf{r} = (\delta x)\mathbf{i} + (\delta y)\mathbf{j}$$

be a small displacement on the path joining A to B. Then the work done in this small displacement is

$$500\mathbf{j} \cdot (\delta\mathbf{r}) = 500\mathbf{j} \cdot [(\delta x)\mathbf{i} + (\delta y)\mathbf{j}]$$
$$= 500(\delta y)$$

Integrating this from A to B (i.e. from $y = 0$ to $y = 20$) gives

$$\text{work done} = \int_0^{20} 500\,dy$$
$$= \left[500y \right]_0^{20}$$
$$= 500 \times 20$$
$$= 1000$$

i.e. the work done by the man on the case is 1000 J.

Repeating this argument for an object of mass m (whose weight therefore is mg), and a displacement whose vertical component is h, we would find that the work done is mgh. The point to note here is that the *shape* of the path from A to B does not matter; it is only the vertical displacement, h, which appears in this expression.

A force for which the work done, in moving from point A to point B, *does not* depend on the form of the path joining A to B, is called a **conservative** force. The work done by this force, returning from B to A has the same magnitude, but it is negative. Consequently the total work done by a conservative force when its point of application moves round any closed path (i.e. a path beginning and ending at the same point) is zero.

For some forces the work done *does* depend on the path chosen, for example the work done against friction would depend on the length of the path; we say that such a force is **non-conservative**. Unlike the previous case, the work done by a non-conservative force *cannot* be recovered.

10.4 Potential energy

The work done by an external conservative force moving a particle from point A to point B is called the **potential energy** of the particle relative to point A. In the case of work done against gravity, the potential energy is usually, but not necessarily, measured from ground level, in which case the potential energy (P.E.) of the particle when it has been raised to a point h m above ground level is mgh, i.e.

P.E. $= mgh$

It must be stressed that gravitational potential energy is always measured relative to a given baseline.

Exercise 10a

Take g to be 9.81 m s^{-2} *throughout this exercise.*

1 Calculate the work done when the following forces undergo the displacements shown:
   (a) a force of 10 N, acting in the direction of a displacement of 20 m;
   (b) a force of 20 N, acting at an angle of 60° to the horizontal, and a horizontal displacement of 5 m;
   (c) a force in newtons represented by the vector $(4\mathbf{i} + 3\mathbf{j})$ and a displacement in metres represented by $(3\mathbf{i} - 2\mathbf{j})$.

2 A conveyor belt, inclined at 20° to the horizontal, carries a package of mass 50 kg. Calculate the work done when the package travels 15 m up the incline.

3 Calculate the work done against gravity when an aircraft, whose mass is 50 tonnes, takes off and climbs to a height of 1500 m.

4 A car of mass 1 tonne travelling at 50 m s^{-1} skids to a halt with constant retardation. In each of the following cases, calculate the magnitude of the work done by the friction force, given that
   (a) the car comes to rest in 4 s;
   (b) the skid marks are 150 m long;
   (c) the skid marks are x metres long.
   What do you deduce from your answers?

5 When a spring whose natural (i.e. unstretched) length is 1 m, is extended by x metres, the force required to hold it in this position is $200x$ newtons. Calculate the work done extending this spring from its unstretched state until the total length is 1.5 m.

6 It is estimated that an escalator in a shop carries 40 people a minute from one floor to the next, a vertical distance of 10 m. Assuming that a typical person has a mass of 60 kg, estimate the average work done per second by the escalator.

7 Two particles, whose masses are 2 kg and 5 kg, are connected by a light inextensible string, which passes over a pulley; the parts of the string not in contact with the pulley hang vertically. Find the overall gain in the potential energy of the system if the 2 kg particle is pulled down 50 cm.

8 A toy car, of mass 50 grams 'loops the loop' on the inside of a plastic track in the form of a circle, radius 25 cm, whose plane is vertical. Denoting the centre

of the circle by O, the lowest point on the track by A, and the position of the car at any instant by P, calculate the change in potential energy of the car between points A and P when:

(a) P is at the highest point on the track;

(b) OP makes an angle of 60° with the downward vertical;

(c) OP makes an angle of 30° with the upward vertical.

9 Writing the mass of the car in question 8 as m kg and the radius of the circle as r metres, express the change in the potential energy of the car between point A and point P, when OP makes an angle α with the upward vertical, in terms of m, g, r and α.

10 A particle A, whose weight is W newtons, is connected by a light inextensible string to a particle B, whose weight is w newtons. The particles are placed on a fixed double wedge, with particle A on one face which is inclined at 60° to the horizontal and particle B on the other face which is inclined at 30° to the horizontal. The string passes over a smooth light pulley fixed to the vertex of the wedge (see Figure 10.3).

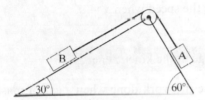

Figure 10.3

The surface of the wedge is rough and the coefficient of friction between each particle and the surface of the wedge is μ. Find an expression for the work done when particle B moves x metres down the inclined plane.

10.5 Kinetic energy

Suppose that a particle of mass m is accelerated from rest, over a distance s, by a constant force F; then the work done by F is Fs. But, from Newton's second law of motion, $F = ma$, where a is the acceleration, so the work done can be written mas. Using the constant acceleration formula $v^2 = u^2 + 2as$, with $u = 0$, we can write $\frac{1}{2}v^2 = as$, and hence the work done can be expressed as follows:

$$\text{work done} = \tfrac{1}{2}mv^2$$

This quantity, $\frac{1}{2}mv^2$, is called the **kinetic energy** (K.E.) of the particle. Since the K.E. is equal to the work done, it is measured in the same units, i.e. joules, and like work done, it is a scalar, not a vector, quantity.

The argument above was based upon the assumption that the force was constant, but this was not actually necessary. Suppose the particle is accelerated, from

$x = a$ to $x = b$, by a *variable force F*, acting along the x-axis (i.e. F is a function of x). Then, from §10.3, the work done is given by

$$\text{W.D.} = \int_a^b F \, dx$$

By Newton's second law of motion $F = m\dfrac{dv}{dt}$. But, in §4.5 we saw that $\dfrac{dv}{dt}$ can be written as $\dfrac{1}{2}\dfrac{d}{dx}(v^2)$. So F can be expressed as $\dfrac{m}{2}\dfrac{d}{dx}(v^2)$. Using this the work done can be written

$$\text{W.D.} = \int_a^b \frac{m}{2}\frac{d}{dx}(v^2)\,dx$$

$$= \left[\tfrac{1}{2}mv^2\right]_a^b$$

$$= \tfrac{1}{2}mv_2^2 - \tfrac{1}{2}mv_1^2$$

where v_2 is the speed when $x = b$ and v_1 is the speed when $x = a$.

This can be expressed in words, as follows:

> The work done is equal to the change in the kinetic energy.

Qu. 4 Neglecting frictional forces, calculate the work done when a car of mass 1 tonne accelerates on level ground from $10\,\text{m s}^{-1}$ to $50\,\text{m s}^{-1}$.

Example 2

A bullet of mass 25 grams is fired horizontally through a fixed block of wood. Its speed on entry is $800\,\text{m s}^{-1}$ and it emerges at $200\,\text{m s}^{-1}$. Calculate the loss of kinetic energy and, given that the block of wood is 10 cm thick, find the average resistance to the motion of the bullet through the block.

$$\text{Initial K.E.} = \tfrac{1}{2} \times 0.025 \times 800^2 \text{ J}$$
$$= 8000 \text{ J}$$
$$\text{Final K.E.} = \tfrac{1}{2} \times 0.025 \times 200^2 \text{ J}$$
$$= 500 \text{ J}$$
$$\therefore \text{ Loss of K.E.} = 7500 \text{ J}$$

Let the average resistance acting on the bullet be R newtons; then the work done as the bullet travels 10 cm through the block is $0.1\,R$ joules. Since the work done by the resistance is equal to the loss of K.E. by the bullet, we have

$$0.1R = 7500$$
$$\therefore \qquad R = 75\,000$$

Hence the average resistance is 75 kN.

10.6 Conservation of energy

In the last section we saw that the work done on an object is equal to the change in its kinetic energy. This principle is especially useful when we are considering motion under the influence of gravity.

The work done when a particle of mass m is raised a distance h vertically is mgh, and, as we saw in §10.4, this is called the potential energy relative to the initial position. If, now, the particle is released, and if its speed is v when it has fallen a distance h, then the work done by gravity, mgh, is equal to the kinetic energy, $\frac{1}{2}mv^2$, gained by the particle. In other words

> the potential energy lost = the kinetic energy gained

This is an example of the important principle of **conservation of energy**. In other words, provided that there are no other external forces which do work on the particle, for example air-resistance, the total amount of energy is constant.

The principle of conservation of energy is especially convenient in problems where we need to examine the initial and final states of the system, rather than the intermediate states.

Example 3

A simple pendulum consists of a particle, P, of mass m, suspended on a light inextensible string of length l. The other end of the string is attached to a point O, and the pendulum swings through an angle of 30° either side of the vertical. Find the speed of P as it passes through the lowest point of its path.

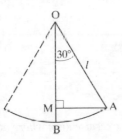

Figure 10.4

At the highest point, A, on the path, the speed of P is zero, so the K.E. at this point is zero. Let the speed at the lowest point, B, of the path be v, then the K.E. at this point is $\frac{1}{2}mv^2$. Hence the K.E. gained between points A and B is $\frac{1}{2}mv^2$. The vertical displacement between these points is MB, and

$$MB = OB - OM$$
$$= l - l\cos 30°$$
$$= l(1 - \tfrac{1}{2}\sqrt{3})$$

Hence the P.E. lost is $mgl(1 - \frac{1}{2}\sqrt{3})$; so, by the principle of conservation of energy, we have

$$\frac{1}{2}mv^2 = mgl(1 - \frac{1}{2}\sqrt{3})$$
$$\therefore \quad v^2 = 2gl(1 - \frac{1}{2}\sqrt{3})$$
$$= gl(2 - \sqrt{3})$$

Hence the speed of P at the lowest point is

$$\sqrt{[gl(2 - \sqrt{3})]}$$

We can use the principle of conservation of energy when there are other forces which do work on the system, provided that these are taken into account.

Suppose an aeroplane, of mass M, flying at $u\,\mathrm{m\,s^{-1}}$ climbs h metres at constant speed. At the end of the climb it has gained potential energy Mgh. In this case

the work done by the plane's engine = the P.E. gained, Mgh

Suppose however that during the climb, the speed of the aeroplane increases to $v\,\mathrm{m\,s^{-1}}$. This means that in addition to the P.E. gained, there is an increase in kinetic energy, namely

K.E. gained $= \frac{1}{2}Mv^2 - \frac{1}{2}Mu^2$

So the total increase in energy is

$$Mgh + \frac{1}{2}Mv^2 - \frac{1}{2}Mu^2$$

[Some readers may find it helpful to regard this expression as the *final* energy, $(Mgh + \frac{1}{2}Mv^2)$, minus the *initial* energy, $\frac{1}{2}Mu^2$.]

Hence, the total amount of work done by the engine is $(Mgh + \frac{1}{2}Mv^2 - \frac{1}{2}Mu^2)$, i.e.

> The work done is equal to the net gain in energy.

Exercise 10b

The principles of work and energy should be used in this exercise even where alternative methods exist.
Take g to be 9.81 m s^{-2} throughout this exercise.

1 A light wheel, of radius 1 m, is free to rotate in a vertical plane, about a smooth axle through its centre. A particle of mass 50 grams is attached to its highest point and slightly displaced. Find the speed of the particle in the subsequent motion, when the radius joining it to the centre of the circle has turned through
   (a) 90°   (b) 180°   (c) 225°   (d) 300°

2 A child on a sledge slides down a hill-side which is inclined to the horizontal at an angle of 30°. Given that the initial speed is zero and that the friction is negligible, calculate the speed attained when the sledge has travelled 100 m.

3   Two particles, each of mass 1 kg, are connected by a light inextensible string, which is then placed symmetrically over two small light pulleys, A and B. The distance between A and B is 2 m, and the line AB is horizontal. A third particle, also of mass 1 kg, is carefully attached to the midpoint of the string and the system is released. Find the distance of the third particle, below the level of AB, when the system first comes instantaneously to rest. (Assume that the particles do not rise to the level of A and B.)

4   A jumbo jet, whose mass is 300 tonnes, descends from an altitude of 12 000 m to an altitude of 2000 m, using its air-brakes to maintain constant speed. Calculate the work done by the air-brakes.

Assuming that the plane's path is a straight line inclined at 15° to the horizontal, calculate the force exerted by the air-brakes.

5   A snooker ball, of mass 100 grams, moving at $5 \, \text{m s}^{-1}$, strikes a stationary ball of the same mass. After the impact the balls move with speeds $1 \, \text{m s}^{-1}$ and $4 \, \text{m s}^{-1}$, respectively. Calculate the net loss of kinetic energy. How could this loss of energy be accounted for?

6   A bullet, of mass 0.050 kg, travelling horizontally at $500 \, \text{m s}^{-1}$, strikes a fixed block of wood and, after travelling 10 cm, it emerges from the block, travelling at $100 \, \text{m s}^{-1}$. Calculate the average resistance of the block to the motion of the bullet.

7   A conker, of mass 40 grams, hangs on a string which is 25 cm long. The other end of the string is fixed and the conker is free to move in a vertical circle about this fixed end. If the conker is given an initial horizontal velocity of $2 \, \text{m s}^{-1}$, calculate its vertical displacement when it comes instantaneously to rest, and find the angle through which the string has turned.

8   A girl slides down a chute into a swimming pool. The chute is 10 m long and it is inclined at 40° to the horizontal. Given that the girl is initially at rest, and that the coefficient of friction is 0.2, calculate her speed when she reaches the end of the chute.

9   A toy car, of mass 40 grams, runs along a plastic track which follows a curved path in a vertical plane. The car starts from rest at a point which is 100 cm above ground level; it runs down the track and then climbs again, coming to rest at a height of 75 cm above ground level. Calculate the work done by the force resisting its motion.

10 A skier travels down a slope which is inclined to the horizontal at an angle of 20°. If she starts from rest, calculate her speed when she has travelled 200 m
    (a) if all resistances are neglected;
    (b) if the friction force is included and the coefficient of friction is taken to be 0.2.
    Discuss the other factors which might affect the calculation.

11 A bob-sleigh team sets its bob in motion by pushing it along a horizontal track and then jumping on board before it starts the descent. The total mass of the team and the bob is 400 kg, and the speed when the whole team is on board is $10 \, \text{m s}^{-1}$. Given that, when the bob has descended 100 m (vertically) its speed is $30 \, \text{m s}^{-1}$, calculate the work done during the descent by the combined resistance forces.

Assuming that the only such forces are the friction and air-resistance, and that the track is a straight line inclined at 30° to the horizontal with a coefficient of friction of 0.1, calculate the work done by the air-resistance.

12 A particle P, of mass m, is suspended from a point O by a light string of length l. The particle is given an initial horizontal velocity u, and it then moves in a circular arc, coming instantaneously to rest when OP makes an angle α with the downward vertical, where $0 < \alpha < \frac{1}{2}\pi$.
Show that

$$u^2 = 2gl(1 - \cos \alpha)$$

13 A particle, of mass m, slides down a rough plane which is inclined to the horizontal at an angle α: the coefficient of friction is μ. Its initial velocity is u, and when it has moved a distance x down the plane, the speed is v, where $v < u$. Show that

$$v^2 = u^2 + 2gx(\sin \alpha - \mu \cos \alpha)$$

Show also that it comes to rest when it has travelled a distance

$$\frac{u^2}{2g(\mu \cos \alpha - \sin \alpha)}$$

14 Water is pumped, at the rate of $12\,\text{m}^3$ per minute, from a reservoir to a height of 25 m, where it is ejected from a pipe whose cross-sectional area is $50\,\text{cm}^2$. Calculate the work done per second. (The mass of $1\,\text{m}^3$ of water is 1000 kg.)

15 When a particle of mass m is at a distance x from the centre of the Earth, (where $x \geqslant R$, the radius of the Earth), the pull of gravity obeys the 'inverse square law', that is, it has the form $\dfrac{\lambda}{x^2}$, where λ is a constant. Given that the weight of the particle when it is on the Earth's surface is mg, show that $\lambda = mgR^2$.
Prove that when the particle is raised from the Earth's surface to a height H, the gain in potential energy is

$$\frac{mgRH}{(R + H)}$$

and show that, if H is small compared with R, this is approximately mgH.

10.7 Power

Imagine two gymnasts, each weighing 800 N, climbing ropes in a gymnasium, and suppose they both climb 10 m vertically upwards; the work done by each of them is 8000 J. Suppose, however, that one does it in 20 seconds while the other takes 50 seconds. Undoubtedly we would think that climbing the rope in 20 seconds is the more impressive performance, because the work was done at an average rate of 400 joules per second, whereas the slower climber only achieved 160 joules per second. In applied mathematics, the rate at which work is done is called the *power*. In other words, for a constant force,

power = work done ÷ time taken

Units

The basic SI unit of power is the watt, and this is defined as follows:

> 1 watt = 1 joule per second   †

The standard abbreviation is W. The watt is quite a small unit and in many practical situations it is more convenient to use the kilowatt (1 kW = 1000 W).

If the point of application of a constant force F newtons moves through a displacement of x metres, along its line of action, in t seconds, then the work done is Fx and the power P watts is given by

$$P = Fx \div t$$

This can also be written $P = F \times (x/t)$ and since (x/t) represents the speed v at which the point of application moves, this formula can be expressed

$$P = Fv$$

If the magnitude of the force F is *not* constant, the work done δW, in a small displacement δx, lies between $F\delta x$ and $(F + \delta F)\delta x$, where δF is the increase in F during this small displacement, i.e.

$$F\delta x \leqslant \delta W \leqslant (F + \delta F)\delta x$$

Dividing this by δt, the time taken, we have

$$F\frac{\delta x}{\delta t} \leqslant \frac{\delta W}{\delta t} \leqslant \frac{(F + \delta F)\delta x}{\delta t}$$

When $\delta t \to 0$, δF also tends to zero, so this becomes

$$\frac{dW}{dt} = F\frac{dx}{dt}$$

but $\dfrac{dW}{dt}$ is the rate of change of W, i.e. the power, and $\dfrac{dx}{dt}$ is v, the velocity, so we can express this equation, in words, as follows:

> power = force × velocity

as before. However, when using vectors, the right-hand side of this expression must be the *scalar product* of the vectors representing the force and the velocity, i.e. if the force is represented by vector **F** and the velocity by **v**, the power is **F . v**.

Qu. 5 The motion of a car, travelling at a constant speed of $20\,\mathrm{m\,s}^{-1}$ along a horizontal road, is resisted by a constant force of 1000 N. Find the power required.

---

† The unit of power is named after the Scottish engineer Sir James Watt (1736–1819), who is famous for his work on the steam engine.

Example 4

The resistance to the motion of a car, when it is travelling at $v\,\mathrm{m\,s^{-1}}$, is $(200 + 0.5v^2)$ newtons. Calculate the power required to drive this car at $30\,\mathrm{m\,s^{-1}}$ along a horizontal road.

Given that the mass of the car is $1000\,\mathrm{kg}$, find the power required to drive it up a hill, inclined at an angle α to the horizontal, where $\sin \alpha = 0.1$, at a speed of $20\,\mathrm{m\,s^{-1}}$.

$$
\begin{aligned}
\text{At } 30\,\mathrm{m\,s^{-1}}, \text{ the resistance} &= (200 + 0.5 \times 900)\,\mathrm{N} \\
&= 650\,\mathrm{N} \\
\text{Power} &= \text{force} \times \text{velocity} \\
&= 650 \times 30\,\mathrm{W} \\
&= 19\,500\,\mathrm{W}
\end{aligned}
$$

The power required is $19.5\,\mathrm{kW}$.

When ascending the hill at $20\,\mathrm{m\,s^{-1}}$, component of the weight acting down the incline

$$
\begin{aligned}
&= 1000g \times \sin \alpha\,\mathrm{N} \\
&= 981\,\mathrm{N} \\
\text{resistance} &= (200 + 0.5 \times 400)\,\mathrm{N} \\
&= 400\,\mathrm{N} \\
\therefore \quad \text{the total force exerted} &= 1381\,\mathrm{N}
\end{aligned}
$$

$$
\begin{aligned}
\text{Hence the power required} &= 1381 \times 20\,\mathrm{W} \\
&= 27\,620\,\mathrm{W}
\end{aligned}
$$

Ascending the hill, the power required is $27.6\,\mathrm{kW}$, correct to three significant figures.

Example 5

The motion of a car is opposed by a resistance of $2000\,\mathrm{N}$. Calculate the maximum speed attainable on a horizontal road with a power output of $25\,\mathrm{kW}$. If the power is now instantaneously increased to $30\,\mathrm{kW}$, find the rate at which the car begins to accelerate given that its mass is $1000\,\mathrm{kg}$.

Using power = force × velocity, we have

$$25\,000 = 2000v$$

where $v\,\mathrm{m\,s^{-1}}$ is the speed. Hence

$$v = 12.5$$

The maximum speed is $12.5\,\mathrm{m\,s^{-1}}$.

$$
\begin{aligned}
\text{The force exerted at increased power} &= 30\,000 \div 12.5\,\mathrm{N} \\
&= 2400\,\mathrm{N}
\end{aligned}
$$

Let the acceleration be $a\,\mathrm{m\,s^{-2}}$, then, using Newton's second law of motion, we have

$$2400 - 2000 = 1000a$$
Hence $\qquad a = 0.4$

The car begins to accelerate at $0.4\,\mathrm{m\,s^{-2}}$.

Example 6

Water is pumped from a reservoir to a height of 5 m, where it is ejected from a pipe with a cross-sectional area of $10\,\mathrm{cm^2}$, at a rate of 120 litres per minute. Calculate the power needed.

$$
\begin{aligned}
120\ \text{litres per minute} &= 2\ \text{litres per second}\\
&= 2000\,\mathrm{cm^3}\ \text{per second}
\end{aligned}
$$

The pipe has a cross-sectional area of $10\,\mathrm{cm^2}$, so the water is ejected at $(2000 \div 10)$ cm per second, i.e. $2\,\mathrm{m\,s^{-1}}$, and since the mass of $1000\,\mathrm{cm^3}$ is 1 kg, the water is pumped at a rate of 2 kg per second.

Hence, in each second

$$
\begin{aligned}
\text{P.E. gained} &= (2 \times g \times 5)\mathrm{J}\\
&= 98.1\,\mathrm{J}\\
\text{K.E. gained} &= (\tfrac{1}{2} \times 2 \times 2^2)\mathrm{J}\\
&= 4\,\mathrm{J}
\end{aligned}
$$

So the total amount of energy gained (i.e. the work done) per second is 102.1 joules, and since the power is equal to the rate of working, the power required is 102 watts, correct to three significant figures.

Exercise 10c

Take g to be $9.81\,\mathrm{m\,s^{-2}}$ throughout this exercise.

1 A man, whose mass is 60 kg, climbs up two flights of stairs, a total height of 10 m, in 90 s. Calculate the power he uses.
2 A ski lift carries people from an altitude of 2500 m to 3000 m. On average it can deliver 12 people per minute. Allowing 60 kg per person, estimate the power required. Discuss briefly some of the other factors which may affect the amount of power needed.
3 A girl and her bicycle have a total mass of 55 kg. Find the power required when she rides at a steady speed of $8\,\mathrm{m\,s^{-1}}$ up a hill inclined at an angle α to the horizontal, where $\sin\alpha = 0.2$, assuming that the wind-resistance is a constant 100 N.
4 A car, of mass 1 tonne, accelerates uniformly from 0 to $90\,\mathrm{km\,h^{-1}}$ in 10 s. Given that a constant force of 200 N opposes its motion, calculate the power when it has been accelerating for 5 seconds.

5 A truck, of mass 2 tonnes, travels at a constant speed of $10\,\mathrm{m\,s^{-1}}$ up a hill which is inclined at an angle α to the horizontal, where $\sin\alpha = 0.25$. There is a constant force of 400 N opposing its motion. Calculate the power used.
If the power is instantaneously increased to 75 kW, calculate the rate at which the truck begins to accelerate.

6 A pump whose power is 20 watts is used to work a fountain. It can pump 30 litres per minute. Calculate the height to which the water rises.

7 A car, of mass 1 tonne, is driven by an engine working at a constant rate of 30 kW. Assuming that forces such as the air-resistance may be neglected, find the speed reached, from a standing start, in a distance of 300 m.

8 The force opposing the motion of a car is $(a + bv^2)$ newtons, where v is the speed of the car in $\mathrm{m\,s^{-1}}$. The power required to maintain a steady speed of $20\,\mathrm{m\,s^{-1}}$ is 6.2 kW, and at $30\,\mathrm{m\,s^{-1}}$, 15.3 kW is required. Find the values of a and b and hence calculate the power required for a steady speed of $40\,\mathrm{m\,s^{-1}}$.

9 A lorry has a maximum power output of P kW, and its motion is opposed by a constant resistance. Its maximum speed when travelling up a certain incline is $u\,\mathrm{km\,h^{-1}}$, and it is $2u\,\mathrm{km\,h^{-1}}$ when going down the same incline. Find its maximum speed, in terms of u, when travelling along a horizontal road.

10 A car, of mass M tonnes, is driven by an engine working at a constant rate of P kW along a horizontal road. Assuming that forces such as the air-resistance may be neglected, show that if its speed increases from $U\,\mathrm{m\,s^{-1}}$ to $V\,\mathrm{m\,s^{-1}}$ while it travels x metres, then $3Px = M(V^3 - U^3)$.

Summary

(a) Work done by a force F N acting at an angle α to a displacement x m

$$= Fx\cos\alpha \text{ newtons} \quad (\text{see }\S10.2)$$

(b) Work done by a variable force F

$$= \int_a^b F\,\mathrm{d}x \quad (\text{see }\S10.3)$$

(c) Potential energy of a particle of mass m kg when raised to a height h m, above a standard base line

$$= mgh \quad (\text{see }\S10.4)$$

(d) Kinetic energy of a particle of mass m kg moving at $v\,\mathrm{m\,s^{-1}}$

$$= \tfrac{1}{2}mv^2 \quad (\text{see }\S10.5)$$

(e) Power = force × velocity   (see §10.7)

Note that in vector notation, if $\mathbf{F}$ represents the force, $\mathbf{r}$ the displacement, and $\mathbf{v}$ the velocity, the formulae for work and power must be amended as follows:

Work done $= \mathbf{F}.\mathbf{r}$ Power $= \mathbf{F}.\mathbf{v}$

using *scalar* products in both cases.

Exercise 10d (Miscellaneous)

In questions which require a numerical value for g, use $10 \, \text{m s}^{-2}$, unless some other value is given.

1 Find the work done by each of the following forces (in newtons) when it undergoes the given displacement (in metres):
   - (a) force 10 N, directed along the positive x-axis, displacement from $x = 0$, to $x = 5$;
   - (b) force 10 N directed along the negative x-axis, displacement from $x = 0$, to $x = 5$;
   - (c) force 10 N, at 60° above the positive x-axis, displacement from $x = 0$, to $x = 5$;
   - (d) force $(3\mathbf{i} + 4\mathbf{j})$ N, displacement from the origin to the point $(2,5)$;
   - (e) a variable force $10x$ directed along the positive x-axis, displacement from $x = 0$, to $x = 5$.

2 A small steel ball of mass 100 grams, moving at $10 \, \text{m s}^{-1}$, collides with an identical ball which is at rest. After the collision their speeds are $2 \, \text{m s}^{-1}$ and $8 \, \text{m s}^{-1}$, respectively. Find the kinetic energy lost in the collision.

3 An athlete carries out a fitness test in which she steps on and off a bench which is 30 cm high. Given that her mass is 60 kg, calculate the power required to do the test at a rate of 40 steps per minute.

4

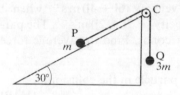

Figure 10.5

The diagram shows a particle P, of mass m, resting on a fixed rough plane inclined at an angle of 30° to the horizontal. A light inextensible string attached to P passes over a small smooth pulley fixed at C, and a particle Q, of mass $3m$, attached to the other end of the string hangs vertically. The portion PC of the string is parallel to a line of greatest slope of the plane.

The coefficient of friction between P and the plane is $\frac{1}{6}\sqrt{3}$. The system is released from rest with Q at C and with the string taut. Show that P will commence to move up the plane, and find the loss in the potential energy of the system when P has moved a distance x up the plane. (Assume that P does not reach C.)

Find also the work done against friction during this motion.

Hence, or otherwise, find, in terms of x, the speed of P.

(C)

5 A fixed plane is inclined at an angle α to the horizontal, where $\tan \alpha = \frac{4}{3}$. A particle of mass m is projected, from the point A on the plane, up a line

of greatest slope. The coefficient of friction between the particle and the plane is $\frac{1}{3}$. The particle has moved a distance d up the plane when it comes instantaneously to rest at the point B.

(a) Find the total work done against the external applied forces during the motion from A to B.
(b) Find the speed of projection from A.
(c) Find the total work done against the external applied forces during the motion from A to B and back to A again.
(d) Find the kinetic energy of the particle when it has passed through A and moved a further distance $4d$ down the plane from A.

(C)

6 The top of a chute whose length is 12 m is 3 m vertically above its lowest point. A parcel of mass 1.6 kg slides from rest from the top of the chute and reaches the lowest point with a speed of $5\,\text{m s}^{-1}$. Calculate, for the parcel,

(a) the gain in kinetic energy;
(b) the loss in potential energy;
(c) the work done in overcoming the frictional resistance;
(d) the average value of this resistance.

After reaching the lowest point of the chute, the parcel slides along a horizontal floor, the resistance to motion being 4 N. Calculate how far the parcel travels before coming to rest.

*(C)

7 A particle of mass 0.5 kg is in motion with velocity $(6\mathbf{i} + 4\mathbf{j})\,\text{m s}^{-1}$ when it strikes a fixed barrier and rebounds with velocity $(-2\mathbf{i} + 2\mathbf{j})\,\text{m s}^{-1}$. The particle is in contact with the barrier for 0.02 seconds. Find the average force exerted by the barrier on the particle.

Calculate also the kinetic energy lost by the particle in the collision.

(JMB)

8 The engine of a car of mass m works at a constant rate S. The car moves on a level road and is subject to a constant resistance R. Show that the time taken for its speed to increase from U to V, where $RU < RV < S$, is

$$\frac{mS}{R^2}\ln\left(\frac{S - RU}{S - RV}\right) - \frac{m}{R}(V - U)$$

(JMB)

9 A train of mass 2×10^5 kg is running, with the engine shut off, down a hill of inclination α to the horizontal, where $\sin\alpha = \frac{1}{100}$. The acceleration of the train is $0.05\,\text{m s}^{-2}$. Find the resistance to motion.

Find the maximum speed of the train when travelling up the same hill against a resistance to motion of 4×10^3 N with the engine working at a power of 3×10^5 W.

(C)

10 A car of mass 1000 kg moves with its engine shut off down a slope of inclination α, where $\sin\alpha = \frac{1}{20}$, at a steady speed of $15\,\text{m s}^{-1}$. Find the resistance, in newtons, to the motion of the car. Calculate the power delivered by the engine when the car ascends the same incline at the same steady speed, assuming that resistance to motion is unchanged.

(L)

11 A cyclist moves against a resistance which is proportional to his speed. When the cyclist is working at the rate of 64 W, his maximum speed along a horizontal road is $4\,\mathrm{m\,s^{-1}}$. Given that the total mass of the cyclist and his machine is 60 kg, find the maximum speed, in $\mathrm{m\,s^{-1}}$, which the cyclist reaches when he travels down a slope of angle θ, where $\sin\theta = \frac{1}{30}$, when he is working at the rate of 56 W.

Show also that, when he is moving along a horizontal road with speed $2\,\mathrm{m\,s^{-1}}$ working at the rate of 46 W, his acceleration is $0.25\,\mathrm{m\,s^{-2}}$.

(L)

12 A car of mass 960 kg has a maximum speed of $50\,\mathrm{m\,s^{-1}}$ on a horizontal road when the power output of the engine is 40 kW. Calculate the frictional resistance.

The car ascends a slope inclined at an angle α to the horizontal where $\sin\alpha = \frac{1}{6}$. The power output of the engine remains the same but the frictional resistance is now 900 N. Calculate
(a) the maximum speed of the car up the slope;
(b) the acceleration of the car up the slope when its speed is $10\,\mathrm{m\,s^{-1}}$.

*(C)

13 The engine of a car has maximum power P kW. The car has mass 1600 kg and a maximum speed of $30\,\mathrm{m\,s^{-1}}$ on a level road. The maximum speed up a hill of inclination α, where $\sin\alpha = \frac{1}{20}$, is $20\,\mathrm{m\,s^{-1}}$. Given that when the speed is $v\,\mathrm{m\,s^{-1}}$ the resistance to motion has magnitude λv^2 N, where λ is a constant, show that the value of λ is $\frac{16}{19}$ and find the value of P, giving three significant figures in your answer.

Find the maximum acceleration of the car at an instant when it is travelling on a level road at a speed of $10\,\mathrm{m\,s^{-1}}$.

(C)

14 A car engine works at a rate of 5.7 kW in driving a car of mass 1100 kg along a level horizontal road at a constant speed of $12\,\mathrm{m\,s^{-1}}$. Find the frictional resistance to motion of the car.

The car descends a slope, inclined at an angle θ to the horizontal, against a frictional resistance of 500 N. Given that $\sin\theta = \frac{1}{20}$, calculate the power required for the car to have an acceleration of $2\,\mathrm{m\,s^{-2}}$ at the instant when the speed is $16\,\mathrm{m\,s^{-1}}$.

*(C)

In question 15 take g to be $9.8\,\mathrm{m\,s^{-2}}$.

15 A lorry of mass 2000 kg is subject to a constant frictional resistance of 2600 N. Find in kW the power at which the engine is working when the lorry is travelling along a level road at a steady speed of $45\,\mathrm{km\,h^{-1}}$.

If the engine continues to work at this rate, find the steady speed in $\mathrm{km\,h^{-1}}$ at which the lorry ascends a hill of inclination $\sin^{-1}(\frac{1}{14})$.

(O & C)

Impulse and momentum

Chapter 11

Impulse and momentum

11.1 Introduction

Few of us (fortunately) will have had the experience of sitting in a Land Rover, while it is charged at by a rhinoceros! But we may have seen such a scene in a wildlife programme on television. It does not take much imagination to realise that a massive object, like a rhinoceros, moving at speed, will give another massive object (the Land Rover) a mighty thump, and the effect of this may well be to change the speed of the second object and its direction of motion. (It may well make a few dents and produce some bruises, but that's another matter!)

In this chapter we shall be considering the properties of massive objects in motion. In particular we shall be looking at what happens when two moving objects collide.

11.2 Impulse

In the last chapter we examined the effect of a force, as it moved through a displacement, and we saw that this led to the concept of the scalar quantities work and energy. We shall now look at the effect of a force over an interval of time. First, we consider a *constant* force. The product of the force and the time for which it acts is called the **impulse** of the force, i.e.

$$\text{impulse} = \text{force} \times \text{time} \qquad \qquad \text{. . . (1)}$$

The units of impulse are newton seconds (abbreviation N s).

If the force is constant in direction, but not in magnitude, we must use the more general definition

$$\text{impulse} = \int_0^T F \, dt \qquad \qquad \text{. . . (2)}$$

where F represents the (variable) force, and $t = 0$ to $t = T$ is the time interval for which it acts.

175

Qu. 1 Show that definition (2) is equivalent to (1) when F is constant.
Qu. 2 A force F newtons increases uniformly from zero to $20\,\text{N}$ in 5 seconds, i.e. $F = 4t$. Show that its impulse is 50 newton seconds.

The reader should note that, since force is a vector quantity, and time a scalar, their product, i.e. the impulse, is a *vector*.

11.3 Impulse and momentum

We have already met momentum, briefly, in Chapter 8 (see §8.5): momentum is the product of the mass of an object and its velocity.

Suppose a constant force $\mathbf{F}$ acts on a particle, of mass m, for t seconds, then the impulse $\mathbf{I}$ is given by

$$\mathbf{I} = \mathbf{F}t$$

By Newton's second law of motion, $\mathbf{F} = m\mathbf{a}$, where $\mathbf{a}$ represents the acceleration of the particle. Consequently

$$\mathbf{I} = m\mathbf{a}t$$

Now suppose that during this time the velocity of the particle changes from $\mathbf{u}$ to $\mathbf{v}$. Using the constant acceleration formula $\mathbf{v} = \mathbf{u} + \mathbf{a}t$, (see §6.4) the impulse equation becomes:

$$\mathbf{I} = m(\mathbf{v} - \mathbf{u})$$
$$\therefore \quad \mathbf{I} = m\mathbf{v} - m\mathbf{u}$$

In words, this says that

> the impulse is equal to the change of momentum

This result is also true for a *variable* force; readers who feel confident about their command of calculus should have little difficulty in proving this; others can take it on trust.

Example 1

A particle of mass $2\,\text{kg}$ is moving along the positive x-axis at a constant speed of $5\,\text{m s}^{-1}$ when a force of $4\,\text{N}$, at angle of $60°$ to this axis, is applied to it for 3 seconds. Find the velocity of the particle at the end of this interval.

Let the final velocity be $v\,\text{m s}^{-1}$, at an angle α to the positive x-axis, then

$$\begin{aligned}
\text{initial momentum} &= 2 \times 5 \quad \text{along the } x\text{-axis} \\
\text{final momentum} &= 2v \quad \text{at angle } \alpha \text{ to the } x\text{-axis} \\
\text{impulse} &= 4 \times 3 \quad \text{at } 60° \text{ to the } x\text{-axis}
\end{aligned}$$

Using the principle that

$$\text{impulse} = \text{final momentum} - \text{initial momentum}$$
or   final momentum = initial momentum + impulse

and remembering that these are *vector* quantities, we can draw the vector triangle shown in Figure 11.1

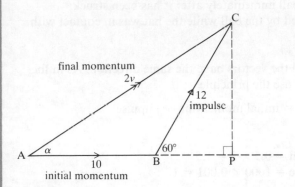

Figure 11.1

In this diagram, P is the foot of the perpendicular from C to AB produced, and hence

$$BP = 12\cos 60° = 6 \quad\text{and}\quad PC = 12\sin 60° = 6\sqrt{3}$$

Then, by Pythagoras' theorem

$$AC^2 = AP^2 + PC^2$$
$$\therefore \quad (2v)^2 = (10 + 6)^2 + (6\sqrt{3})^2$$
$$\therefore \quad 4v^2 = 256 + 108$$
$$= 364$$
$$v^2 = 91$$
$$\therefore \quad v = 9.54 \quad\text{(correct to 3 significant figures)}$$

(Alternatively, this result could be obtained by applying the cosine rule to triangle ABC.)

Also

$$\tan \alpha = \frac{PC}{AP}$$

$$= \frac{6\sqrt{3}}{16}$$

$$\therefore \quad \alpha = 33° \quad\text{(correct to the nearest degree)}$$

Hence the final velocity is $9.54\,\text{m s}^{-1}$, at an angle of 33° to the positive x-axis.

Example 2

A ball, of mass 500 grams, is moving in a straight line with a speed of $3\,\mathrm{m\,s}^{-1}$, when it is struck a blow by a bat in the direction of motion. Assuming that the bat and ball are in contact for 0.001 s, and that during this time the bat exerts a constant force of 1000 N on the ball, find

(a) the speed of the ball immediately after it has been struck;

(b) the distance moved by the ball while the bat was in contact with it.

(a) In this question all the vectors have the same direction. As in the previous example, we use the principle

final momentum = initial momentum + impulse

In this case, we have

$$\text{initial momentum} = 0.5 \times 3 = 1.5$$
$$\text{impulse} = 1000 \times 0.001 = 1$$

Hence

$$\text{final momentum} = 2.5$$
$$\therefore \qquad 0.5v = 2.5 \quad \text{where } v\,\mathrm{m\,s}^{-1} = \text{final speed of the ball}$$
$$\therefore \qquad v = 5$$

The final speed of the ball is $5\,\mathrm{m\,s}^{-1}$.

(b) To find the distance travelled, s, we use $s = ut + \frac{1}{2}at^2$, i.e.

$$s = 3 \times 0.001 + \tfrac{1}{2}a(0.001)^2 \qquad\qquad \text{. . . (1)}$$

The acceleration, a, in this equation, can be found from Newton's second law of motion, as follows:

$$1000 = 0.5a$$
$$\therefore \qquad a = 2000$$

Substituting this into the formula (1) gives

$$s = 0.003 + \tfrac{1}{2} \times 2000 \times 0.000\,001$$
$$= 0.004$$

Hence the ball moves 0.004 m ($= 4\,\mathrm{mm}$) while it is in contact with the bat.

11.4 Impacts

The impulse–momentum principle is especially useful when, as in Example 2, a very large force acts for a very short time. Typical situations which give rise to this are: when an object is struck, when two objects collide, or when a composite body is blown apart by an internal explosion. The very large forces which occur in these situations are often called 'impulsive forces'. In Example 2 we saw that the displacement of the ball during the impact was very small: we shall now show, using the same notation as in the previous sections, that this is true in general.

Suppose a very large constant force $\mathbf{F}$ is applied to a particle of mass m, for a very short time t. (In due course we shall let $\mathbf{F}$ tend to infinity, and t tend to zero, but their product $\mathbf{F}t$, that is the impulse, will remain finite.)

We have already seen in §11.3 that

$$\mathbf{F}t = m\mathbf{v} - m\mathbf{u}$$
$$\text{i.e.} \quad \mathbf{I} = m\mathbf{v} - m\mathbf{u} \qquad \qquad \qquad \text{. . . (1)}$$

or, in words, the impulse is equal to the change in momentum.

Now let us consider the displacement, $\mathbf{s}$, during the impact; using the formula $\mathbf{s} = \mathbf{u}t + \frac{1}{2}\mathbf{a}t^2$, and, from Newton's second law of motion, writing $\mathbf{a} = \mathbf{F}/m$, we have

$$\mathbf{s} = \mathbf{u}t + \frac{1}{2}\left(\frac{\mathbf{F}}{m}\right)t^2$$

$$= t\left[\mathbf{u} + \frac{1}{2}\left(\frac{\mathbf{F}t}{m}\right)\right]$$

$$= t\left[\mathbf{u} + \frac{1}{2}\left(\frac{\mathbf{I}}{m}\right)\right] \qquad \qquad \text{. . . (2)}$$

Now, the factor in square brackets in this equation, is finite (remember that $\mathbf{I}$ is finite), so when t tends to zero, the product on the right-hand side of (2) also tends to zero. Hence we have another important principle:

In an impact, the displacement is negligible.

There are two further points which need special comment:
(a) If there are any *finite* forces acting during the impact, then, when they are multiplied by t, the product will tend to zero when t tends to zero, so they have no effect on equation (1).
(b) The explanation, above, has been based on the assumption that the force $\mathbf{F}$ is constant. It is possible to show that the principles expounded in this section are also valid for a *variable* force, but we shall not do so here.

Example 3

A ball, of mass $0.25\,\text{kg}$, moving horizontally in a straight line at $10\,\text{m s}^{-1}$, is kicked by a girl, back along its original path with a speed of $20\,\text{m s}^{-1}$. Find the magnitude of the impulse of the kick.

Since the motion takes place in a straight line, the vector properties of the impulse and momentum can be taken into account by adopting the direction of the impulse as positive and writing the initial momentum as a negative quantity. Thus

$$\text{initial momentum} = -0.25 \times 10$$
$$= -2.5$$
$$\text{final momentum} = +0.25 \times 20$$
$$= +5.0$$

and since 'impulse = final momentum − initial momentum', we have

$$\text{impulse} = (+5.0) - (-2.5)$$
$$= 7.5$$

Hence the magnitude of the impulse is 7.5 newton seconds.

Example 4

A particle of mass m, moving with a speed $8u$, is struck a blow which deflects it through an angle of 60° and reduces its speed to $5u$. Show that the magnitude of the impulse is $7mu$.

(Since no units are mentioned in the question, we may assume that any consistent set of units may be used.)

The initial momentum is $8mu$ and the final momentum is $5mu$; let the magnitude of the impulse be I. Their directions (but not their magnitudes) are indicated in Figure 11.2a.

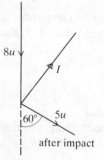
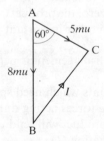

Figure 11.2a   **Figure 11.2b**

Triangle ABC in Figure 11.2b represents the vector equation

$$\text{final momentum } (\overrightarrow{AC}) = \text{initial momentum } (\overrightarrow{AB}) + \text{impulse } (\overrightarrow{BC})$$

Using the cosine rule, we have

$$I^2 = (8mu)^2 + (5mu)^2 - (2 \times 8mu \times 5mu \times \cos 60°)$$
$$= (64 + 25 - 40)m^2u^2$$
$$= 49m^2u^2$$
$$\therefore \quad I = 7mu$$

Hence the magnitude of the impulse is $7mu$.

(Readers who are not familiar with the cosine rule can solve this problem by drawing a line from C perpendicular to the line AB and using elementary trigonometry, as in Example 1.)

In the next example, the impulse takes the form of an 'impulsive reaction' between a moving object and a fixed barrier.

Example 5

A ball of mass 500 grams, moving in a horizontal plane with a velocity $\mathbf{u}\,\mathrm{m\,s^{-1}}$, strikes a vertical wall and rebounds with velocity $\mathbf{v}\,\mathrm{m\,s^{-1}}$. The impulse $\mathbf{I}$ acting on the ball is horizontal and perpendicular to the wall. Taking unit vectors $\mathbf{i}$ perpendicular to the wall, and $\mathbf{j}$ along the wall, (see Figure 11.3a), find the magnitude of the impulses when:
(a) $\mathbf{u} = -4\mathbf{i}$ and $\mathbf{v} = 2\mathbf{i}$;
(b) $\mathbf{u} = (-4\mathbf{i} + 2\mathbf{j})$ and $\mathbf{v} = (2\mathbf{i} + \lambda\mathbf{j})$, and λ is to be found;
(c) $\mathbf{u}$ has magnitude 4 and is at 30° to the normal to the wall, and $\mathbf{v}$ has magnitude 3 and is at α° to the normal (see Figure 11.3c), where angle α is to be found.

(a)

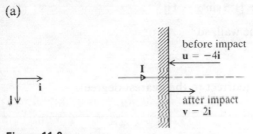

before impact
$\mathbf{u} = -4\mathbf{i}$

after impact
$\mathbf{v} = 2\mathbf{i}$

Figure 11.3a

Using impulse = change of momentum, we have

$$\mathbf{I} = 0.5 \times (2\mathbf{i}) - 0.5 \times (-4\mathbf{i})$$
$$= \mathbf{i} + 2\mathbf{i}$$
$$= 3\mathbf{i}$$

The magnitude of the impulse is 3 newton seconds.

(b)

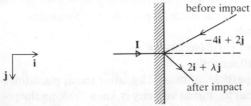

before impact

$-4\mathbf{i} + 2\mathbf{j}$

$2\mathbf{i} + \lambda\mathbf{j}$

after impact

Figure 11.3b

$$I = 0.5 \times (2\mathbf{i} + \lambda\mathbf{j}) - 0.5 \times (-4\mathbf{i} + 2\mathbf{j})$$
$$= \mathbf{i} + 0.5\lambda\mathbf{j} + 2\mathbf{i} - \mathbf{j}$$
$$= 3\mathbf{i} + (0.5\lambda - 1)\mathbf{j}$$

But $\mathbf{I}$ is perpendicular to the wall, i.e. it is a multiple of $\mathbf{i}$ only, so

$$0.5\lambda - 1 = 0$$
$$\therefore \qquad \lambda = 2$$
Hence $\mathbf{I} = 3\mathbf{i}$

The magnitude of the impulse is 3 newton seconds.

(c)

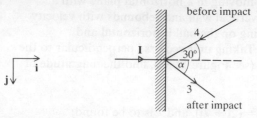

Figure 11.3c

$$I = 0.5[(3\cos\alpha)\mathbf{i} + (3\sin\alpha)\mathbf{j}] - 0.5[(-4\cos 30°)\mathbf{i} + (4\sin 30°)\mathbf{j}]$$
$$= [1.5\cos\alpha + 2\cos 30°]\mathbf{i} + [1.5\sin\alpha - 2\sin 30°]\mathbf{j}$$
$$= [1.5\cos\alpha + \sqrt{3}]\mathbf{i} + [1.5\sin\alpha - 1]\mathbf{j} \qquad \qquad \text{. . .   (1)}$$

But $\mathbf{I}$ is perpendicular to the wall, so

$$1.5\sin\alpha - 1 = 0$$
$$\therefore \qquad \sin\alpha = \tfrac{2}{3}$$
$$\therefore \qquad \alpha = 41.8° \quad \text{(correct to the nearest degree)}$$

Hence, from (1)

$$\mathbf{I} = (1.5\cos\alpha + \sqrt{3})\mathbf{i}$$
$$\therefore \quad = 2.85\mathbf{i} \quad \text{(correct to three significant figures)}$$

Hence the magnitude of the impulse is 2.85 newton seconds.

Exercise 11a

In this exercise, the principles of impulse and momentum should be used, even when alternative methods exist.

1 Find the magnitude of the impulse when the following forces are applied for the given times (the direction of the force in each part is constant):
   (a) 10 newtons, for 5 seconds;
   (b) $12t$ newtons, for $t = 0$ to $t = 5$ (seconds);
   (c) $(3\mathbf{i} + 4\mathbf{j})$ newtons, for 10 seconds.
2 Find the final velocity of a particle, of mass 2 kg, after the application of the following impulses, given that the initial velocity is $3\,\mathrm{m\,s^{-1}}$ along the positive x-axis in each case:

(a) 4 newton seconds along the positive x-axis;

(b) 8 newton seconds parallel to the positive y-axis;

(c) $(4\mathbf{i} - 10\mathbf{j})$ newton seconds.

3 A one tonne spacecraft is travelling in a straight line with a speed of 50 m s^{-1}, when its engine is fired for 10 seconds, in order to alter its course by $20°$, without changing its speed. Calculate (a) the impulse, and (b) the average value of the force exerted.

4 A particle of mass 2 kg, initially moving with a constant velocity of $(3\mathbf{i} + 11\mathbf{j}) \text{ m s}^{-1}$, has a constant force $(5\mathbf{i} + 2\mathbf{j})$ newtons applied to it for 3 seconds. Find (a) the magnitude of the impulse, and (b) the speed of the particle at the end of the 3 seconds.

5 A ship, of mass 3000 tonnes, is towed by a tug-boat, which exerts a constant horizontal force of 100 kN. Neglecting the water-resistance, calculate the time required to reach a speed of 3 km h^{-1}, starting from rest.

6 A bullet, of mass 0.02 kg, moving at 500 m s^{-1}, strikes a fixed wooden block. It passes through the block and emerges with a speed of 200 m s^{-1}. Find the impulse exerted on the bullet. Given that the bullet passed through the block in 0.1 s, calculate the average resistance to its motion through the block.

7 A hosepipe, with a cross-section of 4 cm^2, ejects water at a rate of 5 m s^{-1} (the mass of 1 cm^3 of water is 1 gram). The water is directed horizontally onto a vertical wall, and, after hitting the wall, it runs down the wall at a speed which is negligible. Calculate the momentum lost by the water in one second; hence find the force exerted on the wall by the water.

8 Coal falls vertically, at a rate of 2 tonnes per minute, onto a conveyor belt moving horizontally at 3 m s^{-1}. Calculate the horizontal momentum gained per second by the coal; hence find the average force required to keep the belt moving (neglect other forces).

9 A ball, of mass 0.25 kg, moving in a horizontal straight line with a speed of 2 m s^{-1}, strikes a vertical wall at an angle of $40°$ to the normal. The wall gives it an impulse in the direction of the normal, and the ball rebounds at an angle of $60°$ to the normal. Calculate the magnitude of the impulse, and the speed with which the ball rebounds.

10 A batsman strikes a cricket ball, when it is travelling horizontally at 120 km h^{-1}. Given that the mass of the ball is 0.3 kg, calculate the impulse if the ball is

(a) hit back along its original path at 90 km h^{-1};

(b) turned through $90°$, with its speed unchanged;

(c) deflected by $45°$, and its speed halved.

11.5 Conservation of momentum

When two objects collide, the momentum of each is changed and each of them experiences an impulse.

Imagine two objects, with masses m_1 and m_2, moving with velocities $\mathbf{u}_1$ and $\mathbf{u}_2$. Suppose these objects collide, and let their velocities after the impact be $\mathbf{v}_1$ and $\mathbf{v}_2$.

This information is represented diagramatically in Figure 11.4.

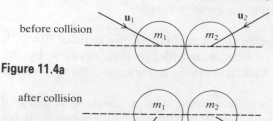

before collision

Figure 11.4a

after collision

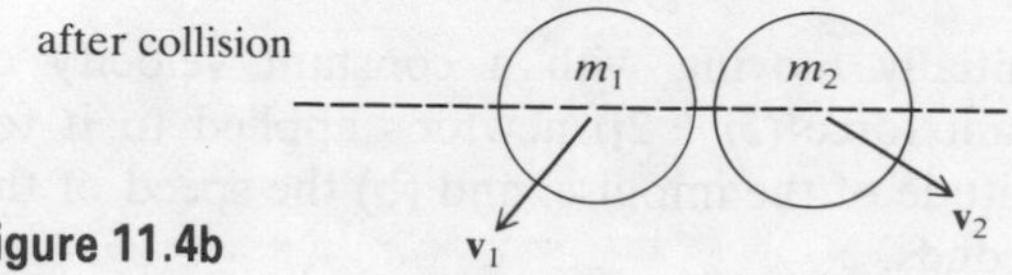

Figure 11.4b

Using the principle that the 'impulse is equal to the change in momentum' we have

$$I_1 = m_1 v_1 - m_1 u_1$$
$$I_2 = m_2 v_2 - m_2 u_2$$

where I_1 and I_2 are the impulses on the two objects.

However while the objects are in contact, the forces they exert on each other are, by Newton's third law of motion, equal in magnitude and opposite in direction. Consequently, the impulses are also equal, but opposite in direction, so their vector sum is zero, i.e. $I_1 + I_2 = 0$.

Hence

$$m_1 v_1 - m_1 u_1 + m_2 v_2 - m_2 u_2 = 0$$
$$\therefore \quad m_1 v_1 + m_2 v_2 = m_1 u_1 + m_2 u_2 \qquad \qquad . \ . \ . \ (1)$$

This equation can be expressed in words, as follows:

> the momentum before the collision is
> equal to the momentum afterwards

or, more briefly,

> momentum is conserved in a collision

This argument could also be applied to an object which is blown apart by an internal explosion.

Notice that, since every term in equation (1) has the form 'mass × velocity', it is not necessary to express these quantities in SI units; any consistent set of units will do.

Example 6

A loaded railway truck, whose total mass is 10 tonnes, moving at $3 \, \text{km h}^{-1}$, collides with an empty truck, whose mass is 5 tonnes, moving towards the first truck at $2 \, \text{km h}^{-1}$. Immediately after the collision the trucks are coupled together; calculate their common velocity.

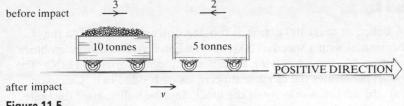

before impact

10 tonnes 5 tonnes

POSITIVE DIRECTION

after impact
v

Figure 11.5

momentum before the collision $= 10 \times 3 + 5 \times (-2)$
$= 20$ units

Let the common velocity after the collision be $v \, \mathrm{km \, h^{-1}}$, then

momentum after the collision $= 15v$ units

Hence, since momentum is conserved,

$$15v = 20$$
$$\therefore \quad v = \tfrac{4}{3}$$

The common velocity of the trucks after the collision is $1\tfrac{1}{3} \, \mathrm{km \, h^{-1}}$.

Example 7

A two stage space rocket is cruising at a speed of $15\,000 \, \mathrm{km \, h^{-1}}$, when the stages are separated by an internal explosive device. The smaller stage, which has a mass of 1 tonne, then moves away from the larger stage, which has a mass of 4 tonnes, with a relative velocity of $100 \, \mathrm{km \, h^{-1}}$ (see Figure 11.6), the direction of motion being unchanged. Find the velocities of the separate stages.

before separation
$15\,000$

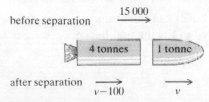

4 tonnes   1 tonne

after separation
$v - 100$ v

Figure 11.6

Let the velocity of the smaller part be $v \, \mathrm{km \, h^{-1}}$; the velocity of the other part is then $(v - 100) \, \mathrm{km \, h^{-1}}$.

momentum before separation $= 5 \times 15\,000 = 75\,000$ units
momentum after separation $= (1 \times v) + 4 \times (v - 100)$
$= (5v - 400)$ units

Hence, using conservation of momentum,

$$5v - 400 = 75\,000$$
$$5v = 75\,400$$
$$\therefore \quad v = 15\,080$$

Hence, after separation, the smaller stage moves at $15\,080 \, \mathrm{km \, h^{-1}}$, and the larger stage at $14\,980 \, \mathrm{km \, h^{-1}}$.

Example 8

A bullet, of mass 100 grams, is moving downwards at 30° to the horizontal with a speed of $500\,\text{m s}^{-1}$, when it strikes a wooden block of mass 1 kg, which is lying at rest on a smooth horizontal table. The bullet penetrates the block and becomes embedded in it. Calculate
(a) the common velocity of the block and the bullet, after the bullet has come to rest relative to the block;
(b) the impulsive reaction of the table on the block.

(a) Before impact, the horizontal component of the bullet's momentum is

$$0.1 \times 500 \times \cos 30° = 50 \cos 30°$$

Let the common (horizontal) velocity of the bullet and the block be $v\,\text{m s}^{-1}$.

After the bullet has come to rest inside the block, the combined object (whose mass is 1.1 kg) has horizontal momentum $1.1v$. Since momentum is a *vector* quantity we may equate these horizontal components, i.e.

$$1.1v = 50 \cos 30°$$

$$v = 50 \cos 30° \div 1.1$$

$$= 39.4 \quad \text{(correct to three significant figures)}$$

The common velocity of the bullet and the block is $39.4\,\text{m s}^{-1}$.

(b) The vertical component of the bullet's momentum $(0.1 \times 500 \times \sin 30°)$ is destroyed in the impact, hence

$$\text{impulsive reaction} = 50 \times \sin 30°$$

$$= 25$$

The impulsive reaction of the table on the block is 25 N s. (N s is the abbreviation for 'newton second'.)

Example 9

A particle A, of mass m, is moving with velocity $3u$, when it collides with a particle B, of mass $2m$, which is moving with velocity u, in the same direction as A. After the collision, the particles separate with relative velocity u. Find
(a) the velocities after the collision;
(b) the kinetic energy lost in the collision.

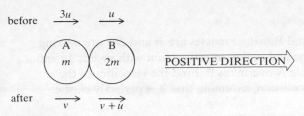

before $\xrightarrow{3u}$ $\xrightarrow{u}$

A _m_ B _2m_ POSITIVE DIRECTION

after $\xrightarrow{v}$ $\xrightarrow{v+u}$

Figure 11.7

(a) Momentum before the collision $= m \times 3u + 2m \times u = 5mu$

Let the velocity of A after the collision be v; the velocity of B is then $(v + u)$. Hence

$$\text{momentum after the collision} = mv + 2m(v + u)$$
$$= 3mv + 2mu$$

Since the momentum is conserved in a collision, we have

$$3mv + 2mu = 5mu$$
$$\therefore \quad 3mv = 3mu$$
$$\therefore \quad v = u$$

Hence the velocity of A after the collision is u, and the velocity of B is $2u$.

(b) Before the collision, the kinetic energy is

$$\tfrac{1}{2}m(3u)^2 + \tfrac{1}{2}(2m)u^2 = 5.5mu^2$$

After the collision, the kinetic energy is

$$\tfrac{1}{2}mu^2 + \tfrac{1}{2}(2m)(2u)^2 = 4.5mu^2$$

Hence the kinetic energy lost in the collision is mu^2.

This last example illustrates a very important point: kinetic energy is lost in collisions (although it is sometimes interesting to consider an *ideal* situation in which the loss is zero; this is called a 'perfectly elastic' collision). Some of the loss can be accounted for in terms of sound energy created by the collision, but most of it goes into work done overcoming internal stresses due to deformation of the objects during a collision. Many modern motorcars are designed so that, in the event of an accident, the energy is absorbed by the structure of the vehicle. The more 'squashy' the object is, the more energy it is likely to absorb in a collision. Strange as it may seem, collisions which come nearest to being 'perfectly elastic', are those between objects made from very hard, rigid materials. Two steel ball bearings are more likely to approach the conditions for a perfectly elastic collision than two balls made from soft rubber.

In Example 9, particle A caught up with B and a collision took place. It is the usual practice, in this subject, to use the word 'overtake' to describe this action; we say that A 'overtakes' B. This is a slightly different use of the word from everyday speech when we say that one vehicle 'overtakes' another!

Example 10

Two particles, A and B, whose masses are m and $2m$, are moving along a straight line and in the same direction with speeds $3u$ and u respectively. Particle A overtakes B. Find the velocities of the particles after the collision, assuming that it is perfectly elastic.

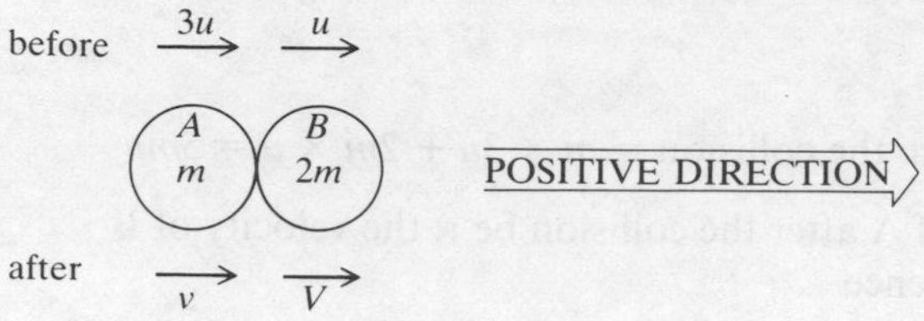

before $\xrightarrow{3u}$ $\xrightarrow{u}$

A (m) B ($2m$) $\boxed{\text{POSITIVE DIRECTION}}\!\!\!\rangle$

after $\xrightarrow{v}$ $\xrightarrow{V}$

Figure 11.8

momentum before the collision $= m \times 3u + 2m \times u = 5mu$

Let the velocity of A and B after the collision be v and V respectively. Conservation of momentum gives

$$5mu = mv + 2mV$$
$$\therefore \quad 5u = v + 2V \qquad \qquad \qquad \cdots \quad (1)$$

Since the collision is perfectly elastic, the kinetic energy is also conserved, so

$$\tfrac{1}{2}m(3u)^2 + \tfrac{1}{2}(2m)u^2 = \tfrac{1}{2}mv^2 + \tfrac{1}{2}(2m)V^2$$
$$\therefore \qquad 4.5u^2 + u^2 = 0.5v^2 + V^2$$
$$\therefore \qquad 11u^2 = v^2 + 2V^2 \qquad \qquad \cdots \quad (2)$$

But from equation (1), $v = 5u - 2V$

Substituting this into equation (2) gives

$$11u^2 = (5u - 2V)^2 + 2V^2$$
$$\therefore \qquad 11u^2 = 25u^2 - 20uV + 4V^2 + 2V^2$$
$$\therefore \qquad 6V^2 - 20Vu + 14u^2 = 0$$
$$\therefore \qquad 3V^2 - 10Vu + 7u^2 = 0$$
$$(3V - 7u)(V - u) = 0 \qquad \qquad \cdots \quad (3)$$

Hence[†]

$$3V - 7u = 0$$
$$\therefore \qquad V = \tfrac{7}{3}u$$

and hence

$$v = 5u - 2(\tfrac{7}{3}u)$$
$$= \tfrac{1}{3}u$$

The velocities of A and B, after the collision, are $\tfrac{1}{3}u$ and $\tfrac{7}{3}u$, respectively.

[†] Note that the alternative solution to equation (3), namely $V = u$, would give $v = 3u$, which is physically impossible because v must be less than V.

Exercise 11b

1 Two railway trucks, whose masses are 30 tonnes and 10 tonnes, are travelling on the same lines, with speeds of $5\,\mathrm{m\,s^{-1}}$ and $3\,\mathrm{m\,s^{-1}}$, respectively, when they collide. In the collision they are coupled together and they then travel with a common velocity. Find this velocity, given that before the collision they were
   (a) travelling in the same direction;
   (b) travelling in opposite directions.

2 A child, whose mass is 25 kg, runs with a speed of $4\,\mathrm{m\,s^{-1}}$ and jumps onto a stationary sledge, which is free to move; the mass of the sledge is 15 kg. Find the resulting speed of the sledge, with the child on it.

3 A gun whose mass is 800 kg fires a shot, of mass 4 kg, horizontally, with a speed of $500\,\mathrm{m\,s^{-1}}$. Initially the gun is at rest, but it is free to move horizontally. Find the velocity with which the gun recoils and find the kinetic energy generated when the gun is fired.

4 Two spacecraft, A and B, have masses m and $2m$, respectively. Initially they are moving in the same direction with speeds v and $(v - u)$, respectively, and B is ahead of A. The spacecraft dock, and after docking they move with a common velocity. Find this velocity in terms of u and v.

5 A particle of mass m, travelling at speed $10u$, collides with a stationary particle, of mass $4m$. After the collision the particles separate with a relative velocity of $5u$. Find their velocities, in terms of u, after the collision and find the percentage of the original kinetic energy lost.

6 A particle P is attached to a fixed point O by a light inextensible string which is 5 metres long. Initially OP is horizontal and the string is taut. The particle is released from rest. Find its speed when it reaches the lowest point on its path. (Take g to be $10\,\mathrm{m\,s^{-2}}$.) At this point P collides with an identical particle and the two particles coalesce. Find their common speed at this instant. Find also the angle which OP makes with the downward vertical when the combined particle comes, instantaneously, to rest.

7 A particle of mass $2m$ is moving with a velocity given by the vector $u\mathbf{i}$, when it collides with another particle, of mass m, moving with velocity $2u\mathbf{j}$. The particles coalesce, and move with a common velocity $\mathbf{V}$. Show that

$$\mathbf{V} = \tfrac{2}{3}u(\mathbf{i} + \mathbf{j})$$

Find also the loss of kinetic energy, expressing it as a percentage of the kinetic energy before the collision.

8 A car travelling due North at 20 mph on an icy road, collides at a crossroads with another car, of equal mass, travelling due East at 10 mph. The cars become entangled in the collision, and slide together on the icy surface. Find their common velocity, giving the speed and the direction.

9 A wooden block of mass M is hanging at rest on a light inextensible string. A bullet of mass m is fired downwards into it at an angle of α to the vertical, striking the block at a speed u. The bullet becomes instantly embedded in the block. Find, in terms of M, m, u and α,
   (a) the common velocity of the bullet and block, immediately after impact;
   (b) the impulsive tension in the string.

10 A projectile of mass $2m$, moving with speed u, is split into two equal parts by an internal explosion, which generates kinetic energy equal to $4mu^2$. Given that after the explosion the parts move along the same line (but not necessarily in the same direction), find their velocities in terms of u.

11.6 Newton's law of restitution[†]

It can easily be shown experimentally that, when a ball is dropped vertically onto a horizontal surface, the ratio of the height from which it is dropped, to the height to which it rebounds, is constant, (i.e. the ratio does not depend on the initial height, but for a different ball or surface the constant may have some other value; it is recommended that the reader should try to verify this). Suppose the ball is dropped from a height H and that it rebounds to height h, then this principle can be expressed

$H:h = $ constant

Suppose that the speed with which the ball strikes the floor is U and that it rebounds with speed V. Using the constant acceleration formula $v^2 = u^2 + 2as$, we have

$$U^2 = 0^2 + 2gH \quad \text{and} \quad 0^2 = V^2 - 2gh$$

i.e. $U^2 = 2gH \qquad$ and $\quad V^2 = 2gh$

Dividing these equations gives

$$\frac{V^2}{U^2} = \frac{h}{H}$$

But the right-hand side of this equation is constant, consequently $\left(\dfrac{V^2}{U^2}\right)$, and hence $\dfrac{V}{U}$ itself, is constant. This constant, $\dfrac{V}{U}$ is called the **coefficient of restitution** and it is always denoted by the letter e. The value of e lies between 0, for a ball which does not rebound, and 1, for a ball which regains its original height. (The latter value, which implies that no energy was lost in the impact, can only be attained in theory; that is, in a perfectly elastic impact.) So we can write $V = eU$. Expressed in words this becomes

> the speed of rebound $= e \times$ the speed of impact

This is called **Newton's law of restitution**.

Qu. 3 A ball is dropped from a height of 1 m onto a horizontal surface and it rebounds to a height of 36 cm. Calculate the coefficient of restitution.

[†] For more information on this subject, see '*Newton's Experimental Law of Impacts*', by Ian Cook, published in the *Mathematical Gazette* dated June 1986.

If the ball strikes the surface obliquely, Newton's law of restitution applies to the normal component of the velocity only. Also, if the surface is smooth there will be no impulsive force parallel to the surface, so the component of the momentum, and hence the velocity, in this direction is unchanged. These facts are summed up in the diagram below, (see Figure 11.9).

Figure 11.9

If the ball strikes the plane at an angle α to the normal, and rebounds at an angle β, we have

$$\tan \alpha = \frac{u}{v}$$

and $$\tan \beta = \frac{u}{ev}$$

$$\therefore \quad e \tan \beta = \frac{u}{v}$$

$$\therefore \quad e \tan \beta = \tan \alpha$$

Example 11

A ball is projected with an initial velocity of $10\mathbf{i} + 10\mathbf{j}$, from a point O on a smooth horizontal plane, whose coefficient of restitution is $\frac{1}{2}$. Taking g to be $10\,\text{m s}^{-2}$, find the horizontal distance the particle has travelled when

(a) it strikes the plane for the first time;
(b) it strikes the plane for the third time;
(c) it stops bouncing.

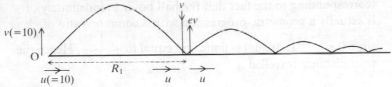

Figure 11.10

(a) Writing the horizontal and vertical components of the initial velocity as u and v respectively, and applying the constant acceleration formula, $s = ut + \frac{1}{2}at^2$, to the vertical motion with $s = 0$, we have

$$0 = vT - \tfrac{1}{2}gT^2$$

where T is the time of flight of the first 'bounce'. Hence

$$T = 2v/g$$

Also, horizontally,

$$R_1 = uT$$
$$= 2uv/g \qquad \qquad \text{. . . . (1)}$$

Using the data, $u = v = 10$, and $g = 10$, we have

$$R_1 = 20$$

The length of the first 'bounce' is 20 m.

(b) Just before the ball strikes the floor for the first time its velocity is $(10\mathbf{i} - 10\mathbf{j})$. When it rebounds the horizontal component is unchanged, but the magnitude of the vertical component is multiplied by e, the coefficient of restitution. Hence, using (1) the range for second 'bounce' is $2euv/g \, (= eR_1)$, and for the next bounce, the range is multiplied by e again.

Consequently the total distance travelled in the first three bounces is

$$(20 + 20e + 20e^2) \text{ metres}$$

and since $e = \frac{1}{2}$, this equals

$$(20 + 10 + 5) \text{ metres}$$

Hence, when the ball strikes the plane for the third time it has travelled 35 m horizontally.

(c) If the ball is allowed to continue bouncing indefinitely, then proceeding as in (b), the total distance travelled horizontally is

$$(20 + 20e + 20e^2 + 20e^3 + 20e^4 \ldots) \text{ metres}$$
$$= 20(1 + e + e^2 + e^3 + e^4 \ldots) \text{ metres}$$

Although the expression in brackets has infinitely many terms (corresponding to the fact that the ball bounces indefinitely), it is actually a geometric progression with a common ratio of e[†]; since $0 < e < 1$, its sum is finite and equal to $\dfrac{1}{(1 - e)}$. Hence the total distance travelled is

$$\frac{20}{(1 - e)} \text{ metres}$$

and putting $e = \frac{1}{2}$, this becomes 40 metres.

[†] See *Pure Mathematics Book 1* by J. K. Backhouse and S. P. T. Houldsworth, revised by P. J. F. Horril, published by Longman, Chapter 13.

The total horizontal distance travelled by the ball while it is bouncing is 40 m.

(After the bouncing stops, the ball continues to move horizontally at $10\,\mathrm{m\,s^{-1}}$.)

11.7 Newton's law of restitution in a direct impact

A *direct* impact is a collision between two objects which are moving along the same straight line. Figure 11.11 represents such a collision between two objects A and B, whose masses are m_1 and m_2 respectively; u_1 and u_2 represent their velocities before the impact, and $u_1 > u_2$ (otherwise there would be no collision). If the objects were moving towards each other, u_2 would be negative.

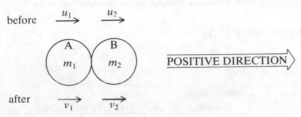

Figure 11.11

The velocities after the impact are represented by v_1 and v_2, and $v_1 < v_2$. (Otherwise A would have to pass through B!). It is possible that, in some circumstances v_1 will turn out to be negative; this would indicate that the direction of A's motion had been reversed by the collision.

As we have already seen, momentum is conserved in such a collision, i.e.

$$m_1u_1 + m_2u_2 = m_1v_1 + m_2v_2 \qquad\qquad\qquad (1)$$

Normally we would expect to know the masses of the objects and their velocities before the collision. However, this leaves us with an important mathematical difficulty: we have *one* equation, but *two* unknown quantities, namely v_1 and v_2. So, with equation (1) only, we cannot find the velocities after the collision. This difficulty can be overcome by using a more general version of the law of restitution which we met in the last section. This more general form says that, in a direct impact,

> the speed with which the objects separate is equal to the speed of approach multiplied by e, the coefficient of restitution

The word 'speed' is used deliberately here. When using this principle it is essential to ensure that both of these speeds are positive quantities, i.e.

the speed of separation $= v_2 - v_1$
the speed of approach $= u_1 - u_2$

Using this notation, Newton's law of restitution can be written

$$v_2 - v_1 = e \times (u_1 - u_2)$$

. . . (2)

(Note carefully the order of the suffices.)

With this second equation at our disposal, the difficulty is overcome; we have two equations, which can be solved to find the two unknowns. The following example will illustrate how to proceed.

Example 12

A railway truck, whose mass is 2 tonnes, moving at $12 \, \text{km h}^{-1}$, collides with another truck, whose mass is 5 tonnes, which is moving in the opposite direction at $2 \, \text{km h}^{-1}$. Given that the coefficient of restitution is $\frac{1}{2}$, find the velocities of the trucks after the collision.

The data is illustrated in Figure 11.12:

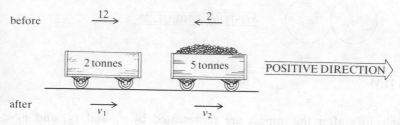

Figure 11.12

Conservation of momentum gives

$$2v_1 + 5v_2 = 2 \times 12 + 5 \times (-2)$$

[Note the use of the minus sign to take account of the fact that the $2 \, \text{km h}^{-1}$ is in the opposite direction to that of the $12 \, \text{km h}^{-1}$.]

Hence

$$2v_1 + 5v_2 = 14$$

. . . (1)

The trucks approach each other with a speed of $(12 + 2) \, \text{km h}^{-1}$, and separate with speed $(v_2 - v_1)$. Consequently, Newton's law of restitution gives

$$(v_2 - v_1) = \tfrac{1}{2} \times 14$$

i.e. $v_2 - v_1 = 7$

. . . (2)

Multiplying this equation by 2 and adding it to equation (1) eliminates v_1, giving

$$7v_2 = 28$$
$$\therefore \quad v_2 = 4$$

Substituting this value in equation (2) gives $v_1 = -3$.

Hence, after the collision, the direction of motion of each truck is reversed, and the speeds are $3 \, \text{km h}^{-1}$, for the 2 tonne truck, and $4 \, \text{km h}^{-1}$ for the 5 tonne truck.

Qu. 4 A particle of mass m moving with speed U collides with an identical particle moving in the same direction with speed u, where $U > u$. If the coefficient of restitution is equal to 1, show that the collision is perfectly elastic (i.e. there is no loss of kinetic energy).

Qu. 5 Prove that if $e = 1$, the kinetic energy lost in a direct impact between two particles is zero. (Do not assume the particles have equal mass.)

In a perfectly elastic collision, $e = 1$.

Exercise 11c

1 A ping-pong ball is dropped from a height of 2 m onto a horizontal floor; the coefficient of restitution is $\frac{3}{4}$. Find the height to which it rises when it rebounds
   (a) for the first time;
   (b) for the second time;
   (c) for the nth time.
2 A stopwatch is started when the ball in question 1 strikes the floor for the first time. Find the time recorded by the watch when the ball
   (a) strikes the floor for the second time;
   (b) stops bouncing.
   (Take g to be $9.81 \, \text{m s}^{-2}$.)
3 A snooker ball strikes the cushion at the edge of the table, making an angle of $30°$ with the normal; the coefficient of restitution is $\frac{1}{3}$. Show that the ball is deflected through $90°$ by the impact.
4 A snooker ball strikes the cushion at the edge of the table, making an angle α with the normal; the coefficient of restitution is e. Show that the kinetic energy lost in the impact is $K(1 - e^2) \cos^2 \alpha$, where K is the kinetic energy before the impact.
5 A railway truck of mass M, moving at $5 \, \text{km h}^{-1}$, collides with a stationary truck, whose mass is $2M$; the coefficient of restitution is 0.8. Find the velocities after the impact.
6 A railway truck of mass M moving at a speed $2u$, collides with another truck, whose mass is $2M$, travelling in the same direction with speed u; the coefficient of restitution is 0.5. Find the velocities after the impact.
7 Two identical particles, each of mass m and moving with speed u, collide head-on. Show that the kinetic energy lost in the collision is $(1 - e^2)mu^2$, where e is the coefficient of restitution.
8 Three identical spheres A, B and C lie in a straight line on a smooth horizontal table. A is projected towards B with velocity u. Show that after the impact between B and C, the latter moves with speed $\frac{1}{4}(1 + e)^2 u$. Show that, if $e < 1$, a second collision between A and B is inevitable.

9 In a 'Newton's Cradle', four identical spheres are suspended on strings of the same length, from a horizontal bar. When they are at rest, the spheres lie in a straight line, almost touching. The first sphere is pulled aside, in the plane of the bar and the other spheres, and released; it strikes the second sphere with speed u. The coefficient of restitution between any pair of spheres is e. Show that, in due course, the fourth sphere begins to move with speed $\frac{1}{8}(1 + e)^3 u$.

10 A similar 'Newton's Cradle' (see question 9) has just two spheres. The first sphere is pulled aside so that its string makes an angle α with the vertical, and released. After the collision, the string supporting the second sphere turns through an angle β, before it comes instantaneously to rest.
Show that

$$(1 - \cos \beta) = \tfrac{1}{4}(1 + e)^2(1 - \cos \alpha).$$

Exercise 11d (Miscellaneous)

1 A smooth sphere, of mass 3 kg, moving with speed $5 \, \text{m s}^{-1}$, collides directly with another smooth sphere, of equal radius, mass 6 kg, and moving in the opposite direction with speed $3 \, \text{m s}^{-1}$. The impact brings the second sphere to rest. Find the speed of the first sphere after the impact and the coefficient of restitution.

(L)

2 Two spheres, A and B, of equal size but different masses m_1 kg and m_2 kg respectively, travel towards each other along the line of centres with speeds $8 \, \text{m s}^{-1}$ and $6 \, \text{m s}^{-1}$ respectively. After the collision A continues to travel in the same direction as before with speed $2 \, \text{m s}^{-1}$, while the direction of motion of B is reversed and its speed reduced to $3 \, \text{m s}^{-1}$. Given that the loss of energy is 9.36 J, calculate the value of m_1 and of m_2.

*(C)

3 Two particles A and B, of masses M kg and m kg respectively, lie at rest inside a smooth horizontal tube. A and B are projected towards each other with speeds of $5 \, \text{m s}^{-1}$ and $2 \, \text{m s}^{-1}$ respectively.

After impact, A continues to move in the same direction as before at $1 \, \text{m s}^{-1}$ while B's direction of motion is reversed and its speed becomes $3 \, \text{m s}^{-1}$.
(a) Calculate the ratio M/m.
(b) If the loss in kinetic energy is 100 J, calculate the values of M and m.

*(C)

4 A space capsule of mass 1050 kg is travelling at $400 \, \text{m s}^{-1}$ when a rocket of mass 50 kg is fired from the capsule forward at an angle of 39° with the direction of flight, so that it has a speed of $500 \, \text{m s}^{-1}$, thereby reducing the mass of the capsule by 50 kg and changing the direction of its flight. By considering the conservation of momentum, calculate the new speed of the capsule after firing the rocket, and the change in the direction of its flight.

*(O & C)

5 A ball of mass 0.4 kg travelling at $20 \, \text{m s}^{-1}$ is struck with an impulse of 6 N s. Calculate the speed with which it continues
(a) if it is struck in the direction opposite to its velocity;

(b) if it is struck in the direction at right-angles to its velocity;

(c) if it is struck so as to deflect it as much as possible.

*(SMP)

6 A particle is released from rest at a height of 20 m above a horizontal plane. The particle falls freely under gravity to strike the plane. The coefficient of restitution between the particle and the plane is e. Find

(a) the height reached after the third bounce;

(b) the time between the first and second bounce.

Given that the total time from the instant of release until the particle comes finally to rest is 4 seconds, find

(c) the value of e;

(d) the total distance travelled by the particle.

(Take g as $10\,\text{m\,s}^{-2}$.)

(L)

7 An elastic sphere of mass M_1, moving with speed $12\,\text{m\,s}^{-1}$, overtakes and collides directly with another elastic sphere of mass M_2 moving in the same straight line with speed $4\,\text{m\,s}^{-1}$. After impact, the spheres continue to move in the same direction as before but with speeds of $6\,\text{m\,s}^{-1}$ and $10\,\text{m\,s}^{-1}$, respectively. Find the coefficient of restitution between the spheres and show that $M_1 = M_2$.

Find the speeds of the spheres after the impact if they had been travelling in opposite directions with speeds of $12\,\text{m\,s}^{-1}$ and $4\,\text{m\,s}^{-1}$.

(L)

8 Sphere A, of mass m, is at rest on a smooth horizontal table, when sphere B, of mass $2m$, and the same radius as A, moving on the table with speed U strikes it directly. Given that the coefficient of restitution between the spheres is $\frac{1}{2}$, show that the speeds of A and B after the collision are U and $\frac{1}{2}U$ respectively.

The velocities of A and B are along the normal to a smooth vertical wall. Sphere A collides with this wall, the coefficient of restitution for this impact being $\frac{2}{3}$. On rebounding, A hits B directly again. Find the speeds of A and B after this collision and show that the total loss of kinetic energy is $\frac{125}{144}mU^2$.

(L)

9 Three identical smooth spheres, A, B and C, each of mass m, lie at rest in a straight line on a smooth horizontal table. Sphere A is projected with speed u to strike B directly. Sphere B then strikes sphere C directly. The coefficient of restitution between any two spheres is e. Find the speeds, in terms of u and e, of the spheres after these two collisions. Show also that the total loss of kinetic energy is

$$\tfrac{1}{16}mu^2(5 + 2e - 4e^2 - 2e^3 - e^4)$$

(L)

10 (a) A sphere of mass m moving along a smooth horizontal table with speed V collides directly with a stationary sphere of the same radius and of mass $2m$. Obtain expressions, in terms of V and the coefficient of restitution e, for the speeds of the two spheres after impact.

Half of the kinetic energy is lost in the impact. Find the value of e.

(b) A particle of mass m moving in a straight line with speed u receives an impulse of magnitude I in the direction of its motion. Show that the increase in the kinetic energy is given by

$$\frac{I(I + 2mu)}{2m}$$

(L)

11 A particle of mass $0.5\,\text{kg}$ is in motion with velocity $(6\mathbf{i} + 4\mathbf{j})\,\text{m s}^{-1}$ when it strikes a fixed barrier and rebounds with velocity $(-2\mathbf{i} + 2\mathbf{j})\,\text{m s}^{-1}$. The particle is in contact with the barrier for 0.02 seconds. Find the average force exerted by the barrier on the particle.

Calculate also the kinetic energy lost by the particle in the collision.

(JMB)

12 A particle moving along a smooth horizontal floor hits a smooth vertical wall and rebounds in a direction at right-angles to its initial direction of motion. The coefficient of restitution is e. Find, in terms of e, the tangent of the angle between the initial direction of motion and the wall. Prove that the kinetic energy after the rebound is e times the initial kinetic energy.

(JMB)

13 Two particles, A of mass $2m$ and B of mass m, moving on a smooth horizontal table in opposite directions with speeds $5u$ and $3u$ respectively, collide directly. Find their velocities after the collision in terms of u and the coefficient of restitution e. Show that the magnitude of the impulse exerted by B on A is $\frac{16}{3}mu(1 + e)$.

Find the value of e for which the speed of B after the collision is $3u$. Moving at this speed B subsequently collides with a stationary particle C of mass km, and thereafter remains attached to C. Find the velocity of the combined particle and find the range of values of k for which a third collision will occur.

(JMB)

14 Two points A and B lie on a smooth horizontal plane and are such that the line AB is perpendicular to a vertical wall standing on the plane, A being between B and the wall. A particle P of mass $2m$ is at rest at A and a particle Q of mass m is projected with velocity u from B directly towards A. Find the velocities of the particles immediately after the impact, given that the coefficient of restitution between them is $\frac{1}{3}$.

P hits the wall and rebounds, hitting Q again. The coefficient of restitution between P and the wall is $\frac{1}{6}$. Does P hit the wall a second time? Justify your answer.

(O & C)

15 Three particles A, B and C are lying in a straight line in that order on a smooth horizontal table. The particles have masses $3m$, $2m$ and m respectively, and the coefficient of restitution between any pair of particles is $e\,(<1)$. A is projected towards B with velocity u. Show that the velocity of B immediately after the second collision is

$$\tfrac{1}{5}u(1 + e)(2 - e)$$

and find the velocities of the other two particles at this instant. Show that there will be no more collisions if

$$e \geqslant \tfrac{1}{2}(3 - \sqrt{5})$$

(O & C)

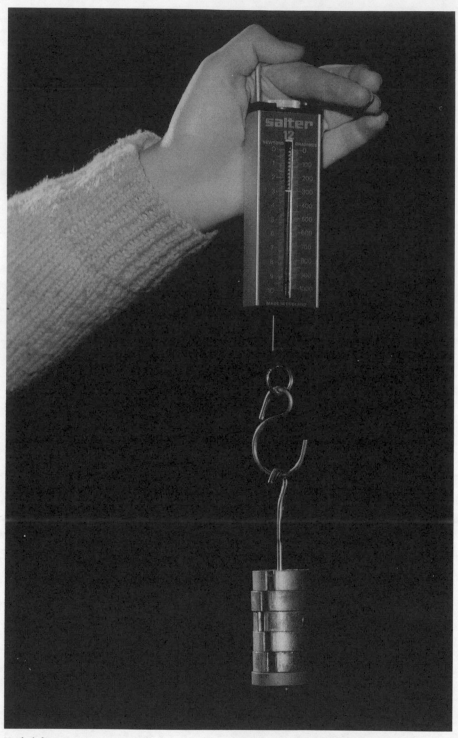

Hooke's law

Chapter 12

Hooke's law–simple harmonic motion

12.1 Springs and elastic strings

Most mechanical devices, from the humblest sewing machine to the most complicated jet engine, depend at some stage of their operation on the action of a spring. In this chapter we shall be looking at the action of elastic strings and simple springs (i.e. springs designed to act along a straight line and usually made from wire, or other suitable material, arranged in a helix).

In order to avoid the mathematical difficulties which would follow, we shall assume that the springs and strings are 'light', i.e. their mass is negligible. As we saw in Chapter 9 (§9.1), the effect of this assumption is to make the tension the same throughout the length of the spring or string. This is also true for a spring when it is compressed, (we then say the spring exerts a 'thrust'); an elastic string cannot be compressed.

It is a matter of common experience that, if an excessive load is applied to a spring, it may be permanently damaged; it may fail to return to its natural shape, or in the case of an elastic string, it may break. When permanent damage is caused, we say it has passed its **elastic limit**. In this book, we shall assume that this stage is not reached; in other words, all the loads we shall be considering will be such that, when they are removed, the string or spring will return to its 'natural' (i.e. unstretched) length. It is usual to represent the natural length by l.

When a string, or a spring, hangs vertically, supporting a load, W, (see Figure 12.1), the tension, T, is equal to the load. If this were not so, there would be a non-zero resultant force and the load would accelerate.

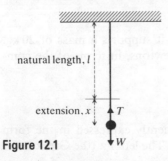

natural length, l

extension, x T

Figure 12.1 W

The relationship between the tension T, and the extension x, can be investigated experimentally by applying different loads to a given spring. The reader should do this, if the necessary equipment is available.

Qu. 1 The results of an experiment to investigate the relationship between the tension, T newtons, and the extension, x mm, in a stretched spring, are shown below:

load (= tension) (in newtons)	1	2	3	4	5	6
extension (in mm)	14	34	50	62	82	95

Plot these results on graph paper. Draw a straight line which, as nearly as possible, fits the points. From the graph,
(a) estimate the tension when the extension is 40 mm;
(b) estimate the extension which would be produced by a load of 3.5 N;
(c) express T in terms of x.

12.2 Hooke's law

Experiments, like that described in §12.1, were conducted by the English physicist Robert Hooke (1635–1703), and he concluded that, subject to experimental error and provided the elastic limit is not exceeded, the tension T is proportional to the extension x, i.e.

$$T \propto x$$

This can be expressed as an equation of the form

$$T = kx$$

where k is (for a given string or spring) a constant. For the spring in Qu. 1, k is approximately $\frac{1}{16}$.

Although only an ideal spring would obey Hooke's law exactly, it is a very simple principle and consequently it makes a convenient basis for our study of elastic strings and springs; we say it provides a good 'mathematical model' of the action of a spring or string. Hooke's law may also be applied to a spring under compression.

Qu. 2 A certain spring is compressed 8 cm when it supports a mass of 20 kg. Taking g to be 10 m s^{-2}, express the thrust, T newtons, in terms of the compression, x cm. Find:
(a) the thrust when $x = 2$;
(b) the compression when $T = 300$.

The constant k in the equation $T = kx$, is frequently expressed in the form $k = \lambda/l$, where l is the natural length of the spring. The letter λ (the Greek letter lambda) is called the **modulus of elasticity**. (An important reason for using λ is that it does not depend on the natural length of the spring.) Hooke's law can then be written in the form

$$T - \frac{\lambda}{l} x$$

i.e. $T = \dfrac{\lambda x}{l}$

Notice that if $x = l$, then $T = \lambda$. Consequently, λ can be interpreted as the tension when the extension is equal to the natural length of the spring, i.e. when the natural length is doubled. (However, most real springs would have passed their 'elastic limit' before this stage is reached.) Since λ can be interpreted as a tension, it is expressed in the same units as the tension.

Example 1

Two elastic strings, AP and BP, are attached to a particle of mass 2 kg at P. The ends A and B are attached to fixed points which are 135 cm apart, with A vertically above B. Each string has a natural length of 50 cm and the moduli of elasticity are 60 N and 50 N, respectively. The system hangs in equilibrium with the strings in the same vertical line. Taking g to be $10\,\text{m s}^{-2}$, calculate the extensions in the strings.

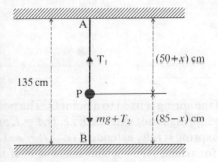

Figure 12.2

Let the extension of the string AP be x cm. Then, since $AB = 135$ cm, the extension of string BP is

$$[(85 - x) - 50]\,\text{cm} = (35 - x)\,\text{cm}$$

The tensions in the strings are:

in AP $T_1 = \dfrac{60x}{50} = 1.2x$ . . . (1)

in BP $T_2 = \dfrac{50(35 - x)}{50} = 35 - x$ . . . (2)

Using Newton's second law of motion, with the acceleration equal to zero, we have

$$T_1 - (T_2 + mg) = 0$$

and hence, (putting $m = 2$ and $g = 10$)

$$T_1 - T_2 = 20$$

Using equations (1) and (2), we have

$$1.2x - (35 - x) = 20$$
$$1.2x - 35 + x = 20$$
$$\therefore \qquad 2.2x = 55$$
$$\therefore \qquad x = 25$$

Hence the extensions in AP and BP are 25 cm and 10 cm, respectively.

12.3 The work done in stretching a light spring

Suppose a light spring, whose modulus of elasticity is λ, is stretched from its natural length, l, until its total length is $l + e$.

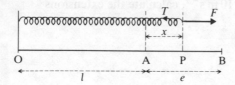

Figure 12.3

In the diagram (Figure 12.3), one end of the spring is fixed to a point O. The point A is the position of the free end when the spring is not extended, i.e. $OA = l$, and B is the position of the free end when the spring is fully extended, i.e. $AB = e$. The point P represents a general position, for which $AP = x$ and $0 \leqslant x \leqslant e$. If the force exerted on the spring at P is represented by F, the work done by this force as its point of application moves from A to B is given by

$$\int_0^e F \, dx \qquad \text{(see §10.3)}$$

Since we are considering an *ideal* spring (one whose mass is negligible) and since there is no particle attached to P, there is no mass to be accelerated as P moves, so the force F is equal in magnitude to the tension T in the spring. This in turn is given by Hooke's law, i.e.

$$T = F = kx$$

where $k = \lambda/l$. Substituting for F in the integral opposite, we have

$$\text{work done} = \int_0^e kx\,dx$$

$$= \left[\frac{1}{2}kx^2\right]_0^e$$

$$= \tfrac{1}{2}ke^2$$

It is frequently convenient to write this expression in the form

$$\text{work done} = \tfrac{1}{2}(ke)e$$

i.e.   work done = $\frac{1}{2}$ × tension (at **B**) × extension

(The reader may find it easier to remember this result in words, rather than in symbols.)

Qu. 3 Using the notation above, prove that the work done in stretching the spring from $x = a$ to $x = b$ (where $a < b$) is $\tfrac{1}{2}k(b^2 - a^2)$, and show that this can be expressed as follows:

work done = the average of the tensions × the extension

The results above may also be applied to a *compressed* spring (although in this case it is usual to call the force exerted by the spring the 'thrust').

If the spring is held in this state, the work done in stretching it, or, in the case of a compressed spring, the work done in compressing it, is called the **elastic potential energy**; this may then be converted into some other form of energy, as the following examples illustrate.

Example 2

In a toy gun, a projectile, of mass 5 grams, is propelled by a spring whose modulus of elasticity is 20 N. Before firing, the spring, whose natural length is 10 cm, is compressed by 6 cm. The gun is then held horizontally and fired. Neglecting frictional losses, find the speed given to the projectile.

By Hooke's law, the initial thrust in the spring is

$$\frac{20 \times 0.06}{0.10} = 12\,\text{N}$$

The 'elastic potential energy' stored

$= \frac{1}{2}$ × initial thrust × compression
$= \frac{1}{2} \times 12 \times 0.06\,\text{J}$
$= 0.36\,\text{J}$

When the gun is fired, this elastic potential energy is converted into kinetic energy (of the projectile). Let the final speed of the projectile be $v \, \mathrm{m\,s^{-1}}$, then its kinetic energy is

$$\tfrac{1}{2} \times 0.005v^2 \, \mathrm{J} = 0.0025v^2 \, \mathrm{J}$$

Equating the elastic potential energy lost and the kinetic energy gained, we have

$$
\begin{aligned}
0.0025v^2 &= 0.36 \\
\therefore \quad v^2 &= 144 \\
\therefore \quad v &= 12
\end{aligned}
$$

The final speed of the projectile is $12 \, \mathrm{m\,s^{-1}}$.

Example 3

A particle of mass m is attached to one end of a light elastic string, of natural length l, whose modulus of elasticity is $2mg$. The other end of the string is attached to a fixed point O. Initially the particle is held at rest at O; it is then released and allowed to fall vertically. It comes to rest instantaneously at a point A, vertically below O. Find the distance OA.

Find also the acceleration of the particle when it is at A.

Suppose the particle comes instantaneously to rest when the extension of the string is x, i.e. $OA = l + x$.

The tension in the string at A is $\lambda x/l$, where $\lambda = 2mg$, i.e.

$$\text{the tension in the string} = \frac{2mgx}{l}$$

Hence at A,

$$
\begin{aligned}
\text{the elastic potential energy} &= \frac{1}{2} \times \frac{2mgx}{l} \times x \\
&= \frac{mgx^2}{l}
\end{aligned}
$$

The gravitational potential lost by the particle in falling from O to A is $mg(l + x)$. Inbetween O and A the particle has kinetic energy, but at A it is instantaneously at rest, so its kinetic energy *at this instant* is zero. Consequently, using the principle of conservation of energy, we have

$$\frac{mgx^2}{l} = mg(l + x)$$

Hence

$$x^2 = (l + x)l$$
$$\therefore \quad x^2 - lx - l^2 = 0$$

Regarding this as a quadratic equation in x and using the quadratic formula, we have

$$x = \frac{l \pm \sqrt{(l^2 + 4l^2)}}{2}$$

$$= \tfrac{1}{2}(1 \pm \sqrt{5})l$$

But x, the extension, must be positive, so we must use the positive square root, i.e.

$$x = \tfrac{1}{2}(1 + \sqrt{5})l \qquad\qquad\qquad\qquad (1)$$

$$\therefore \quad OA = l + x = l + \tfrac{1}{2}(1 + \sqrt{5})l$$
$$= \tfrac{1}{2}(3 + \sqrt{5})l$$

Although the particle is at rest at A, the forces acting on it, at this point, are not in equilibrium. In fact

the downward force $= mg$

the upward force (i.e. the tension in the string)

$$= \frac{2mgx}{l}$$

$$= \frac{2mg \times \tfrac{1}{2}(1 + \sqrt{5})l}{l} \qquad \text{(from equation (1))}$$

$$= mg(1 + \sqrt{5})$$

Hence

the resultant upward force $= mg(1 + \sqrt{5}) - mg$
$$= mg\sqrt{5}$$

Using Newton's second law of motion gives

$$mg\sqrt{5} = ma$$

where a is the (upward) acceleration of the particle, i.e.

$$a = g\sqrt{5}$$

When the particle is at A its acceleration is $g\sqrt{5}$, upwards.

Exercise 12a

1 An elastic string, whose natural length is 50 cm, is stretched 10 cm by a force of 25 N. Find
(a) the modulus of elasticity;

(b) the extension when a force of 20 N is applied;

(c) the tension in the string when the extension is 35 cm.

2 A vertical spring, of natural length l metres is compressed by $\frac{1}{4}l$ metres when a particle of weight W newtons is placed on it. Find its modulus of elasticity, in terms of W. Find the amount by which the spring is compressed if the particle is replaced by another of weight kW newtons.

3 Two identical springs each have natural length l and modulus λ. Find the extension in each spring if they are suspended vertically, side by side, and jointly support a weight W.

Find also the total extension if the two springs are joined together, making a single spring of length $2l$, supporting the same weight.

4 Find the work done when an elastic string of natural length 2 metres and modulus of elasticity 20 newtons, is stretched to a total length of 2.5 metres.

5 An elastic string, of natural length 2 metres and modulus of elasticity 100 newtons, is placed on a smooth horizontal table, with one end fixed to a point on the table. The string is maintained in a stretched state by a force of 20 newtons. Calculate the extension and find the elastic potential energy stored in the string.

6 A simple pop-gun consists of a spring, of natural length 10 cm and modulus of elasticity 0.04 N. It is used to propel a small projectile, whose mass is 100 grams. Initially the spring is compressed by 5 cm and the gun is held with its barrel horizontal. Assuming that the projectile remains in contact with the end of the spring as it is released, find

(a) the initial acceleration of the projectile;

(b) the acceleration at the instant when the spring returns to its natural length;

(c) the acceleration when the spring has a compression of x cm, where $0 \leqslant x \leqslant 5$, giving the answer in terms of x;

(d) the velocity when $x = 0$.

7 A light spring is placed on a smooth horizontal table with one end fixed; the other end has a small particle of mass m attached to it. The spring is stretched by an external force and it is then released. Describe in words the subsequent motion of the particle, paying special attention to its acceleration

(a) when the particle is first released,

(b) when the spring returns to its unstretched state,

(c) when the particle first comes to rest.

8 Repeat question 7 for a light elastic *string*. (Note that when the string is slack it exerts no force.)

9 A particle of mass 2 kg is attached to one end of a light elastic string of natural length 1 metre. The other end of the string is attached to a fixed point A. Initially the particle is held at A, and then it is released; it falls vertically and comes to rest at a point 1.5 metres below A. Using the principle of conservation of energy, find the modulus of elasticity of the string. (Take g to be $9.81 \, \text{m s}^{-2}$.)

10 As an alternative to Hooke's law, it is suggested that the tension in newtons, in a spring of natural length l metres, is given by a function $T(x)$ of the form

$$T(x) = ax + bx^2 + cx^3$$

where a, b and c are constants, and the extension of the spring is x metres. Find expressions for a, b and c, given that the law must satisfy the three following conditions:

(i)   For a given extension, the tension must be equal in magnitude to the thrust for a compression of the same amount, i.e. $T(x) = -T(-x)$.

(ii) For small values of x the new law must be the same as Hooke's law for a spring of modulus λ newtons.

(iii) The maximum value of the tension in the spring is $\frac{2}{3}\lambda$ newtons. For this law, sketch a graph showing the tension in the spring plotted against the displacement. Could such a law be used for *all* values of x? Would it be more realistic than Hooke's law? Give reasons for your answers.

12.4 Simple harmonic motion

In this section, we shall examine the acceleration which is given to a particle by a spring.

Suppose a light spring is lying in a straight line on a smooth horizontal table, with one end attached to a fixed point O and with a particle of mass m attached to the free end. (Having the spring on a smooth horizontal table simplifies the problem, since this avoids the extra difficulties arising from friction, or from the effect of gravity; these can be considered at a later stage.) Suppose that the natural length of the spring is l, and its modulus of elasticity λ.

Throughout this section it will be convenient to use x to represent the extension of the spring, and, as usual, t will represent the time which has elapsed. The reader is reminded that $t = 0$ represents the instant at which we start to record the time, i.e. we can imagine starting a stopwatch at this instant; $t = 0$ does not necessarily represent the instant at which the particle starts to move.

In this particular case, let us suppose that the particle is given an initial velocity u in the positive direction, in other words, away from O, that is, when $t = 0$, $x = 0$, and $\dfrac{\mathrm{d}x}{\mathrm{d}t} = u$. Figure 12.4a shows this information, and Figure 12.4b shows the particle P in a general position on its path, with the tension T acting on it.

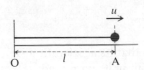

Figure 12.4a

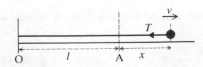

Figure 12.4b

As always the velocity, v, at any instant, is given by $\dfrac{dx}{dt}$ and the acceleration by $\dfrac{dv}{dt}$.

Using Newton's second law of motion, we have

$$-T = m\frac{dv}{dt}$$

(Note the minus sign, indicating that T acts in the *negative* direction, that is, towards O.)

This equation can be written

$$-T = m\frac{d^2x}{dt^2}$$

The tension, T, is given by Hooke's law, i.e. $T = \lambda x/l$. Hence

$$\frac{-\lambda x}{l} = m\frac{d^2x}{dt^2}$$

$$\frac{d^2x}{dt^2} = -\frac{\lambda x}{ml} \qquad \qquad . \quad . \quad . \quad (1)$$

This is an extremely important equation and we shall study it in more detail in the next section. But first let us examine its main features in general terms.

Notice that the acceleration is *not* constant; in fact it is proportional to x, the displacement. Consequently, *the constant acceleration formulae must not be used.*

The velocity when $x = 0$ is u. Immediately, the spring begins to stretch, and the tension begins to slow the particle down. As the displacement increases, the tension increases in magnitude and consequently the magnitude of the retardation increases. Eventually the particle comes to rest. Although the particle is instantaneously at rest, there is a large force acting on it towards O, namely, the tension. Consequently the particle now starts to move back towards O, with increasing speed. When it reaches A, the point where $x = 0$, the tension, and hence the acceleration, becomes zero. But the particle is now moving quite quickly in the negative direction, so the spring will start to compress.

Qu. 4 Describe the subsequent motion in words.

Definition

Motion in which a point moves in a straight line, so that its acceleration is
(a) proportional to its displacement from a fixed point in its path, and
(b) directed towards the fixed point, is called
Simple Harmonic Motion (SHM).

Writing the displacement as x, this definition can be expressed as an equation, thus:

$$\frac{d^2x}{dt^2} = -n^2x \qquad \qquad . \quad . \quad . \quad (2)$$

The minus sign ensures that the acceleration is directed towards O. The constant of proportionality is written as n^2, because the square of a real number cannot be negative, so this factor cannot reverse the effect of the minus sign. Using n^2 also saves us, at a later stage, from having to make repeated use of the square root signs. The equation is usually called **the SHM equation**.

Equation (1), above, is an example of the SHM equation in which

$$n^2 = \frac{\lambda}{ml}$$

12.5 The SHM equation

We shall now look in more detail at the SHM equation

$$\frac{d^2x}{dt^2} = -n^2x \qquad \qquad \cdots \quad (1)$$

which was introduced in the last section.

First we shall write the left-hand side, which represents the acceleration, as $\dfrac{dv}{dt}$ and this in turn we shall write as $v\dfrac{dv}{dx}$ (see §4.5).

Equation (1) can now be expressed as follows:

$$v\frac{dv}{dx} = -n^2x$$

Integrating with respect to x gives

$$\tfrac{1}{2}v^2 = -\tfrac{1}{2}n^2x^2 + c$$
i.e. $v^2 = -n^2x^2 + 2c$

where c is the constant of integration. Let the velocity at the origin be u, i.e. when $x = 0$, $v = u$; then

$$u^2 = 2c$$

Substituting for $2c$ in the expression for v^2, we have

$$v^2 = -n^2x^2 + u^2$$
i.e. $v^2 = u^2 - n^2x^2$ $\qquad \qquad \cdots \quad (2)$

[Readers are reminded that the constant acceleration formulae must not be used in this context; equation (2) should be used in circumstances which, in motion with constant acceleration, would require the equation $v^2 = u^2 + 2as$.]

In equation (2) the velocity, v, becomes zero when

$$n^2x^2 = u^2, \quad \text{i.e. when } x = \pm u/n$$

The positive quantity u/n is called the **amplitude** of the motion, and it is always represented by a, i.e. $u^2 = n^2a^2$. Using this in equation (2) we have

$$v^2 = n^2a^2 - n^2x^2$$

$$\therefore \quad v^2 = n^2(a^2 - x^2) \qquad \qquad \qquad . \ . \ . \ (3)$$

Notice that, since the square of a real number cannot be negative, x^2 cannot exceed a^2, that is

$$-a \leqslant x \leqslant +a$$

so the displacement from the origin never exceeds a in magnitude.

Equation (3) can be written

$$v = \pm n\sqrt{(a^2 - x^2)}$$

or $\quad \dfrac{\mathrm{d}x}{\mathrm{d}t} = \pm n\sqrt{(a^2 - x^2)} \qquad \qquad \qquad . \ . \ . \ (4)$

This equation can be solved by a technique known as 'separating the variables', which is beyond the scope of this book; details may be found in any standard text-book which deals with 'Differential Equations'[†].

It is sufficient for our purposes to note that the general solution of equation (4), and hence of the original SHM equation (equation (1)), is

$$x = A \cos nt + B \sin nt \qquad \qquad \qquad . \ . \ . \ (5)$$

where A and B are arbitrary constants, i.e. they can take any value, but they remain constant as t varies.

It is easy to verify that this expression for x does indeed satisfy equation (1). Differentiating (5) with respect to t gives

$$\frac{\mathrm{d}x}{\mathrm{d}t} = -nA \sin nt + nB \cos nt$$

and differentiating again gives

$$\frac{\mathrm{d}^2x}{\mathrm{d}t^2} = -n^2A \cos nt - n^2B \sin nt$$

$$= -n^2 (A \cos nt + B \sin nt)$$
$$= -n^2x$$

So expression (5) for x does satisfy equation (1) for all values of A and B.

The sine and cosine functions are 'periodic', that is, their values are repeated at regular intervals. The period of these functions is 2π (remember that whenever the trigonometrical functions are differentiated or integrated, it is essential to

[†] See *Pure Mathematics Book 2*, by J. K. Backhouse and S. P. T. Houldsworth, revised by P. J. F. Horril, published by Longman, Chapter 14.

measure the 'angle' in radians). Consequently the function given by (5), above, is also periodic. Its values are repeated whenever nt is increased by 2π, i.e. every $2\pi/n$ seconds. The period of the SHM is always denoted by T, and so we write

$$T = \frac{2\pi}{n}$$
                                                                    . . . . (6)

Example 4

A particle is performing simple harmonic motion with a period of 12 seconds and an amplitude of 2 m. Find an expression for its displacement, x metres, from the origin, O, in terms of the time, t seconds, given that $x = 0$ when $t = 0$. Hence find (a) its displacement, (b) its speed and (c) its acceleration, when $t = 1$.

We are given that T, the period, is 12 s, hence

$$12 = \frac{2\pi}{n}$$

so $n = \frac{1}{6}\pi$

Using this in the general solution of the SHM equation (see equation (5)) gives

$$x = A\cos\left(\tfrac{1}{6}\pi t\right) + B\sin\left(\tfrac{1}{6}\pi t\right)$$

We know that when $t = 0$, $x = 0$, therefore

$$0 = A\cos 0 + B\sin 0$$

but $\cos 0 = 1$ and $\sin 0 = 0$, so we have $A = 0$. Hence

$$x = B\sin\left(\tfrac{1}{6}\pi t\right)$$

This function oscillates between $\pm B$, i.e. its amplitude is B, so $B = 2$. Consequently the displacement, x, at time t is given by

$$x = 2\sin\left(\tfrac{1}{6}\pi t\right)$$

Differentiating this with respect to t gives the velocity v, i.e.

$$v = 2 \times \left(\tfrac{1}{6}\pi\right) \times \cos\left(\tfrac{1}{6}\pi t\right)$$
$$= \tfrac{1}{3}\pi \cos\left(\tfrac{1}{6}\pi t\right)$$

and differentiating again gives the acceleration, $\dfrac{dv}{dt}$, i.e.

$$\frac{dv}{dt} = -\left(\tfrac{1}{18}\pi^2\right)\sin\left(\tfrac{1}{6}\pi t\right)$$

Putting $t = 1$, we have

(a) $x = 2\sin\left(\tfrac{1}{6}\pi\right) = 1$

(b) $v = \left(\tfrac{1}{3}\pi\right)\cos\left(\tfrac{1}{6}\pi\right) = \tfrac{1}{6}\pi\sqrt{3}$

(c) $\dfrac{dv}{dt} = -(\tfrac{1}{18}\pi^2)\sin(\tfrac{1}{6}\pi) = -\tfrac{1}{36}\pi^2$

Hence when $t = 1$
(a) the displacement is 1 m;
(b) the speed is $\tfrac{1}{6}\pi\sqrt{3}$ m s^{-1};
and
(c) the acceleration is $-\tfrac{1}{36}\pi^2$ m s^{-2}.

Although $x = A\cos nt + B\sin nt$ is the *general* solution of the SHM equation, it is only rarely necessary to use it. For many purposes two special cases are more convenient, namely:

CASE 1, when $t = 0$, $x = 0$.
This appeared in Example 4, and as we saw there, the general solution can be reduced to the form

$$x = a\sin nt \qquad\qquad\qquad \text{. . . (7)}$$

CASE 2, when $t = 0$, $x = a$ (and hence $v = 0$).
In this case $v = 0$, when $t = 0$, but

$$v = -nA\sin nt + nB\cos nt$$

and putting $t = 0$ and $v = 0$ gives

$$0 = nB$$

hence $B = 0$, leaving $x = A\cos nt$. The amplitude of this function is A, so $A = a$. Therefore

$$x = a\cos nt \qquad\qquad\qquad \text{. . . (8)}$$

Summary

(The numbers used below are those which appear earlier in this section.)

The SHM equation

$$\dfrac{d^2x}{dt^2} = -n^2x \qquad\qquad\qquad \text{. . . (1)}$$

has a general solution

$$x = A\cos nt + B\sin nt \qquad\qquad\qquad \text{. . . (5)}$$

and its period is given by

$$T = \dfrac{2\pi}{n} \qquad\qquad\qquad \text{. . . (6)}$$

If $x = 0$, when $t = 0$, then

$x = a \sin nt$ where $a =$ the amplitude   . . . (7)

If $x = a$, when $t = 0$, then

$x = a \cos nt$ . . . (8)

The velocity, v, can be expressed in terms of x as follows:

$v^2 = n^2(a^2 - x^2)$ . . . (3)

Exercise 12b

1 A particle P is performing SHM, about a fixed point O, with a period of 4 s and amplitude 0.5 m. Find
   (a) the maximum value of its acceleration;
   (b) the maximum value of its speed;
   (c) its velocity when its displacement from O is 0.3 m.
2 A particle P is performing SHM about a fixed point, O. The amplitude of the motion is 0.2 m and the maximum speed is $0.05 \, \mathrm{m \, s^{-1}}$. Find the period of the motion.
   Write down, in terms of t, an expression for the displacement from O, t seconds after the particle passes O.
3 A particle P is performing SHM and making 10 oscillations per second. The largest allowable value of its acceleration is $5 \, \mathrm{m \, s^{-2}}$. Find the largest possible value of its amplitude.
4 A particle of mass 0.04 kg is moving along the x-axis and performing SHM about the origin O, with a period of 0.1π s and amplitude 0.05 m. Find an expression for its kinetic energy, in terms of its displacement, x, from the origin.
5 Find the maximum value of the magnitude of the force acting on the particle in question 4.
6 A particle P is moving along the x-axis and its displacement, in metres, from the origin, at time t seconds is given by

$$x = 5 + 3 \sin (\tfrac{1}{2}t)$$

Show that P is performing SHM. Find the period of the motion and the coordinates of
   (a) the centre of the oscillation;
   (b) the extreme points of the oscillation.
   Sketch the displacement–time graph.
7 A particle P is moving along the x-axis and its displacement, in metres, from the origin, at time t seconds is given by

$$x = 4 \sin^2 t$$

Show that it is performing SHM and find its maximum speed.
Sketch the displacement–time graph.

8 A particle P is moving along the x-axis and its displacement, in metres, from the origin, at time t seconds is given by

$$x = 3\cos 2t + 4\sin 2t$$

Show that it is performing SHM. Find its period and amplitude. Find also its maximum speed. Sketch the displacement–time graph.

9 A particle P, of mass 0.05 kg, lies on a smooth horizontal table, and it is attached by a light elastic string, of modulus 5 N and natural length 1 m, to a fixed point O on the table. The particle is pulled aside until the string's total length is 1.5 m, and then it is released. Show that, until the string becomes slack, P performs SHM. Find the speed of the particle at the instant when the string slackens. Find also the time taken from the instant when the particle is released, until it reaches O.

10 A light spring, of modulus λ and natural length l, lies in a straight line on a smooth horizontal table. One end of it is attached to a fixed point O on the table, and at the other end there is a particle P, of mass m. Initially the spring is straight, but not extended, i.e. $OP = l$. The particle is then given a velocity u, in the direction of the spring, away from O. By considering the equation of motion when the particle is at a general point, where the extension of the spring is x, show that the motion is simple harmonic. Find the amplitude (assuming it to be less than l) and the period.

Write down, in terms of t, an expression for the total length of the spring t seconds after the particle is set in motion.

12.6 Notation

At this point it will be useful to introduce a new piece of mathematical notation. When we differentiate a variable x (usually, but not always, representing a length), with respect to t, the time, we write:

$$\frac{dx}{dt} = \dot{x}$$

(read 'x dot'), and

$$\frac{d^2x}{dt^2} = \ddot{x}$$

(read 'x double dot')

For example, if $x = t^3$, then $\dot{x} = 3t^2$ and $\ddot{x} = 6t$.

In this notation, the SHM equation becomes

$$\ddot{x} = -n^2 x$$

It is important to remember that x, $\dot{x}$ and $\ddot{x}$, must all obey the same sign convention, i.e. their positive values must all be in the same direction, and their negative values must all be in the opposite direction.

This is also a convenient place to introduce the reader to two more terms which are occasionally used in the context:

(i)   the **stiffness** of an elastic spring (or string) is its modulus of elasticity divided by its natural length; the unit of stiffness is newtons per metre;

(ii)   the **frequency** of an oscillation is the number of oscillations per second ($=1/T$, where T is the period in seconds). While one might reasonably measure this in 'oscillations per second', the correct SI name for this unit is 'hertz', (named after Heinrich Hertz, a German physicist who lived from 1857 to 1894). The abbreviation for 'hertz' is Hz.

12.7 The motion of a particle on a vertical spring

In §12.4 we examined the motion of a particle attached to a spring and supported by a horizontal table, and this led to the discussion of SHM in §12.5.

We shall now consider the motion of a particle moving in a vertical line under the action of a spring. In view of the preceding sections, it will not be surprising to find that this also is simple harmonic, but we must consider the equation of motion carefully, before jumping to this conclusion.

Consider a particle, of mass m, hanging on a vertical spring which is attached to a fixed point O. As usual, let the modulus of elasticity of the spring be λ, and its natural length be l; let A be the point vertically below O such that $OA = l$ (see Figure 12.5). Also, let E be the point where the particle hangs in equilibrium, that is, the point where the resultant force on the particle is zero. We shall write $AE = e$, as shown in Figure 12.5.

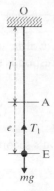

Figure 12.5

At point E there are two forces acting on the particle, its weight mg, acting downwards, and the tension, T_1, upwards; by Hooke's law

$$T_1 = \frac{\lambda e}{l}$$

Since the resultant force when the particle is at E is zero, we have

$$mg = T_1$$

and hence

$$mg = \frac{\lambda e}{l} \qquad \qquad \cdots \quad (1)$$

Now let us examine the forces acting on the particle when it is moving and passing through a general point P on its path. It is very important that P should be a *general* point, and not some particular point such as the point E, or an extreme point on the path (see Figure 12.6). We shall assume that in the course of the particle's motion the spring does not exceed its elastic limit (and, in exercises involving strings, the particle must not rise above point A, because, above this point a string would be slack).

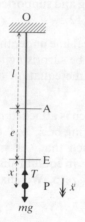

Figure 12.6

Let $EP = x$; this variable, x, is measured from E *downwards*, in other words, we shall take the downward direction to be positive. The tension, T, corresponds to a total extension $(e + x)$, and so, by Hooke's law

$$T = \frac{\lambda(e + x)}{l} \qquad \qquad \cdots \quad (2)$$

The resultant downward force at P is $mg - T$. Consequently Newton's second law of motion gives

$$mg - T = m\ddot{x}$$

and substituting for T from equation (2) gives

$$mg - \frac{\lambda(e + x)}{l} = m\ddot{x}$$

i.e. $mg - \dfrac{\lambda e}{l} - \dfrac{\lambda x}{l} = m\ddot{x}$

But from equation (1) we know that $mg = \dfrac{\lambda e}{l}$. Consequently

$$-\frac{\lambda x}{l} = m\ddot{x}$$

i.e. $\ddot{x} = -\left(\dfrac{\lambda}{ml}\right)x$ . . . (3)

Using the notation of §12.5, this is the SHM equation with

$$n^2 = \frac{\lambda}{ml}$$

The centre of the oscillation is the point where $x = 0$, that is, point E, and the period is $2\pi\sqrt{\left(\dfrac{ml}{\lambda}\right)}$.

(It is not recommended that the formulae in this section should be memorised, but the reader should understand the method and should be able to reproduce it.)

In Examples 5 and 6, refer to the abbreviations and diagrams from the preceding text.

Example 5

A particle of mass 0.4 kg is suspended by a light spring of modulus 20 N and natural length 50 cm. Find its extension when it is in equilibrium.

Show that, when it is disturbed in a vertical direction, the motion is simple harmonic, and find its period. Find also the amplitude when the particle is set in motion
(a) by giving it an initial downward velocity of 0.5 m s^{-1}, from the equilibrium position;
(b) by pulling it down 4 cm below the equilibrium position and releasing it.
In each case, write down an expression for the displacement t seconds after the particle is released.
(Take g to be 10 m s^{-2}.)

In equilibrium:

$$\text{tension, } T_1 = \left(\frac{20}{0.5}\right)e\,\text{N}$$
$$= 40e\,\text{N}$$
$$\text{weight}\ (= mg) = 0.4 \times 10\,\text{N}$$
$$= 4\,\text{N}$$

Since the resultant force is zero, $T_1 = mg$

i.e. $40e = 4$
$e = 0.1$

In equilibrium, the extension of the spring is 0.1 m (= 10 cm).

When the particle is at a general point P, below the equilibrium point, E, (with $EP = x$), the equation of motion is (see Figure 12.6)

$$mg - T = m\ddot{x}$$

which, in this example becomes

$$0.4 \times 10 - \left(\frac{20}{0.5}\right)(0.1 + x) = 0.4\ddot{x}$$

$$\therefore \qquad 4 - 4 - 40x = 0.4\ddot{x}$$

$$\therefore \qquad -40x = 0.4\ddot{x}$$

$$\therefore \qquad \ddot{x} = -100x$$

This is the SHM equation, with $n = 10$.

Hence the motion is simple harmonic, the centre of the oscillation is E, (where the extension is 10 cm) and the period is $\frac{2\pi}{10}$ (≈ 0.628) seconds.

(a) Since this is SHM, we may use the formula $v^2 = n^2(a^2 - x^2)$. In this case, $n = 10$, and we are given that when $x = 0$, the velocity is $0.5\,\mathrm{m\,s^{-1}}$, i.e. when $x = 0$, $v = 0.5$, hence

$$0.5^2 = 10^2 \times a^2$$

$$\therefore \qquad a = 0.5 \div 10$$

$$= 0.05$$

The amplitude is 0.05 m (= 5 cm).

Since the particle is at the *centre* of the oscillation when $t = 0$, the solution to the SHM equation can be written in the form $a \sin nt$, i.e.

$$x = 0.05 \sin (10t)$$

(b) In this case the particle is at rest when it is 4 cm below E, so the amplitude is 4 cm (= 0.04 m).

Since the particle is at an extreme point of the oscillation when $t = 0$, the solution to the SHM equation can be written in the form $a \cos nt$, i.e.

$$x = 0.04 \cos (10t)$$

Example 6

When a particle of mass m is suspended on a light spring of stiffness k it oscillates with frequency f. Express k in terms of m and f.

In this example Hooke's law takes the form

tension = k × extension

The extension, e, when the particle is in equilibrium, is given by

$ke = mg$

At a general point P, the equation of motion is

$mg - T = m\ddot{x}$

i.e. $mg - k(e + x) = m\ddot{x}$

∴ $mg - ke - kx = m\ddot{x}$

But $mg = ke$, so we have

$-kx = m\ddot{x}$

i.e. $\ddot{x} = -\left(\dfrac{k}{m}\right)x$

This is SHM of period $2\pi\sqrt{\left(\dfrac{m}{k}\right)}$. Now the frequency is the reciprocal of the period, so

$$f = \frac{1}{2\pi\sqrt{(m/k)}}$$

$$= \frac{1}{2\pi}\sqrt{\left(\frac{k}{m}\right)}$$

∴ $\sqrt{\left(\dfrac{k}{m}\right)} = 2\pi f$

∴ $\dfrac{k}{m} = 4\pi^2 f^2$

Hence

$k = 4m\pi^2 f^2$

Exercise 12c

In the exercises which follow, the reader is reminded that most examining boards expect candidates to know the formulae which appear in the summary of section 12.5. The other formulae which appear in the text above are not normally quoted; they should be proved from first principles when required.
In numerical questions take g to be $10\,\mathrm{m\,s^{-2}}$.

1   A heavy object of mass 10 kg is supported vertically by a spring whose natural length is 1 m and whose modulus of elasticity is 200 N. The object is pulled down 20 cm below its equilibrium position and released. Show that the motion is simple harmonic and find the period and amplitude of the oscillation. Find also the speed of the object when it is 10 cm from the equilibrium position.

2 A spring is required to make a mass of 0.2 kg oscillate in a vertical line with a frequency of 0.5 hertz. Find the stiffness needed.

3 A spring AB, of length 0.5 m and modulus of elasticity 50 N, is suspended with the end A attached to a fixed point. The other end B is attached to a particle of mass 2 kg. Initially the particle is held at a point 0.5 m vertically below A (i.e. the spring is not extended). The particle is released from this position and allowed to fall vertically.

Show that the subsequent motion is simple harmonic and find its amplitude and period. Find also the maximum speed.

4 A light spring, of stiffness 40 N m^{-1} is arranged so that it can cause a particle to oscillate in a vertical line. Find the mass of a particle which, when it is attached to the free end of the spring, will oscillate with a period of 2 seconds.

5 Two identical springs AB and CD have natural length 0.5 m and modulus of elasticity 50 N. They are used to support a mass of 2 kg which can oscillate in a vertical line. Find the period of the oscillation when they are joined
   (a) in series (i.e. B and C are joined together so that a single spring AD is formed);
   (b) in parallel (i.e. with AB parallel to CD).

6 A light spring has natural length 1 m and modulus of elasticity 4 N. One end is attached to a fixed point O and the other is attached to a particle of mass 0.1 kg, which can oscillate in a vertical line. Calculate the period of the oscillation.

If the particle is set in motion by pulling it down 50 cm below its equilibrium position and releasing it, find the speed when the spring first returns to its natural length.

7 A light spring of modulus of elasticity $4mg$ and natural length l has one end attached to a fixed point and the free end has a particle of mass m attached to it. The particle performs simple harmonic motion in a vertical straight line with period T. Show that

$$T = \pi \sqrt{\left(\frac{l}{g} \right)}$$

The particle is replaced by another of mass M and for this the period of oscillation is $2T$. Find M in terms of m.

8 Two springs AB and CD each have natural length l but unequal stiffnesses, k_1 and k_2. They are used to support a particle of mass m which can oscillate in a vertical line. Find the period of the oscillation when they are joined (a) in series, (b) in parallel.

9 A light elastic string has natural length l and, when a particle of mass m is suspended on it, the extension is $\frac{1}{5}l$. Write down its modulus of elasticity.

The particle is now pulled down a further distance $\frac{1}{5}l$ from its equilibrium position and released. Prove that the subsequent motion is SHM and find expressions for
   (a) the amplitude;
   (b) the period;
   (c) the maximum speed.

10 Using the principle of conservation of energy in question 9, find an expression for v, the velocity of the particle, in terms of x, its displacement from the equilibrium position. By differentiating this expression obtain the SHM equation

$$\ddot{x} = -\left(\frac{5g}{l}\right)x$$

12.8 The simple pendulum

A simple pendulum is a pendulum which consists of a heavy bob, small enough to be treated as a particle, supported by a light inelastic string, which swings to and fro in a vertical plane; as it does so, the bob moves along the arc of a circle. Figure 12.7 shows such a pendulum at time t, with the particle at a general point, P, on its path. One end of the string is attached to a fixed point O; OY is the downward vertical and angle YOP is θ radians. When $t = 0$, the angle θ is equal to α, and the particle is released from this position (i.e. when P is at A in Figure 12.7). Point C is the lowest point of the path.

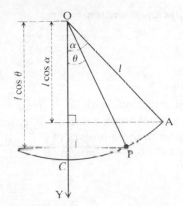

Figure 12.7

We shall apply the principle of conservation of energy to the motion of the particle. The length of the arc CP is $l\theta$, where l is the length of the string. Differentiating this with respect to time, we can see that the speed at point P is $l\dot{\theta}$. Consequently the kinetic energy gained by the particle is $\frac{1}{2}m(l\dot{\theta})^2$.

Initially, the particle is at A and, in this position, its depth below the fixed point O is $l\cos\alpha$. When it reaches point P the depth of the particle below O is $l\cos\theta$. So the vertical displacement of the particle, as it moves from A to P, is

$(l\cos\theta - l\cos\alpha)$

The P.E. lost, then, as the particle moves from A to P, is

$mg(l\cos\theta - l\cos\alpha)$

Equating the K.E. gained and the P.E. lost, we have

$$\tfrac{1}{2}m(l\dot\theta)^2 = mg\,(l\cos\theta - l\cos\alpha)$$

Dividing both sides by ml and removing brackets gives

$$\tfrac{1}{2}l\dot\theta^2 = g\cos\theta - g\cos\alpha \qquad \ldots \ (1)$$

When we differentiate with respect to t, $\dot\theta^2$ becomes $2\dot\theta\ddot\theta$, and $\cos\theta$ becomes $-(\sin\theta)\dot\theta$. Consequently differentiating equation (1) with respect to t gives

$$l\dot\theta\ddot\theta = -g(\sin\theta)\dot\theta$$

and dividing both sides by $\dot\theta$ reduces this to

$$l\ddot\theta = -g\sin\theta \qquad \ldots \ (2)$$

As it stands, this equation is difficult to solve, but if the oscillations are small in magnitude, we can use the approximation $\sin\theta \approx \theta$, where θ is small.

(Reference to a calculator will show that if $\theta = 0.1$, i.e. about 6°, $\sin\theta = 0.0998$, correct to three significant figures, so this is a good approximation for a simple pendulum which is swinging through about five or six degrees either side of the vertical.)

Consequently, for small oscillations equation (2) can be written

$$l\ddot\theta = -g\theta$$

or $\quad \ddot\theta = -\left(\dfrac{g}{l}\right)\theta \qquad \ldots \ (3)$

Now this equation has the form of the SHM equation, with $n^2 = (g/l)$, so although the motion of the bob of the pendulum is not along a straight line, we may make use of the standard results which appear in the summary of §12.5. In particular, the formula for the period of the oscillation, $T = 2\pi/n$, gives

$$T = 2\pi\sqrt{\left(\dfrac{l}{g}\right)} \qquad \ldots \ (4)$$

Qu. 5 Obtain equation (3) above, by applying Newton's second law of motion in the direction of the tangent in Figure 12.7.

Qu. 6 Taking g to be $9.81\ \text{m s}^{-2}$, calculate the length of a simple pendulum which would 'beat' in seconds, i.e. a pendulum with a period of 2 seconds.

12.9 SHM–a useful model

Imagine a point P, moving around a circle, centre O radius a, with uniform angular velocity n radians per second (see Figure 12.8). Suppose the particle is at A $(a, 0)$ when $t = 0$, so that t seconds later

angle AOP $= nt$ radians

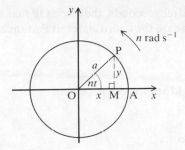

Figure 12.8

The coordinates (x, y) of P at time t are

$$x = a \cos nt \qquad y = a \sin nt$$

If we could view this circle from a point in its plane, at a large distance from O, and looking along the x-axis (the distance being large enough to enable us to ignore the effects of perspective and regard all the rays of light as being parallel to the axis), we would see P moving up and down in a straight line and its displacement from the centre of its path would be the y-coordinate of P, i.e. $a \sin nt$. Now this is the standard formula for the displacement of a particle performing SHM with period $2\pi/n$, with the time measured from the instant when the particle is moving through the centre point of its path. In other words we would be observing a point moving with simple harmonic motion. (Readers who have access to a microcomputer should arrange this to be displayed on the visual display unit, so that they can see what SHM looks like.)

If we were to transfer the y-coordinate to a displacement–time graph, that is a graph with y on the vertical axis and t, the time, on the horizontal axis, we would obtain a curve like that in Figure 12.9.

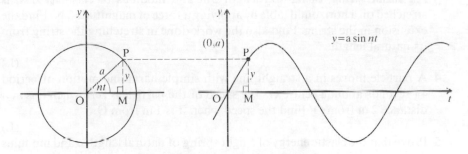

Figure 12.9

On the other hand, looking along the y-axis, one would see the displacement parallel to the x-axis, i.e. $a \cos nt$, and this is the standard formula for the displacement in the case $x = a$ when $t = 0$.

We can use this model to demonstrate the *general* solution of the SHM equation (see §12.5), by taking the initial angular displacement of OP to be α with the positive x-axis, i.e. when $t = 0$, $\angle$ POA $= \alpha$. After t seconds, the angle OP makes with the positive x-axis is then $(nt + \alpha)$. In this case the coordinates of P at time t are

$$x = a\cos(nt + \alpha) \qquad y = a\sin(nt + \alpha)$$

Viewing this motion along the y-axis, we would see the displacement

$$x = a\cos(nt + \alpha)$$

and applying the 'addition formula'† to this gives

$$x = a[\cos(nt)\cos\alpha - \sin(nt)\sin\alpha]$$

Writing $A = a\cos\alpha$ and $B = -a\sin\alpha$, this becomes

$$x = A\cos nt + B\sin nt$$

which is the general solution to the SHM equation we used in §12.5.

Exercise 12d (Miscellaneous)

1 Verify that $x = A\cos 5t + B\sin 5t$ satisfies the equation

$$\ddot{x} = -25x$$

for all values of A and B. Find the values of A and B for which $x = 3$ and $\dot{x} = 20$ when $t = 0$. Find the maximum and minimum values of x.

2 Sketch the graph of l against T given by the formula for the period of a simple pendulum (see §12.8). In order to verify this result experimentally, the period of a simple pendulum is measured, for various values of the length. What variables should be used in order to produce a *straight line* graph which would confirm the formula?

3 An elastic string, of natural length $2\,\text{m}$ and modulus of elasticity $6\,\text{N}$, is stretched on a horizontal table by applying a force of magnitude $2\,\text{N}$. Find the extension in the string. Find also the work done in stretching the string from its natural length.

(L)

4 A particle moves in a straight line with simple harmonic motion of period 4π seconds about a centre O. The speed of the particle is zero when it is at a distance $2\,\text{m}$ from O. Find the speed when it is $1\,\text{m}$ from O.

(L)

5 Prove that the elastic energy of a light spring of natural length a and modulus of elasticity λ, stretched by an amount x, is $\dfrac{\lambda x^2}{2a}$.

A trolley of mass m runs down a smooth track of constant inclination $\frac{1}{6}\pi$ to the horizontal, carrying at its front a light spring of natural length a and

---

†See *Pure Mathematics Book 1* by J. K. Backhouse and S. P. T. Houldsworth, revised by P. J. F. Horril, published by Longman, Chapter 17.

modulus mga/c, where c is a constant. When the spring is fully compressed it is of length $\frac{1}{4}a$, and it obeys Hooke's law up to this point. After the trolley has travelled a distance b from rest the spring meets a fixed stop. Show that when the spring has been compressed a distance x, where $x < \frac{3}{4}a$, the speed of the trolley is given by

$$cv^2/g = c(b + x) - x^2$$

Given that $c = \frac{1}{10}a$ and $b = 2a$, find the total distance covered by the trolley before it momentarily comes to rest for the first time.

(L)

6 One end of a light elastic string, of natural length $2a$ and modulus $4mg$, is attached to a fixed point A of a smooth horizontal table. A particle of mass m is attached to the other end of the string. The particle is released from rest from a point C of the table, where $AC = 3a$. Show that the particle reaches C again after time

$$(4 + \pi)\sqrt{(2a/g)}$$

(L)

7 A light elastic spring, of modulus $4mg$ and natural length l, has one end attached to a fixed point and carries a scale pan, of mass m, at the other end. Show that the period T of small oscillations of the pan about its equilibrium position is $\pi\sqrt{(l/g)}$.

When the pan is hanging in equilibrium, a particle of mass km is gently placed in the pan. Given that the periodic time of the ensuing motion is $2T$, find the value of k and the amplitude of the oscillation.

(L)

8 A particle of mass m is attached to one end of a light elastic string of natural length l and modulus $2mg$. The other end of the string is fixed to a point A. The particle rests on a support B vertically below A, with $AB = \frac{5}{4}l$. Find the tension in the string and the reaction exerted on the particle by the support B.

The support B is suddenly removed. Show that the particle will execute simple harmonic motion and find
(a) the depth below A of the centre of the oscillation;
(b) the period of the motion.

(JMB)

9 A particle of mass 5 kg executes simple harmonic motion with amplitude 2 m and period 12 seconds. Find the maximum kinetic energy of the particle, leaving your answer in terms of π.

Initially the particle is moving with its maximum kinetic energy. Find the time that elapses until the kinetic energy is reduced to one quarter of the maximum value, and show that the distance moved in this time is $\sqrt{3}$ m.

(JMB)

10 A particle P of mass m is attached to one end of a light elastic string of length l and modulus $3mg$. The other end of the string is fastened to a fixed point A. The particle is projected from A vertically upwards with a speed $\sqrt{(3gl)}$. Find its speed when it reaches the point B which is at height l above A. (*continued*)

At time t after passing through B, but before it first returns to B, the particle is at a distance x above B. Find $\dfrac{d^2x}{dt^2}$ in terms of g, l and x, and verify that the equation you have found is satisfied if

$$x = -\tfrac{1}{3}l + H\cos\omega t + K\sin\omega t$$

where $\omega^2 = 3g/l$ and H and K are constant.

Find H and K in terms of l.

(JMB: part)

11 A particle of mass m is attached to one end of a light elastic string of length l and modulus $\tfrac{3}{2}mg$. The other end of the string is attached to a fixed point O and the particle hangs in equilibrium under gravity at the point E. Find the distance OE.

If the particle is a *further* distance x below E show that the resultant force acting on the particle is proportional to x.

The particle is pulled down to the point at a distance a below E and released from rest. Show that, in the subsequent motion and while the string is taut, the particle executes simple harmonic motion and that its distance below E at time t after being released is $a\cos\omega t$, where $\omega^2 = \tfrac{3}{2}g/l$.

(O & C)

12 One end of a light elastic string of length a and modulus mg is attached to a fixed point O. To the other end A are attached two particles P and Q, P having mass $2m$ and Q having mass m. The particles hang down in equilibrium under gravity. If Q falls off, show that P subsequently performs simple harmonic motion and state the period and amplitude of this motion.

If on the other hand P falls off, find the distance from O of the highest point reached by Q.

(O & C)

13 Two fixed points A and B on a smooth horizontal table are at a distance $10a$ apart. A particle of mass m lies between A and B. It is attached to A by means of a light elastic string of modulus λ and natural length $2a$ and to B by means of a light elastic string of modulus 2λ and natural length $5a$. M is the midpoint of AB and O is the point at which the particle would rest in equilibrium. Prove that $OM = \tfrac{5}{3}a$.

The particle is released from rest at P, the midpoint of OM. Prove that it moves in simple harmonic motion of amplitude $\tfrac{5}{6}a$ and period $\dfrac{2\pi}{n}$, where

$$n^2 = \dfrac{9\lambda}{(10am)}$$

Show that the speed of the particle when it passes through Q, the midpoint of OP, is $\tfrac{5}{12}\sqrt{3}an$, and find the time taken by the particle to move the distance from Q to O.

(O & C)

14 A particle of mass m moves in simple harmonic motion about O in a straight line, under the action of a restoring force of magnitude proportional to the distance from O. At time $t = 0$ the speed of the particle is zero. After 1 second the speed of the particle is $2\,\mathrm{m\,s^{-1}}$; after a further second the speed is $2\sqrt{3}\,\mathrm{m\,s^{-1}}$ and subsequently the particle passes through O for the first time. Show that the speed of the particle when it passes through O is $4\,\mathrm{m\,s^{-1}}$ and find

(a) the period and amplitude of the motion;

(b) the time at which the speed of the particle is equal to $2\,\mathrm{m\,s^{-1}}$ after it first passes through O.

A stationary particle of mass m is placed at O and the moving particle collides and coalesces with this particle. The combined particle moves under the action of the same restoring force as before. Find the period of the subsequent motion and the speed when the displacement is $12/\pi$ from O.

(C)

15 (*Take g to be* $10\,\mathrm{m\,s^{-2}}$ *in this question.*)

A light elastic spring has natural length a and modulus of elasticity λ. Prove that the energy stored in the spring when it is stretched is $\lambda x^2/(2a)$, where x is the extension.

A light elastic spring of natural length 0.2 m and modulus of elasticity 50 N hangs vertically with one end attached to a fixed point and with a particle of mass 2 kg attached to the lower end.

(a) Calculate the extension of the spring when the particle is in equilibrium.

(b) The particle is pulled down below its equilibrium position until the *total* extension is 0.2 m and it is then released from this position. Calculate the speed of the particle when it passes through the equilibrium position, and find the maximum compression of the spring in the resulting motion.

(C)

Circular motion

Chapter 13

Circular motion

13.1 Introduction

Most readers will have seen, at least on television, an athlete 'throwing the hammer', and some will have attempted to do this themselves. Those who have tried to do it will know, and the others will appreciate, that this event calls for great strength and considerable skill. In this chapter we shall be considering some of the mechanical forces involved in throwing the hammer.

In order to study its mechanical properties, we must first simplify the problem. The forces acting throughout the athlete's body are, of course, enormously complicated; we shall only concern ourselves with the force acting along the radius as the athlete swings the hammer around in a circle prior to releasing it. In order to reduce the motion to a manageable problem, we shall have to consider that the hammer is moving in a circle whose centre, the athlete, is stationary.

Some of the questions one might want to ask about the motion of the hammer are:

(a) What initial speed is needed to achieve a throw of, say 80 m?

(b) At what rate, in revolutions per second, should the athlete be turning the hammer in order to achieve the linear speed required?

(c) At what angle to the horizontal should the path of the hammer be inclined when it is released by the thrower?

(d) What force must the athlete be able to exert in order to produce such a throw?

The reader may well be able to think of other pertinent questions. We can provide answers to *some* of the questions on the basis of work covered in earlier chapters; in this chapter we shall be especially concerned with (d).

Before we embark on this, the reader should give some thought to Qu. 1 on the next page.

Qu. 1 Figure 13.1 represents the path of a hammer. Initially it is moving in a circle (the plane of the circle would normally be inclined to the horizontal), and when the hammer is at point A it is released.

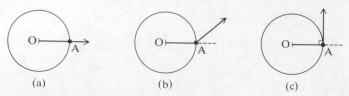

(a)   (b)   (c)

Figure 13.1

Which diagram represents the initial path of the hammer immediately after it is released? What is the direction of the force exerted by the athlete just before the hammer is released?

13.2 Angular velocity

Consider the motion of a particle P which is moving round a circle, centre O, radius r metres. There are several ways in which we can describe the rate at which the radius OP is turning. One way is to state the number of revolutions it makes per minute (r.p.m.). If OP is turning through N r.p.m., then the distance travelled by the particle P in one minute is $2\pi r N$ metres. In other words, the linear speed of P is $(2\pi r N \div 60)$ metres per second.

Qu. 2 The minute hand of a clock is 10 cm long. Find the speed of its tip in mm s^{-1}.

In mathematics and science, however, it is usual to measure angles in radians (because this makes it easier to use calculus methods), and so, in these subjects, it is usual to express angular velocity in radians per second (rad s^{-1}). (Remember that one complete revolution is 2π radians.) In Figure 13.2 the radius OP makes an angle θ radians with the positive x-axis. If θ is changing with time, the instantaneous angular velocity of the radius OP is $\dfrac{d\theta}{dt}$.

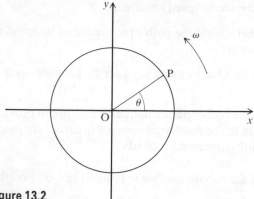

Figure 13.2

The Greek letter ω (omega) is normally used to represent angular velocity in radians per second. For the radius OP in Figure 13.2 we would write

$$\omega = \frac{d\theta}{dt}$$

Sometimes it is necessary to distinguish between clockwise and anticlockwise rotation. To do this, we use the normal sign convention for measuring angles; that is, *anticlockwise* rotations are written as *positive* numbers, and *clockwise* rotations as *negative* numbers. The direction associated with the angular velocity is the axis about which the rotation occurs. In Figure 13.2 it would be a line through O, perpendicular to, and coming out of, the page. In this book, however, we shall only be concerned with two-dimensional problems and it will be sufficient to record the appropriate plus or minus sign.

Qu. 3 Express the angular velocity of the minute hand of a clock in radians per second.

Qu. 4 A particle P is moving in a circle at a rate of $10\,\text{rad}\,\text{s}^{-1}$. Express this in r.p.m. Find the speed of a particle, given that the radius of the circle is $2\,\text{m}$.

If a particle P is moving in a circle of radius r, centre O, and if the radius OP is rotating at a constant rate of $\omega\,\text{rad}\,\text{s}^{-1}$, then, in each second, P will move a distance $r\omega$, so v, the speed of P, is given by

$$v = r\omega$$

(Notice the use of the word *speed* in the last sentence: the particle P is moving with constant speed $r\omega$, but we must be careful not to use the word *velocity* here, because this word combines the concepts of speed and direction. Although the speed of P is constant, its direction is changing continuously, so its velocity is variable. This has important consequences when we come to look at the acceleration of P.)

13.3 Angular acceleration

The rate at which angular velocity changes, i.e. the angular acceleration, is measured in $\text{rad}\,\text{s}^{-2}$ (read 'radians per second squared'). If this rate is constant, we may use the constant acceleration formulae which we met in §3.2, replacing s, the displacement, by θ, the angle turned through in radians. As usual t is used to represent the time elapsed, normally but not necessarily in seconds. The angular velocity in $\text{rad}\,\text{s}^{-1}$, when $t = 0$, is represented by ω, and the angular velocity after t seconds by Ω. The angular acceleration, in $\text{rad}\,\text{s}^{-2}$ is represented by α. With this notation, the constant acceleration formulae, for rotational motion, are:

$$\Omega = \omega + \alpha t \qquad\qquad\qquad\qquad \text{. . . (1)}$$

$$\theta = \tfrac{1}{2}(\omega + \Omega)t \qquad\qquad\qquad\qquad \text{. . . (2)}$$

$$\theta = \omega t + \tfrac{1}{2}\alpha t^2 \qquad\qquad\qquad . \;\; . \;\; . \quad (3)$$

$$\theta = \Omega t - \tfrac{1}{2}\alpha t^2 \qquad\qquad\qquad . \;\; . \;\; . \quad (4)$$

$$\Omega^2 = \omega^2 + 2\alpha\theta \qquad\qquad\qquad\; . \;\; . \;\; . \quad (5)$$

Example 1

The drum of a spin-drier, which is initially at rest, reaches its top speed of 1000 r.p.m. in 5 seconds. It then runs at this speed for 10 seconds and finally comes to rest in 20 seconds. Assuming that its acceleration and retardation are constant, find their magnitudes, and the total number of revolutions between the beginning of the cycle and the end.

In the notation above, $\omega = 0$, $\Omega = (1000 \times 2\pi) \div 60 = \tfrac{100}{3}\pi$ and $t = 5$.

Using formula (1), we have

$$\tfrac{100}{3}\pi = 0 + 5\alpha$$
$$\alpha = \tfrac{20}{3}\pi$$
$$ = 20.9 \quad \text{(correct to three significant figures)}$$

The drum accelerates at 20.9 rad s^{-2}.

For the final phase of the motion, $\omega = \tfrac{100}{3}\pi$, $\Omega = 0$ and $t = 20$, hence

$$0 = \tfrac{100}{3}\pi + 20\alpha$$
$$\alpha = -\tfrac{5}{3}\pi$$
$$ = -5.24 \quad \text{(correct to three significant figures)}$$

The $-$ sign indicates that this is a retardation.

The drum slows down at a rate of 5.24 rad s^{-2}.

Applying formula (2) to the initial phase

$$\theta = \tfrac{1}{2}(0 + \tfrac{100}{3}\pi) \times 5$$
$$ = \tfrac{250}{3}\pi$$

So the drum turns through $\tfrac{250}{3}\pi$ radians while it is accelerating.

While the drum is rotating at top speed, the angle turned through is

$$1000 \times 2\pi \times 10 = 20\,000\pi \text{ radians}$$

Applying formula (2) to the final phase, we have

$$\theta = \tfrac{1}{2}(\tfrac{100}{3}\pi + 0) \times 20$$
$$ = \tfrac{1000}{3}\pi$$

So the drum turns through $\tfrac{1000}{3}\pi$ radians while it is slowing down.

Hence the total angle turned through is

$$(\tfrac{250}{3}\pi + 20\,000\pi + \tfrac{1000}{3}\pi) \text{ radians}$$
$$= (\tfrac{1250}{3}\pi + 20\,000\pi) \text{ radians}$$

To convert this to revolutions we divide by 2π, giving $(\tfrac{625}{3} + 10\,000)$ revolutions.

Over the complete cycle, the drum turns through 10 208 revolutions, correct to the nearest whole number.

Exercise 13a

1 A long-playing record rotates at $33\tfrac{1}{3}$ r.p.m. Express this in radians per second. Find the speed of the needle along the groove relative to the record when it is
  (a) at the start, 14 cm from the centre;
  (b) at the end of the record, 6 cm from the centre.
2 The drum of a spin-drier rotates at 1000 r.p.m. Express this in radians per second. Find the speed of the washing if the drum has a radius of 15 cm, and the washing is in contact with the drum's cylindrical surface, and at rest relative to it.
3 The radius of a bicycle wheel is 36 cm. Find the angular velocity of the wheel in rad s^{-1}, when the bicycle is travelling at 30 km h^{-1}.
4 A hammer thrower wishes to release the hammer when it is travelling at 25 m s^{-1}. Just before it is released, the hammer is moving in a circle of radius 2.5 m. Find the angular velocity required in r.p.m.
5 A compact disc is read by a laser which is designed to read the disc at a rate of 1.2 m s^{-1}. Find the angular velocity required when the reading head is 5 cm from the centre of the disc, giving the answer (a) in rad s^{-1}, (b) in r.p.m.
6 A helicopter is moving horizontally in a straight line at 200 km h^{-1}. Its rotor blades are 3 m long and they rotate at 100 r.p.m. Assuming that they rotate in a horizontal plane, find the greatest and least values of the speed of the tip of a rotor blade, as seen by a stationary observer.
7 When a turbine is switched on, it accelerates uniformly up to 2000 r.p.m. in one minute. The turbine is then turned off and it slows down uniformly, coming to rest two minutes later. Express its acceleration and retardation in rad s^{-2}, and find the total angle through which it turns.
8 A spin-drier accelerates uniformly from rest up to a speed of 1500 r.p.m., at an angular acceleration of 40 rad s^{-2}. Find the total angle through which it turns.
9 A cyclist accelerates uniformly from 10 km h^{-1} to 30 km h^{-1} in 30 seconds. If the bicycle's wheels have a radius of 36 cm, find their angular acceleration.
10 A point P moves round a circle, centre O. Starting from rest, the radius OP accelerates uniformly up to an angular velocity of ω rad s^{-1}. It then immediately begins to slow down uniformly, coming to rest after a total of t seconds from the instant when it started. Find an expression for the total angle through which OP turns, and show that this does *not* depend on the magnitudes of the acceleration and retardation.

13.4 Motion in a circle with constant angular velocity

We shall now consider the linear velocity and acceleration of a point P moving in a circle with constant angular velocity.

It may seem odd to talk about the acceleration of an object whose speed is constant. But remember that velocity and acceleration are *vector* quantities, and so we must take account of their directions as well as their magnitudes. Remember also that acceleration is the rate of change of velocity. A particle moving in a circle, even if its speed is constant, has changing direction and consequently it has (non-zero) acceleration. Our task is to discover the relationship between the acceleration, the speed of the particle and the radius of the circle. At this point, readers may find it helpful to revise Chapters 4 and 5.

13.5 Differentiating unit vectors

Before we embark on the task outlined in the previous section, we shall establish an important result which will prove to be very useful.

Figure 13.3 shows a circle, centre O and radius one unit. The point P moves around the circle and, at time t, OP makes an angle θ with the positive x-axis.

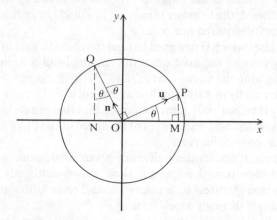

Figure 13.3

Let **u** represent the unit vector $\overrightarrow{OP}$. In the diagram, $OM = \cos \theta$ and $MP = \sin \theta$, so, using **i** and **j** to represent the unit vectors parallel to the x- and y-axes, respectively, we can express **u** as follows:

$$\mathbf{u} = (\cos \theta)\mathbf{i} + (\sin \theta)\mathbf{j} \qquad \qquad \text{. . . . (1)}$$

The vector **n** $(= \overrightarrow{OQ})$ in Figure 13.3, is perpendicular to vector **u**. (Think of **u** and **n** as two hands on a dial which are free to rotate, but they are rigidly connected so that **n** is always at right-angles to **u**.)

The radius OQ makes an angle θ with the y-axis, as shown. From this it follows that angle OQN equals θ, and consequently $NQ = \cos\theta$ and $ON = \sin\theta$. Hence

$$\mathbf{n} = (-\sin\theta)\mathbf{i} + (\cos\theta)\mathbf{j} \qquad \cdots \quad (2)$$

If we differentiate $\mathbf{u}$ (equation (1) above) with respect to θ, we get

$$\frac{d\mathbf{u}}{d\theta} = (-\sin\theta)\mathbf{i} + (\cos\theta)\mathbf{j} \qquad \cdots \quad (3)$$

Now the right-hand side of this is also a unit vector; in fact, it is the unit vector $\mathbf{n}$. Consequently we can write

$$\frac{d\mathbf{u}}{d\theta} = \mathbf{n} \qquad \cdots \quad (4)$$

We can express this in words by saying that 'differentiating a unit vector $\mathbf{u}$ with respect to θ, gives another unit vector perpendicular to $\mathbf{u}$'.

Qu. 5 Prove that $\dfrac{d\mathbf{n}}{d\theta} = -\mathbf{u}$.

However, we shall need to differentiate these unit vectors with respect to t, the time. This is simple to arrange, because, by the chain rule

$$\frac{d\mathbf{u}}{dt} = \frac{d\mathbf{u}}{d\theta}\frac{d\theta}{dt}$$

Hence from (4)

$$\frac{d\mathbf{u}}{dt} = \mathbf{n}\frac{d\theta}{dt}$$

$$= \mathbf{n}\omega \qquad \cdots \quad (5)$$

where, as in the previous section, $\omega = \dfrac{d\theta}{dt}$, the angular velocity of the radius vector $\overrightarrow{OP}$. Applying the same method to the result of Qu. 5, we have

$$\frac{d\mathbf{n}}{dt} = -\mathbf{u}\omega \qquad \cdots \quad (6)$$

Qu. 6 By combining results (5) and (6), above, show that

$$\frac{d^2\mathbf{u}}{dt^2} = -\omega^2\mathbf{u}$$

With these useful results at our disposal, we can proceed to our main task, finding the velocity and acceleration of a point R, moving in a circle of radius r, at constant angular velocity ω (see Figure 13.4 on the next page).

Using the notation above, we can write the position vector of R as

$$\mathbf{r} = r\mathbf{u}$$

We find the velocity, **v**, of R, by differentiating its position vector with respect to time, i.e.

$$\mathbf{v} = \frac{d\mathbf{r}}{dt}$$

$$= r\frac{d\mathbf{u}}{dt}$$

$$= r\omega\mathbf{n} \quad \text{(from equation (5) on page 237)}$$

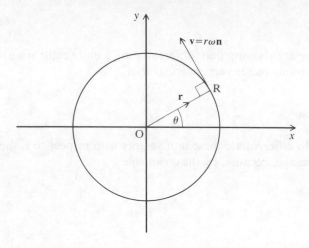

Figure 13.4

This result, however, is not very surprising; it merely says that the velocity of R has magnitude $r\omega$ and it is at right angles to **u**, i.e. it is directed along the tangent to the circle at R, (see Figure 13.4).

Now we differentiate the velocity, **v**, with respect to t, to find the acceleration, **a**, of the point R.

$$\mathbf{a} = \frac{d\mathbf{v}}{dt}$$

$$= \frac{d}{dt}(r\omega\mathbf{n})$$

$$= r\omega\frac{d\mathbf{n}}{dt}$$

$$= -r\omega^2\mathbf{u} \quad \text{(from equation (6) on page 237)}$$

Unlike the previous result, this one could not have been foreseen. It tells us that the acceleration of point R has magnitude $r\omega^2$ and that its direction is that of the unit vector, $-\mathbf{u}$, that is, towards the centre of the circle.

The *speed* of point R could be written as $v = r\omega$. If we substitute $\omega = v/r$ in the formula for the magnitude of the acceleration, we obtain

$$r\omega^2 = r\left(\frac{v}{r}\right)^2$$

$$= \frac{rv^2}{r^2}$$

$$= \frac{v^2}{r}$$

Summary

(a) The velocity of a point R moving in a circle of radius r with constant angular velocity ω has magnitude $r\omega$, and its direction is along the tangent to the circle at R.

(b) The acceleration of R has magnitude $r\omega^2$ (or alternatively v^2/r, where $v = r\omega$, the speed of R), and its direction is towards the centre of the circle.

Example 2

A car goes round a roundabout at a constant speed of 20 km h^{-1}, moving in a circle of radius 20 m. Calculate the magnitude of its acceleration.

Car's speed $= v$
$$= (20 \times 1000 \div 3600) \text{ m s}^{-1}$$
$$= \tfrac{50}{9} \text{ m s}^{-1}$$
Car's acceleration $= v^2/r$
$$= (\tfrac{50}{9})^2 \div 20 \text{ m s}^{-2}$$
$$= 1.54 \text{ m s}^{-2} \quad \text{(correct to three significant figures)}$$

Example 3

On a fairground ride, the passengers rotate in a horizontal circle of radius 10 m. It is decided that the maximum acceleration allowed should be 30 m s^{-2}. Find the fastest permissible rate of rotation in r.p.m.

Since the maximum acceleration is to be 30 m s^{-2}, we have

$$r\omega^2 = 30$$

where ω is the angular velocity in rad s^{-1}, and $r = 10$.

$$\therefore \quad \omega^2 = 3$$
$$\therefore \quad \omega = \sqrt{3}$$

The angular velocity is $\sqrt{3} \text{ rad s}^{-1}$.

Number of revolutions per minute $= \sqrt{3} \times 60 \div (2\pi)$
$$= 16.5 \quad \text{(correct to three significant figures)}$$

Exercise 13b

1 A particle is moving in a circle of radius 2 m, at constant speed. Calculate the magnitude of its acceleration when it is moving with
 (a) a linear speed of $4\,\mathrm{m\,s^{-1}}$;
 (b) an angular speed of 5 rad s^{-1};
 (c) an angular speed of 12 r.p.m.

2 A 'wall of death' rider is moving in a horizontal circle of radius 15 m, at a constant speed of $40\,\mathrm{km\,h^{-1}}$. Calculate the acceleration.

3 A racing car has to negotiate a bend which is in the form of an arc of a circle of radius 400 m. Find the maximum possible (constant) speed, in km h^{-1}, given that its acceleration must not exceed $9\,\mathrm{m\,s^{-2}}$.

4 An aircraft makes a horizontal turn at $400\,\mathrm{km\,h^{-1}}$ around part of a circle. Find the minimum radius allowed if the acceleration towards the centre of the circular path must not exceed $40\,\mathrm{m\,s^{-2}}$.

5 A big wheel at a fun fair turns at a steady rate of 3 r.p.m.; its radius is 20 m. Calculate the acceleration of a point on the circumference towards its centre.

6 Taking the axis of the Earth to be fixed in space, calculate the acceleration towards the centre of the Earth due to rotation, of a point on the equator. (Take the radius of the Earth to be 6400 km.)

7 A jet plane used for training astronauts simulates 'weightlessness' by flying at a constant speed of $540\,\mathrm{km\,h^{-1}}$ in a vertical circle. Taking g to be $9.81\,\mathrm{m\,s^{-2}}$, calculate the radius required for the astronaut to appear to be weightless when the plane is at the highest point of its path. Explain briefly how you interpret the word 'weightless' in this context.

8 A roundabout at a funfair turns at constant speed in a horizontal circle, with the passengers at a distance of 10 m from the centre. Calculate the rate of rotation required in r.p.m. if the acceleration experienced by the passengers is to be $20\,\mathrm{m\,s^{-2}}$.

9 A point P is moving in a circle of radius r with constant angular velocity ω rad s^{-1}. When $t = 0$, it is at the point $(r, 0)$, relative to coordinate axes through the centre of the circle. Using $\mathbf{i}$ and $\mathbf{j}$ to represent unit vectors parallel to these axes, show that the position vector $\mathbf{r}$ of P can be expressed as

$$\mathbf{r} = r(\mathbf{i}\cos\omega t + \mathbf{j}\sin\omega t)$$

By differentiating this twice with respect to t, show that the acceleration vector, $\mathbf{a}$, of P is given by

$$\mathbf{a} = -\omega^2\mathbf{r}$$

Express this result in words.

10 A particle P is moving in a plane and its polar coordinates (r, θ) at time t are given by

$$r = 5t^2 \qquad \theta = 2t$$

Sketch its path.
Using the Unit vectors $\mathbf{u}$ and $\mathbf{n}$, as in §13.5, find in terms of t
 (a) the position vector of P;

(b) the velocity vector of P;
(c) the acceleration vector of P.

Hence find, when $t = 2$,

(i) the distance of P from the origin O;
(ii) its speed;
(iii) the magnitude of its acceleration.

13.6 The force acting on a particle moving in a circle

Newton's first law of motion tells us that a particle moves with constant velocity, that is, in a straight line and with constant speed, *unless it is acted on by a force.* Consequently motion in a circle is only possible if there is a force causing it. To the hammer thrower we met in §13.1 this would be fairly obvious; he has to exert a considerable force while he is 'winding up' the hammer prior to releasing it. The washing in a spin-drier moves in a circular path because the reaction of the drum acts upon the washing. A motorcyclist can travel around a circular track, provided the surface of the track is sufficiently rough to exert a force towards its centre; if he hits a greasy patch and loses this force, then he goes off the track, quite literally, 'at a tangent'.

Qu. 6 Give three further examples of circular motion, stating, in each case, the force which causes the motion.

Newton's second law tells us how to calculate the force. For a particle of constant mass:

force = mass × acceleration

Remember that force and acceleration are vectors, and since this equation tells us that the force is a scalar multiple of the acceleration, we can deduce that the force has the same direction as the acceleration it causes.

From the work we have done in the preceding sections, we can calculate the acceleration of a particle moving in a circle with constant angular velocity, and we know that this acceleration is directed towards the centre of the circle. The following examples show how to use these facts.

Example 4

A particle, of mass 0.5 kg, moves in a circular path on a smooth horizontal table. The particle is attached to one end of a light inextensible wire, of length 2 m, the other end of which is fixed to a point O on the table. It is known that the wire will break if the tension in it exceeds 25 N. Calculate the maximum possible speed of the particle.

Let the maximum allowable speed of the particle be $v\,\text{m}\,\text{s}^{-1}$. The acceleration of the particle at this speed is given by v^2/r, where r is the radius of the circle. This particle is moving in a circle of radius 2 m, so its acceleration has magnitude $\frac{1}{2}v^2$, and this is directed towards O, the centre of the circle.

By Newton's second law of motion, the force equals the mass multiplied by the acceleration. Consequently the force acting on the particle

$$= 0.5 \times \tfrac{1}{2}v^2$$
$$= \tfrac{1}{4}v^2$$

and, like the acceleration, it acts towards O.

This force is provided by tension in the wire; it is the only horizontal force acting on the particle, and when the wire is on the point of breaking, it is equal to 25 N. Hence

$$\tfrac{1}{4}v^2 = 25$$
$$\therefore \quad v^2 = 100$$
$$\therefore \quad v = 10$$

The maximum possible speed is $10\,\text{m}\,\text{s}^{-1}$.

Example 5

A button, of mass 2 grams, is left in a spin-drier when the washing is removed. If the drier is then switched on and spins at 1000 r.p.m., calculate the force acting on the button, given that it moves in a horizontal circle of radius 15 cm.

The angular velocity in rad s^{-1} is $1000 \times (2\pi) \div 60 = \frac{100}{3}\pi$

The magnitude of the acceleration ($= r\omega^2$)

$$= 0.15 \times (\tfrac{100}{3}\pi)^2\,\text{m}\,\text{s}^{-2}$$

The force acting on the button (which is provided by the reaction of the drum on the button)

$$= 0.002 \times 0.15 \times (\tfrac{100}{3}\pi)^2\,\text{N}$$
$$= 3.29\,\text{N} \quad \text{(correct to three significant figures)}$$

13.7 The conical pendulum

A conical pendulum consists of a small heavy bob, attached to the end of a light inextensible string. The other end of the string is attached to a fixed point and the bob moves in a horizontal circle, under the action of gravity and the tension in the string (see Figure 13.5).

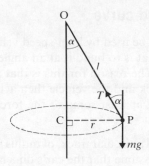

Figure 13.5

Let the mass of the bob be m, the length of the string be l, the radius of the circle in which the bob moves be r, and let the string make an angle α with the vertical, as shown in the diagram.

The tension, T, in the string can be resolved into two components, $T \sin \alpha$ horizontally, and $T \cos \alpha$ vertically. It is the horizontal component, acting towards C, the centre of the circle, which causes the circular motion. When the bob is rotating at ω rad s^{-1}, the acceleration towards C is $r\omega^2$. Consequently, by Newton's second law of motion, we have

$$T \sin \alpha = mr\omega^2$$

but $\qquad r = l \sin \alpha$, so

$$T \sin \alpha = m(l \sin \alpha)\omega^2$$

$\therefore \qquad\qquad T = ml\omega^2 \qquad\qquad\qquad\qquad\qquad\qquad \cdots \quad (1)$

Resolving vertically gives

$$T \cos \alpha = mg \qquad\qquad\qquad\qquad\qquad\qquad \cdots \quad (2)$$

and substituting for T, from (1), we have

$$(ml\omega^2) \cos \alpha = mg$$

$\therefore \qquad\qquad \cos \alpha = \dfrac{g}{l\omega^2} \qquad\qquad\qquad\qquad\qquad \cdots \quad (3)$

This equation can only be solved for α if the right-hand side is less than or equal to one, i.e. $g \leqslant l\omega^2$, that is, provided $\omega^2 \geqslant g/l$. In other words, as one might expect on common sense grounds, the motion is impossible unless the bob is given a sufficiently large angular velocity. With $\omega = \sqrt{(g/l)}$, the angle α is 0°, i.e. the string will be vertical. For larger values of ω, the bob moves in a horizontal circle with the string inclined at an angle given by equation (3), that is

$$\cos^{-1}\left(\frac{g}{l\omega^2}\right)$$

13.8 The motion of a vehicle on a banked curve

If a race track includes bends which are intended to be used by high-speed vehicles, it is a common practice to 'bank' the track; that is to build it at an angle, tilting the vehicle towards the centre of the curve. The reason for this is that a component of the normal reaction between the track and the vehicle then acts towards the centre of the curve, thereby supplying at least some of the force required in this direction.

Consider a car of mass m, travelling at speed v around a circular track, of radius r. (As with all the examples in this chapter, we must assume that the car's dimensions are small compared with the radius of the circle, so that we can treat the car as a particle.)

Suppose that the track is banked at an angle α to the horizontal. Figure 13.6 shows a vertical cross-section of the car as it travels round the track. The force, F, shown in the diagram, represents the sideways friction force between the wheels of the car and the road, and N represents the normal reaction of the track acting on the car. The radius of the circle is shown in the diagram as a broken line, to indicate that it is large compared with the other dimensions.

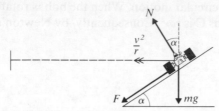

Figure 13.6

Resolving vertically gives

$$N \cos \alpha = mg + F \sin \alpha \qquad \qquad . \; . \; . \; (1)$$

The acceleration towards the centre of the circle is v^2/r, so, resolving horizontally, we have

$$F \cos \alpha + N \sin \alpha = m(v^2/r) \qquad \qquad . \; . \; . \; (2)$$

Now, the ideal arrangement is one in which there is no tendency for the car to side-slip, that is, for F to be zero. In this case we have

$$N \cos \alpha = mg$$
and $$N \sin \alpha = m(v^2/r)$$

Eliminating N from these equations gives

$$\tan \alpha = \frac{v^2}{gr}$$

If the track is banked at the angle given by this last formula, a car could travel at speed v without any sideways friction force between the tyres and the track; at other speeds, the force F required could be calculated from equations (1) and (2).

Example 6

A racing car, of mass 1 tonne, travels round a bend in the form of a circular arc of radius 100 m. Given that the track is banked at 20° to the horizontal, calculate

(a) the speed at which the bend can be taken with no tendency to side-slip;

(b) the friction force required if the driver wishes to take the bend at a speed 50% above this 'ideal' speed.

(Take g to be $9.81 \, \mathrm{m \, s^{-2}}$.)

[In the work which follows, results from the preceding section will be quoted. However, the reader should bear in mind that, when answering a question on this topic in a public examination, the candidate would be expected to derive these from first principles.]

(a) Since there is no tendency to side-slip

$$\tan \alpha = \frac{v^2}{gr}$$

and hence

$$v^2 = gr \tan \alpha$$

Putting $\alpha = 20°$, $r = 100$ and $g = 9.81$

$$v^2 = 9.81 \times 100 \times \tan 20°$$
$$\therefore \quad v = \sqrt{(981 \tan 20°)} \qquad \qquad \qquad \ldots \quad (3)$$
$$\therefore \quad v = 18.9 \quad \text{(correct to three significant figures)}$$

The vehicle should take the bend at $18.9 \, \mathrm{m \, s^{-1}}$ ($= 68.0 \, \mathrm{km \, h^{-1}}$).

(b) In this part of the question, the speed is 50% higher than that given by (3); let this speed be V, where $V = 1.5v$. Using equations (1) and (2) in the preceding text

$$N \cos \alpha = mg + F \sin \alpha$$

$$F \cos \alpha + N \sin \alpha = m(V^2/r)$$

Eliminating N gives

$$F \cos \alpha + \left(\frac{mg + F \sin \alpha}{\cos \alpha} \right) \sin \alpha = m(V^2/r)$$

and multiplying both sides by $\cos \alpha$

$$F \cos^2 \alpha + (mg + F \sin \alpha) \sin \alpha = m(V^2/r) \times \cos \alpha$$

Hence

$$F(\cos^2 \alpha + \sin^2 \alpha) + mg \sin \alpha = m(V^2/r) \times \cos \alpha$$

and since $\cos^2 \alpha + \sin^2 \alpha = 1$, we have

$$F = m(V^2/r) \times \cos \alpha - mg \sin \alpha \qquad \qquad \text{. . . .} \quad (4)$$

Now putting $V = 1.5v$, we have (from (3))

$$V = 1.5 \times \sqrt{(981 \tan 20°)} \qquad \dagger$$

Substituting this, together with the data, $m = 1000$, $r = 100$ and $\alpha = 20°$, into equation (4) gives

$$F = 4190 \quad \text{(correct to three significant figures)}$$

The friction force required is 4190 N.

Qu. 7 Obtain equation (4) in the last example by resolving parallel to the track.

Exercise 13c

1 A small button is placed on a turntable at a distance of 20 cm from the axis of rotation; the turntable can rotate in a horizontal plane. Given that the coefficient of friction between the button and the turntable is 0.5, find the fastest rate, in r.p.m., at which the turntable can rotate without the button slipping. (Take g to be $9.81 \, \text{m s}^{-2}$.)

2 A car, of mass 2 tonnes, is travelling round a bend in the form of the arc of a circle of radius 40 m, on a road whose surface is horizontal. Given that the maximum possible sideways friction force between the tyres and the road is 5000 N, find the greatest speed, in km h^{-1}, at which the car can take the bend without losing its grip.

3 On a fairground ride, the passengers revolve in a large cylindrical drum, whose axis is vertical. Initially the drum is at rest, and the passengers, who stand with their backs to the cylindrical surface, are supported by a horizontal floor. The drum then begins to turn until it is rotating at a speed at which the friction force between the surface of the drum and the passengers is sufficient to support their weight. The floor is then lowered, leaving the passengers apparently sticking to the side of the drum. Given that the drum's radius is R and the coefficient of friction is μ, show that the minimum angular velocity ω which will achieve the desired effect is given by $\mu R\omega^2 = g$.
Choose some reasonable values for R and μ and estimate the rate of rotation required.

4 A big wheel in a fairground rotates in a vertical plane at a constant rate of N r.p.m. Show that the vertical reaction, R, of the seat on a passenger at the highest point of the circle is given by

$$R = mg - \tfrac{1}{900} mr\pi^2 N^2$$

where r is the radius of the wheel, and m is the mass of the passenger.
Choose some suitable values for m and r, and estimate the smallest value of N, for which the passenger will lose contact with the seat at the highest point of the path.

---

†When using a calculator, it is preferable to use the uncorrected value of v, not that corrected to a number of significant figures.

5 A motorcyclist travels on a horizontal surface at 60 km h^{-1}, round a bend in the form of an arc of a circle. The direction of the track changes by 30° as the machine travels 100 m. Given that the combined mass of the machine and its rider is 750 kg, calculate the sideways force required to keep the machine on its intended track.

6 A jet plane pulls out of a dive and flies, at a constant speed of v m s^{-1}, along a path forming an arc of a vertical circle of radius r metres. The reaction between the pilot and the plane must not exceed nW, where W is the weight of the pilot in newtons. Show that v^2 must not be greater than $(n - 1)gr$.

7 Assuming that the pull of gravity on a particle of mass m, at a distance x from the Earth's centre (where $x > R$, the radius of the Earth) obeys the inverse square law, show that this force can be expressed in the form $\dfrac{mgR^2}{x^2}$.

Prove that the period, T seconds, of a satellite moving in a circular orbit, of radius x metres, with its centre at the centre of the Earth, is given by

$$T^2 = \frac{4\pi^2 x^3}{gR^2}$$

(The period is the time taken to complete a 360° orbit.)
A satellite is 'parked' in a stationary orbit, that is, it remains above a fixed point on the Earth's surface. Taking g to be 9.81 m s^{-2} and R to be 6400 km, calculate the height of the satellite above the Earth's surface.

8 A satellite moves in a circular orbit about the Earth with a period of 90 minutes. Using the data of question 7, calculate the radius of its orbit in km, and its speed in m s^{-1}.

9 A small particle, of mass m, is attached by a light inelastic string of length 13l to a fixed point A. The particle rotates, at constant speed v, in a horizontal circle, of radius 5l, centre B which is vertically below point A. Show that
  (a) $AB = 12l$
  (b) the tension in the string is $\frac{13}{12} mg$
  (c) $v^2 = \frac{25}{12} gl$.

10 An aircraft flies in a horizontal circle, of radius 2 km, at a constant speed of 540 km h^{-1}, with its wings banked at an angle α to the horizontal. Assuming that the only forces acting on the aircraft are (i) its weight, and (ii) the aerodynamic lift perpendicular to the wings, and taking g to be 9.81 m s^{-2}, find the value of α.

11 In an attempt to estimate the force exerted by the hammer thrower, the athlete is regarded as the vertical axis of a conical pendulum. The hammer is treated as a small heavy particle, of mass 6 kg, attached to a light inextensible wire of length 1 m. The length of the wire together with the athlete's arms is estimated to be 2 m. Taking g to be 10 m s^{-2}, calculate the force exerted when the particle has an angular speed of 5 rad s^{-1}. Find also the angle between the wire and the vertical at this speed.
Discuss briefly the assumptions made in this 'model' of the hammer thrower.

12 A particle, of mass m, is suspended from a fixed point by a light elastic string of natural length a, and modulus of elasticity λ. The particle moves in a

horizontal circle with constant angular velocity ω, with the string inclined at an angle α to the vertical. Prove that

$$\lambda a\omega^2 \cos\alpha = g(\lambda - ma\omega^2)$$

13 A small particle moves at constant speed v in a horizontal circle on the smooth inner surface of a hollow cone. The axis of the cone is vertical and its vertex is downwards. The semi-vertical angle of the cone is α. Prove that the height h of the centre of the circle above the vertex of the cone is given by $v^2 = gh$.

14 A railway track in the form of a circular arc of radius 100 m, is to be designed so that a train can travel along it at a speed of 60 km h^{-1}, with no sideways force between the wheels and the track. Taking g to be 9.81 m s^{-2}, find the angle at which the track should be inclined to the horizontal.

Calculate the sideways force when a truck of mass 5 tonnes travels round this curve at 120 km h^{-1}.

15 A small particle moves at constant speed around a horizontal circular path on the smooth inner surface of a hollow sphere. The centre of the circle is at a depth H below the centre of the sphere. Prove that its period, T seconds, is given by

$$T^2 = \frac{4\pi^2 H}{g}$$

13.9 Motion in a circle with variable angular velocity

So far in this chapter, we have confined our attention to motion in a circle with *constant* angular velocity. We shall now consider what happens if the angular velocity is variable. We shall be using the notation and methods employed in §13.5, especially the results labelled (5) and (6) in that section, i.e.

$$\frac{d\mathbf{u}}{dt} = \omega\mathbf{n} \quad \text{and} \quad \frac{d\mathbf{n}}{dt} = -\omega\mathbf{u}$$

where $\mathbf{u}$ is the unit vector along the radius, and $\mathbf{n}$ is the unit vector perpendicular to it, (see Figure 13.3). As before, the position vector of point R in Figure 13.4 is $\mathbf{r}$, where

$$\mathbf{r} = r\mathbf{u}$$

and its velocity vector $\mathbf{v}$ is found by differentiation, i.e.

$$\mathbf{v} = r\omega\mathbf{n}$$

That much is the same as before, but now comes the important difference: the angular velocity ω is not constant, so when we differentiate again to find the acceleration vector, $\mathbf{a}$, we must use the product rule, giving

$$\mathbf{a} = r\frac{d\omega}{dt}\mathbf{n} + r\omega\frac{d\mathbf{n}}{dt}$$

and using the results from §13.5 (quoted earlier) we have

$$\mathbf{a} = r\dot{\omega}\mathbf{n} + r\omega \times (-\omega\mathbf{u})$$
$$\mathbf{a} = r\dot{\omega}\mathbf{n} - r\omega^2\mathbf{u}$$

where $\dot{\omega} = \dfrac{d\omega}{dt}$

This tells us that the acceleration vector has two components:

(i) $r\dot{\omega}$ in the direction of $\mathbf{n}$ (i.e. $r\dot{\omega}$ parallel to the tangent)
(ii) $-r\omega^2$ in the direction of $\mathbf{u}$ (i.e. $r\omega^2$ along the radius, but towards the centre)

The first of these can be written $\dfrac{d}{dt}(r\omega)$, and in this expression we can replace $r\omega$

by v, the speed of R, giving $\dfrac{dv}{dt}(= \dot{v})$, as the component of acceleration along the

tangent: not a very surprising conclusion!

The second is the same result we had for motion with *constant* angular velocity.

These components of the acceleration vector are shown in Figure 13.7.

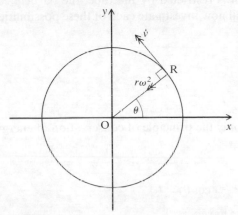

Figure 13.7

13.10 Motion in a vertical circle

We shall first consider the motion of a particle, of mass m, moving inside a smooth circular tube, of radius r, standing in a vertical plane, whose cross-section is sufficiently small to restrict the particle to a circular path (see Figure 13.8 on the next page). Suppose the particle is set in motion from the lowest point A of the circle, by giving it an initial velocity u.

The diagram shows the particle P at time t, with the radius OP making an angle θ with the downward vertical; let the speed of the particle at this point be v. For the subsequent motion, we shall consider three possible cases which depend on the magnitude of u, the initial speed.

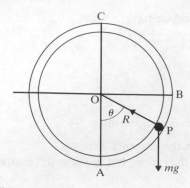

Figure 13.8

(i) The particle may fail to reach the level of O, coming to rest instantaneously before $\theta = \frac{1}{2}\pi$, and then falling back.

(ii) It may rise above the level of O, but come to rest instantaneously before reaching the highest point C on the circle.

(iii) If u is sufficiently large, it may go all the way round the circle.

Remember that, because its motion is restricted by the tube, the particle can never leave its circular path. We shall now investigate each of these possibilities.

Case (i)

Since the tube is smooth, we may apply the principle of conservation of energy.

$$\text{Initial kinetic energy} = \tfrac{1}{2}mu^2$$
$$\text{Kinetic energy at P} = \tfrac{1}{2}mv^2$$

$$\text{Potential energy gained} = mg(r - r\cos\theta)$$

Hence by conservation of energy, we have

$$\tfrac{1}{2}mu^2 = \tfrac{1}{2}mv^2 + mg(r - r\cos\theta) \qquad \qquad \cdots \quad (1)$$

The particle will just reach the point B if $v = 0$ when $\theta = \frac{1}{2}\pi$, i.e.

$$\tfrac{1}{2}mu^2 = mg[r - r\cos(\tfrac{1}{2}\pi)]$$
$$= mgr$$

i.e. $\quad u^2 = 2gr$

$$u = \sqrt{(2gr)}$$

For any value of u less than $\sqrt{(2gr)}$, the particle will come to rest, instantaneously, before it reaches B; it will then oscillate about point A. Let the angle at which it comes to instantaneous rest be α_1.

Putting $v = 0$ in equation (1) of this section, this angle is given by

$$\tfrac{1}{2}mu^2 = mg(r - r\cos\alpha_1)$$

i.e. $gr \cos \alpha_1 = gr - \frac{1}{2}u^2$

$$\cos \alpha_1 = \frac{(gr - \frac{1}{2}u^2)}{gr}$$

$$= \frac{(2gr - u^2)}{2gr}$$

Notice that, since $u^2 < 2gr$, the top line of this expression is positive and less than $2gr$, so $\cos \alpha_1 < 1$, confirming that a value for α_1 can be found with $\alpha_1 < \frac{1}{2}\pi$.

Throughout the motion the reaction, R, of the tube on the particle can be found by resolving along PO, giving

$$R - mg \cos \theta = mv^2/r$$

i.e. $R = mg \cos \theta + mv^2/r$ . . . (2)

But, from equation (1),

$$\frac{1}{2}mv^2 = \frac{1}{2}mu^2 - mg(r - r\cos \theta)$$
$$v^2 = u^2 - 2g(r - r\cos \theta)$$
$$= u^2 - 2gr + 2gr \cos \theta$$

Substituting this in equation (2) gives

$$R = mg \cos \theta + m(u^2 - 2gr + 2gr \cos \theta)/r$$
$$= mg \cos \theta + mu^2/r - 2mg + 2mg \cos \theta$$
$$= 3mg \cos \theta - 2mg + mu^2/r \qquad \qquad . . . (3)$$

Case (ii)

We now consider the motion when $u^2 > 2gr$, i.e. the initial velocity is sufficiently large to take the particle beyond point B. Although Figure 13.8 shows θ as an acute angle, we may continue to use equations (1) and (2), even though θ is now obtuse.

The particle will just reach the point C if $v = 0$ when $\theta = \pi$. Using equation (1) on page 250,

$$\frac{1}{2}mu^2 = mg(r - r\cos \pi)$$
$$= 2mgr$$

i.e. $u^2 = 4gr$
$$u = 2\sqrt{(gr)}$$

For any value of u less than $2\sqrt{(gr)}$, the particle will come to rest, instantaneously, before it reaches C, the highest point of the circle. Let the angle at which it comes to instantaneous rest be α_2. Putting $v = 0$ in equation (1), this angle is given by

$$\frac{1}{2}mu^2 = mg(r - r\cos \alpha_2)$$

i.e. $gr \cos \alpha_2 = gr - \frac{1}{2}u^2$

$$\cos \alpha_2 = \frac{gr - \frac{1}{2}u^2}{gr}$$

$$= \frac{(2gr - u^2)}{2gr}$$

Notice that, since $4gr > u^2 > 2gr$, the top line of this expression is negative, and the value of $\cos \alpha_2$ lies between -1 and 0, confirming that a value for α_2 can be found with $\frac{1}{2}\pi < \alpha_2 < \pi$.

Throughout the motion the reaction, R, of the tube on the particle can be found from equation (3), i.e.

$$R = 3mg \cos \theta - 2mg + mu^2/r$$

However, when θ is obtuse, $\cos \theta$ is negative, and consequently it is possible for R to become zero; this will happen for an angle θ_1 given by

$$0 = 3mg \cos \theta_1 - 2mg + mu^2/r$$

$$3g\cos \theta_1 = 2g - u^2/r$$

$$\cos \theta_1 = \frac{2gr - u^2}{3gr}$$

With $u^2 > 2gr$, the top line is negative, confirming that $\cos \theta_1$ is negative and hence that an obtuse value of θ_1 can be found (provided $\cos \theta_1$ is not less than -1).

Case (iii)

We saw in Case (ii) that if $u^2 = 4gr$, the particle will just reach point C, the highest point on the circle. For larger values of u the particle will make complete circles.

Qu. 8 Describe in general terms the nature of the motion, if the initial speed is sufficiently great to take the particle over the highest point, but the tube is rough.

Example 7

A small particle P, whose mass is m, is free to move in a vertical circle on the smooth inner surface of a cylinder, centre O of radius r, whose axis is horizontal. When the particle is at the lowest point, A, of the circle, it is given a velocity u. Assuming that it remains in contact with the cylinder throughout the motion, find an expression for the reaction, R, of the cylinder on the particle when the radius OP has turned through θ radians, in terms of m, r, u, g and θ.

Show that this motion is possible provided $u^2 > 5gr$.

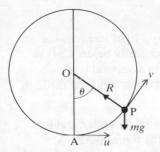

Figure 13.9

The diagram shows the particle when the radius OP has turned through an angle θ; let the speed of the particle at this instant be v.

Using the principle of conservation of energy, we have

K.E. lost = P.E. gained

i.e. $\frac{1}{2}mu^2 - \frac{1}{2}mv^2 = mg(r - r\cos\theta)$

where m is the mass of the particle, and r is the radius of the circle. Hence

$$\frac{1}{2}u^2 - \frac{1}{2}v^2 = g(r - r\cos\theta)$$

$$\therefore \quad v^2 = u^2 - 2gr + 2gr\cos\theta \qquad \qquad \cdots \quad (1)$$

Resolving towards O, the equation of motion at this instant is

$$R - mg\cos\theta = m(v^2/r)$$

where R is the reaction of the cylinder on the particle.

$$\therefore \quad R = m(v^2/r) + mg\cos\theta$$

and substituting the value of v^2 given by (1), we have

$$R = m(u^2 - 2gr + 2gr\cos\theta)/r + mg\cos\theta$$
$$\therefore \quad R = m(u^2/r) - 2mg + 3mg\cos\theta$$

Since the particle is to remain in contact with the surface of the cylinder at all times, the reaction must be positive for all values of θ; in particular it must be positive when $\theta = \pi$.

When $\theta = \pi$, the reaction is, from the result above,

$$m(u^2/r) - 2mg + 3mg\cos\pi$$
$$= m[(u^2/r) - 2g - 3g]$$
$$= m[(u^2/r) - 5g]$$

Hence $R > 0$ for all values of θ, provided $u^2/r > 5g$,

i.e. $u^2 > 5gr$

(If $u^2 = 5gr$, the reaction becomes instantaneously zero at the highest point, but the particle will continue to move in a circle.)

Example 8

A particle P is attached to a fixed point C, by a light inextensible string of length a. Initially P is at point A, vertically below C. It is then given an initial velocity u and it moves in a vertical circle about C. Given that the string becomes slack when OP has turned through $\frac{2}{3}\pi$ radians, find u^2 in terms of a and g.

In the subsequent motion, the particle moves freely under gravity. Show that its path meets the circle again at the initial point, A.

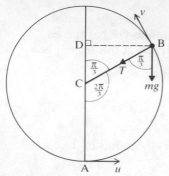

Figure 13.10

The diagram shows the forces acting on the particle when it has reached the point B, where $\angle ACB = \frac{2}{3}\pi$. Notice that $\angle DCB = \frac{1}{3}\pi$, so $CD = a\cos(\frac{1}{3}\pi) = \frac{1}{2}a$, and consequently $AD = \frac{3}{2}a$. This length is required in the working below.

The tension T is shown in the diagram, but at this instant it is actually zero.

Resolving along BC, we have

$$T + mg\cos(\tfrac{1}{3}\pi) = m(v^2/a)$$

where v is the speed of the particle at B.

But at this point $T = 0$, so

$$\tfrac{1}{2}mg = mv^2/a$$
$$\therefore \quad 2v^2 = ga \qquad\qquad \cdots \quad (1)$$

Between A and B, the particle undergoes a vertical displacement equal to AD, and we have already noted that this equals $\frac{3}{2}a$, so the potential energy gained is $\frac{3}{2}mga$. The kinetic energy lost between these points is $(\frac{1}{2}mu^2 - \frac{1}{2}mv^2)$. By the principle of conservation of energy, we have

$$\tfrac{1}{2}mu^2 - \tfrac{1}{2}mv^2 = \tfrac{3}{2}mga$$
$$\therefore \qquad\qquad v^2 = u^2 - 3ga$$

Substituting this in equation (1), we have

$$2(u^2 - 3ga) = ga$$
$$2u^2 - 6ga = ga$$
$$\therefore \qquad u^2 = \tfrac{7}{2}ga$$

When the string becomes slack, the particle is free to move in a parabolic path. We shall take axes through B, with the y-axis vertically upwards, and the x-axis *to the left* (see Figure 13.11).

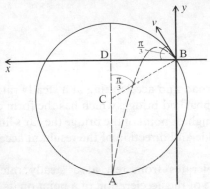

Figure 13.11

The initial velocity for the particle when it is moving as a projectile is v, given by (1), above. Since the velocity v was inclined to the horizontal at an angle of $\tfrac{1}{3}\pi$ at this instant, the horizontal and vertical components of the initial velocity are $v\cos(\tfrac{1}{3}\pi)$ and $v\sin(\tfrac{1}{3}\pi)$ respectively, that is $\tfrac{1}{2}v$ and $\tfrac{1}{2}v\sqrt{3}$.

The coordinates of the particle, relative to B, t seconds later are given by

$$x = \tfrac{1}{2}vt \qquad\qquad\qquad\qquad\qquad (2)$$
$$y = (\tfrac{1}{2}v\sqrt{3})t - \tfrac{1}{2}gt^2 \qquad\qquad\qquad (3)$$

The coordinates of A relative to B are $(a\sin(\tfrac{1}{3}\pi), -a - a\cos(\tfrac{1}{3}\pi))$, i.e. $(\tfrac{1}{2}a\sqrt{3}, -\tfrac{3}{2}a)$. So, using this x-coordinate in equation (2), the time taken for the particle to travel the horizontal distance from B to A is given by

$$\tfrac{1}{2}a\sqrt{3} = \tfrac{1}{2}vt$$
i.e. $$t = \frac{a\sqrt{3}}{v}$$

Using this value of t in equation (3), we find that the y-coordinate at this instant is

$$y = \left(\frac{v\sqrt{3}}{2}\right)\left(\frac{a\sqrt{3}}{v}\right) - \frac{g}{2}\left(\frac{3a^2}{v^2}\right)$$
$$= \frac{3}{2}a - \frac{3ga^2}{2v^2}$$

But, from (1) $2v^2 = ga$,   so

$$y = \frac{3}{2}a - \frac{3ga^2}{ga}$$
$$= \tfrac{3}{2}a - 3a$$
$$= -\tfrac{3}{2}a$$

Hence the particle passes through the point whose coordinates relative to B are $(\tfrac{1}{2}a\sqrt{3}, -\tfrac{3}{2}a)$, i.e. it passes through point A.

Exercise 13d

Where necessary, take g to be $9.81\,\mathrm{m\,s^{-2}}$.

1 A car is travelling along a straight road and accelerating at a steady rate of $3\,\mathrm{m\,s^{-2}}$, when it passes over a humpbacked bridge, which has the form of a circular arc of radius 100 m. At the highest point of the bridge the car's linear speed is $20\,\mathrm{m\,s^{-1}}$. Find the magnitude and direction of the resultant acceleration at this point.

2 A flywheel, of radius 0.5 m, accelerates from rest at a steady rate of $0.2\,\mathrm{rad\,s^{-2}}$. Calculate the magnitude of the acceleration of a point on its rim at the end of five seconds.

3 A particle, of mass m, moves around the inside of a smooth circular tube of radius r, which stands in a vertical plane. Initially the particle is slightly displaced from the highest point, i.e. its initial speed is zero. Show that when it reaches the lowest point, its speed is $2\sqrt{(gr)}$, and that the reaction of the tube on the particle is $5mg$.

4 A car, travelling at a constant speed of $v\,\mathrm{m\,s^{-1}}$, passes over a humpbacked bridge in the form of a circular arc of radius r m. Show that, at the highest point, the reaction of the road on the car is $mg - mv^2/r$. Taking r to be 40, find the greatest speed at which the car can travel without 'lifting off' at the highest point.

5 A boy swings a bucket of water in a vertical circle of radius 1 m at a constant speed. Find the rate, in r.p.m. at which he must swing it, to avoid spilling the water.

6 A toy car runs down a plastic track, arranged in a vertical plane and standing on a horizontal floor, as shown in Figure 13.12. At the lowest point, A, it starts to 'loop-the-loop', travelling in a vertical circle, centre O, radius r, as indicated in the diagram. Friction is negligible.

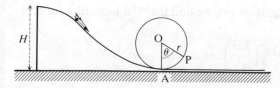

Figure 13.12

The car starts from rest at a height H above the level of the floor. Find the reaction of the track on the car when it is at point P, with the radius OP making an angle θ with the downward vertical. Show that if the car subsequently leaves the track, it does so when at a height $\frac{1}{3}(r + 2H)$ above the floor. Hence show that it will complete the loop without leaving the track if $H > \frac{5}{2}r$.

7 A particle P is suspended on a light inextensible string of length a from a point O. When the string is vertical, the particle is given an initial horizontal velocity u; it then moves in a circular path in a vertical plane. Show that when OP makes an angle θ with the downward vertical, the tension in the string is

$$m\left(3g\cos\theta + \frac{u^2}{a} - 2g\right)$$

Given that $u^2 = \frac{7}{2}ga$, find the height of the particle above the initial position, when the string first becomes slack.

8 Show that the particle in question 7 will complete its circular path without the string becoming slack if $u^2 \geqslant 5ga$.

9 A particle P, of mass m, is slightly displaced from the highest point on the surface of a smooth sphere, centre O, radius r. Show that when OP makes an angle θ with the upward vertical, the reaction of the sphere on the particle is $mg(3\cos\theta - 2)$. Hence show that the particle will leave the surface of the sphere when $\cos\theta = \frac{2}{3}$.

10 A simple pendulum consists of a small bob P, of mass m, attached by a light inextensible string to a point O. The pendulum swings through an angle α (less than $\frac{1}{2}\pi$) either side of the vertical through O. Show that when OP makes an angle θ with the vertical, the tension in the string is $mg(3\cos\theta - 2\cos\alpha)$.

Exercise 13e (Miscellaneous)

1 A particle, of mass 0.4 kg, is tied to one end of a taut light inextensible string, of length 0.5 m. The other end of the string is tied to a fixed point A on a smooth horizontal plane. Given that the particle moves with speed 8 m s^{-1} on the plane and in a circular path with centre A, find, in newtons, the tension in the string. Given that the string snaps when its tension exceeds 60 N, find, in radians per second to one decimal place, the greatest angular speed with which the string can rotate.

(L)

2 A particle P, of mass m, is suspended from a fixed point O by a light elastic string of natural length l. In equilibrium, P hangs at a depth $\frac{3}{2}l$ below O. Find the modulus of elasticity of the string.

The particle is set in motion and describes a horizontal circle with uniform angular speed and with the string inclined at 60° to the vertical. Find the tension in the string, the radius of the circle, and the angular speed.

(L)

3 A particle, of mass 2 kg, is attached to one end of a light inextensible cord, the other end of which is fixed at a point O. The particle moves with speed 5 m s^{-1}

in a horizontal circle, which is of radius 6 m and whose centre is vertically below O. Calculate
(a) the magnitude of the tension in the cord;
(b) the length of the cord.

The path of the particle is at a height of 20 m above level ground when the cord breaks. Find the tangent of the angle which the velocity of the particle makes with the horizontal when the particle strikes the ground.
(Take g to be $10 \, \text{m s}^{-2}$.)

(L)

4 A particle, of mass m, is suspended from a fixed point O, by a light elastic string of natural length l and modulus λ. The particle moves with constant angular speed ω in a horizontal circular path, with the string at a constant angle θ with the downward direction of the vertical. Show that x, the extension of the string, is given by

$$\lambda x = \omega^2 lm(l + x)$$

Deduce that the motion described cannot take place unless $\omega^2 < \dfrac{\lambda}{lm}$.

Show further that, for a given value of ω, the depth of the horizontal circle below O is independent of λ.

(L)

5 An artificial satellite of mass m moves under the action of a gravitational force which is directed towards the centre, O, of the Earth and is of magnitude F. The orbit of the satellite is a circle of radius a and centre O. Obtain an expression for T, the period of the satellite, in terms of m, a and F.

Show that, if the gravitational force acting on a body of mass m at a distance r from O is $m\mu/(r^2)$, where μ is a constant, then $T^2\mu = 4\pi^2 a^3$.

Assuming that the radius of the Earth is 6400 km and that the acceleration due to gravity at the surface of the Earth is $10 \, \text{m s}^{-2}$, show that $\mu = 6.4^2 \times 10^{13} \, \text{m}^3 \, \text{s}^{-2}$.

Hence, or otherwise, find the period of revolution, in hours to 2 decimal places, of the satellite when it travels in a circular orbit 600 km above the surface of the Earth.

(L)

6 A pendulum consists of a small ball of mass m attached to a fixed hook by a long string of length l. The ball is pulled aside until the string makes an angle of 30° with the vertical, and is then released.

Draw separate diagrams, indicating the forces on the ball
(a) immediately after it has been released;
(b) when it is at the lowest point of the swing.
Mark clearly in each diagram, the direction of the acceleration of the ball at that instant.

Find the magnitude of the acceleration, in terms of g, in each case.

(SMP)

7 (a) One end of a light rigid rod is attached to a fixed point O and a particle P of mass m is fixed to the other end. The rod, which is of length l, is made to rotate in a vertical plane with uniform angular velocity ω. Find expressions, in terms of m, l, ω and g, for the tension in the rod when P is (i) at the highest point of its path, (ii) at the lowest point of its path.

(b) A particle is attached to one end of a light inextensible string of length 1 m, the other end of which is fixed at a point O. The particle moves with uniform speed in a horizontal circle whose centre is directly below O, and the time for one revolution is 1 s. Find, in terms of g, the cosine of the angle that the string makes with the vertical.

(C)

8 A fixed smooth hollow sphere, of external radius a and centre O, has a small hole at the highest point. A light inextensible string passes through the hole and carries a particle of mass $2m$ which hangs freely, at rest, inside the sphere. A particle P, of mass $3m$, is attached to the other end of the string and moves on the outer surface of the sphere in a circle of radius $\frac{1}{2}a$, with angular velocity ω. Show that

$$\frac{a\omega^2}{g} = \frac{2\sqrt{3}}{9}$$

and find, in terms of m and g, the magnitude of the reaction of the sphere on the particle P.

(C)

9 The diagram (Figure 13.13) shows a light inextensible string APR where $AP = PR = l$

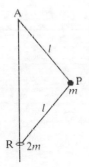

Figure 13.13

The end A of the string is attached to a fixed point. A particle of mass m is attached to the string at P and a ring of mass $2m$ is attached at R. The ring is free to move on a fixed smooth vertical wire passing through A. The plane APR rotates with constant angular velocity ω about the vertical through A, and P moves in a fixed horizontal circle. Find the tension in AP, and show that $AR = 10g/\omega^2$. Prove that $\omega^2 > 5g/l$.

(C)

10 The diagram (Figure 13.14) shows a particle of mass m which is attached to fixed points A and B by means of two light inextensible strings of lengths l and $l\sqrt{3}$, respectively. B is a distance l vertically above A.

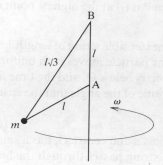

Figure 13.14

The system rotates with constant angular speed ω about AB, and both strings are taut.

(a) Show that the strings make angles of 60° and 30° with the vertical;

(b) Find expressions in terms of m, g, l and ω for the tensions in the strings.

(c) Show that $\dfrac{2g}{3l} < \omega^2 < \dfrac{2g}{l}$

(C)

Revision Exercise 3

In questions 1 to 5 take g to be 9.8 m s^{-2}.

1 A car of mass 1000 kg is travelling along a level road at a steady speed of 45 km h^{-1} with its engine working at 22.5 kW. Calculate in newtons the total resistance (assumed constant) due to friction, etc.

The engine is disconnected and simultaneously the brakes are applied, bringing the car to rest in 30 metres. Find the force, assumed constant, exerted by the brakes. (O & C)

2 The frictional resistance to the motion of a car of mass 1000 kg is kv newtons, where v m s^{-1} is its speed and k is constant. The car ascends a hill of inclination sin^{-1} ($\frac{1}{10}$) at a steady speed of 8 m s^{-1}, the power exerted by the engine being 9.76 kW. Prove that the numerical value of k is 30.

Find the steady speed at which the car ascends the hill if the power exerted by the engine is 12.8 kW.

When the car is travelling at this speed, the power exerted by the engine is increased by 2 kW. Find the immediate acceleration of the car.

(O & C)

3 A car of mass 900 kg moving with speed v m s^{-1} is subject to a frictional resistance of R newtons where $R = av + b$, a and b being constants. With the engine working at 37 280 W it can ascend a hill of inclination sin^{-1} ($\frac{1}{7}$) at a steady speed of 20 m s^{-1}. With the engine working at 37 080 W it can ascend a hill of inclination sin^{-1}($\frac{1}{14}$) at a steady speed of 30 m s^{-1}. Calculate a and b.

When the car is moving along a horizontal road at a speed of 25 m s^{-1} the engine is working at 26 375 W. Calculate its acceleration at this instant.

(O & C)

4 A car of mass 3×10^3 kg travels down a hill of inclination sin^{-1} ($\frac{1}{15}$) to the horizontal at a constant speed of 40 m s^{-1} with the engine shut off. If the resistance is proportional to the square of the speed show that the resistance at a speed of 20 m s^{-1} is 490 N.

If the engine is working at a power of 9×10^4 W find the acceleration at an instant when the car is moving up the hill at a speed of 20 m s^{-1}.

(C)

5 When a tractor of mass M kg is ascending a hill of inclination α at a steady speed of u m s^{-1} its engine is working at P W. With the engine working at the same power the tractor can descend the same hill at a steady speed of $2u$ m s^{-1}. If the frictional resistance is kv^2 N, where v m s^{-1} is the speed, show that

$$k = 4.2\left(\frac{M \sin \alpha}{u^2}\right)$$

and find P in terms of M, u and α.

If $u = 5$, find the steady speed at which the tractor can move along a level road when the engine is again working at P W. (C)

6 Two particles of mass 5 kg and 15 kg are moving towards each other in the same straight line with speeds of $4 \, \text{m s}^{-1}$ and $2 \, \text{m s}^{-1}$ respectively.

(a) If there is no loss of energy in the collision find their speeds immediately afterwards. Find also the magnitude of the impulse between the particles, stating the units in which it is measured.

(b) If the particles coalesce in the collision, find the loss of energy in the collision, stating the units in which it is measured.

(C)

7 Three particles A, B and C, of masses m, $2m$ and $3m$ respectively, are at rest, in that order, in a straight line on a smooth horizontal table. The particles A and B are joined by a light inelastic string. Initially the string is slack and the particle B is projected away from A and towards C with speed V. The string becomes taut before B strikes C. Show that the loss of kinetic energy when the string becomes taut is $\frac{1}{3}mV^2$ and find the magnitude of the impulse in the string.

The coefficient of restitution between B and C is $\frac{1}{2}$. Show that immediately after the collision between B and C the relative velocity of A and B has magnitude $\frac{3}{5}V$.

(C)

8 On a smooth horizontal table, n identical particles lie at rest in a straight line. Each particle has mass m and the distance between each pair of particles is d. The particle at the end of the line is projected with speed u directly towards the others. In each collision that occurs, the two particles that are involved coalesce to form a single particle. Find expressions for the time that elapses between the instant of projection and the last collision, and for the total loss of kinetic energy in the $(n-1)$ collisions.

(C)

9 Two small smooth spheres P and Q, of equal radii and having masses $4m$ and m respectively, are placed at rest on a smooth horizontal plane. The line PQ is perpendicular to a fixed vertical barrier with Q between P and the barrier. The coefficient of restitution between Q and the barrier is $\frac{1}{2}$ and between P and Q is $\frac{2}{3}$.

The sphere Q is projected, with speed u, directly towards the barrier. Show that the kinetic energy lost when Q collides with the barrier is $\frac{3}{8}mu^2$.

Determine how many collisions take place altogether and, for the second collision, find the loss of kinetic energy of the system and find the magnitude of the impulse on Q.

(C)

10 Three beads A, B and C, of masses $3m$, $2m$ and m respectively, are threaded in that order on a smooth horizontal straight wire. Initially A, B and C are separated and at rest, and A is then projected towards B with a speed $10V$. The coefficient of restitution between A and B is $\frac{1}{2}$. Show that the speed of B after the collision is $9V$.

Find the velocity of A after the collision and find also the kinetic energy lost in the collision.

After A and B have collided, C is projected towards B with speed $6V$. Given that the collision between B and C is perfectly elastic ($e = 1$), find the velocities of B and C after this collision.

(C)

11 A small bead B of mass m is free to slide on a fixed smooth vertical wire, as indicated in the diagram.

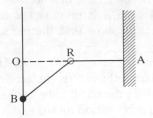

Figure 13.15

One end of a light elastic spring, of unstretched length a and modulus of elasticity $2mg$, is attached to B. The string passes through a smooth fixed ring R and the other end of the string is attached to a fixed point A, AR being horizontal. The point O on the wire is at the same horizontal level as R, and $AR = RO = a$.
(a) Prove that, in the equilibrium position, $OB = \frac{1}{2}a$.
(b) The bead B is raised to a point C of the wire above O, where $OC = a$, and is released from rest. Find the speed of the bead as it passes O, and find the greatest depth below O of the bead in the subsequent motion.

(C)

12 *Take g to be 10 m s^{-2} in this question*
A particle P of mass 2 kg is free to move in a straight groove in a rough horizontal table. P is attached to one end of a light elastic string of natural length 0.4 m and modulus of elasticity 30 N, and the other end of the string is fixed to a point O in the groove. The coefficient of friction between P and the groove is $\frac{1}{4}$.
(a) Find the furthest distance from O at which P can rest in equilibrium.
(b) The particle P is released from rest at a point where the extension of the string is 0.05 m. Find the extension of the string when P subsequently comes to rest.

(C)

13 Using the binomial approximation $(1 + x)^2 \approx 1 + 2x$, where x is small, or otherwise, show that for the period of a simple pendulum to be increased by $p\%$, where p is small, the length of the pendulum must be increased by about $2p\%$.

A clock which is operated by a simple pendulum gains 100 seconds a day. Find the percentage by which the length of the pendulum must be increased or decreased in order that the clock should give the correct time.

Show that the height above the Earth's surface to which the incorrect clock should be taken in order that it should give the correct time is about 7.4 km.

(Assume that the radius of the Earth is 6371 km and that gravity varies inversely as the square of the distance from the centre of the Earth.)

(C)

14 A light inextensible string of length $5a$ has one end fixed at a point A and the other end fixed at a point B which is vertically below A and at a distance $4a$ from it. A particle P of mass m is fastened to the midpoint of the string and moves with speed u, and with the parts AP and BP both taut, in a horizontal circular path whose centre is the midpoint of AB. Find, in terms of m, u, a and g, the tensions in the two parts of the string, and show that the motion described can take place only if $8u^2 \geqslant 9ga$.

(JMB)

15 At any point outside the Earth, at a distance r from the centre, the acceleration f due to gravity is proportional to r^{-2}. Taking the Earth to be a sphere of radius a and given that g is the gravitational acceleration on the surface, find f in terms of a, g and r.

Show that the speed, v, of an artificial satellite in a circular orbit at a height h above the surface of the Earth is given by

$$v^2 = \frac{ga^2}{a + h}$$

Find, in terms of a, g and h, the time taken to complete one orbit.

(JMB)

16 A particle P is projected horizontally with speed u from the lowest point A of the smooth inside surface of a fixed hollow sphere of internal radius a.
(a) In the case when $u^2 = ga$, show that P does not leave the surface of the sphere. Show also that, when P is halfway along its path from A towards the point at which it first comes to rest, its speed, v, is given by $v^2 = ga(\sqrt{3} - 1)$.
(b) Find u^2 in terms of ga in the case when P leaves the surface at a height $\frac{3}{2}a$ above A, and find, in terms of a and g, the speed of P as it leaves the surface.

(JMB)

17 A light rod OP of length a is free to rotate in a vertical plane about a fixed point O, and a mass m is attached to it at P. Denote by θ the angle between OP and the downward vertical. When $\theta = 0$, P is moving horizontally with speed u. Find the tension in the rod as a function of θ, and show that the value of the function is positive throughout the motion if $u^2 > 5ag$.

Find the least value of the tension when $u^2 = 6ag$.

Show that, if $u^2 < 2ag$, P remains below the level of O and deduce that, in this case also, the value obtained for the tension is positive throughout the motion.

(JMB)

18 One end of a light inextensible string of length a is attached to a particle of mass m. The other end is attached to a point O which is at a height $\frac{5}{2}a$ above the horizontal ground. Initially the string is taut and horizontal. The particle

is projected vertically downwards with velocity $\sqrt{(2ag)}$. Find the velocity v of the particle when the string has turned through an angle θ ($\theta < \pi$), and show that the tension is then $mg(2 + 3\sin\theta)$.

(a) If the string can withstand a tension of at least $5mg$, prove that the string will not break.

(b) If the string can withstand a tension of at most $\frac{7}{2}mg$, find the values of θ and v when the string breaks. In this case show that the particle strikes the ground at the point vertically below O.

(O & C)

Forces in equilibrium

Chapter 14

Forces in equilibrium

14.1 Introduction

Some of the most important questions which engineers have to answer are concerned with the stability of structures or machines. For example:

What is the largest load a tower crane can carry?

How far can a double-decker bus be tilted before it topples over?

What load of freight can a plane carry and where should it be placed?

Up to this stage in this book we have been concerned with motion, and the forces which cause objects to move. (This branch of the subject is usually called **Dynamics**.) We shall now turn our attention to forces which are in equilibrium. If such a system of forces acts on a stationary object, it remains at rest. This branch of the subject is usually called **Statics**.

14.2 The equilibrium of two forces

Two forces can only be in equilibrium if they are

(a) equal in magnitude;

(b) opposite in direction;

(c) acting in the same straight line.

The reader will no doubt think that this is obvious; one only has to think of two perfectly matched tug of war teams, pulling equally hard in opposite directions, to see that it makes sense.

We may also regard this as a consequence of Newton's second law of motion. Suppose two forces, represented by vectors F_1 and F_2, act upon a particle of mass m, but produce zero acceleration, then, from Newton's law, we have

$$F_1 + F_2 = m \times 0$$
$$F_1 + F_2 = 0$$

i.e. $\qquad F_1 = -F_2$

In other words the forces represented by F_1 and F_2 are equal in magnitude, but opposite in direction.

If two forces which are equal in magnitude, but opposite in direction do *not* act in a straight line (as, for example, when one applies such forces to the arms of a wheel brace in order to remove a nut from the wheel of a car), they will not produce equilibrium. Instead they produce a 'turning effect'. Such a pair of equal, but opposite, forces is called a **couple**. We shall return to this later.

14.3 The equilibrium of three forces

It is easily verified experimentally, that two forces which are not parallel are 'added' by the **parallelogram law** for adding vectors. That is, if **P** and **Q** are two forces, then their resultant, or sum, is given by the diagonal **R** of the parallelogram shown in Figure 14.1a.

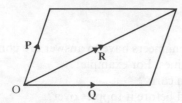

Figure 14.1a

We write

$$P + Q = R$$

(The reader is reminded that when we write **P**, **Q** and **R** in bold, i.e. heavy, type, these symbols are intended to represent the forces in magnitude *and direction*.)

Notice that the lines of action of the forces, **P**, **Q** and **R**, all pass through point O and are all in the same plane. The side of the parallelogram opposite **Q**, also represents the magnitude and direction of the force **Q**, but *not* its line of action. It is usually sufficient to draw the triangle shown in Figure 14.1b, but when we do this, it is important to remember that the sides of the triangle are not the actual lines of action of the forces

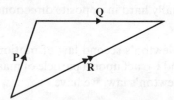

Figure 14.1b

If **S** represents a force which combines with **P** and **Q** to produce equilibrium, then, by the rule for equilibrium of two forces we have **R** = − **S**, so

$$P + Q = -S$$

Hence

P + Q + S = 0

This means that we can represent the forces **P**, **Q** and **S** by the sides of a triangle, taken in order, as in Figure 14.2.

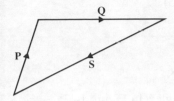

Figure 14.2

Also, **S** and **R** must have the same line of action, otherwise they will form a couple (see §14.2). The line of action of **R** passes through the common point (i.e. O in Figure 14.1a) of **P** and **Q**, so **S** must also do this. So the three forces which are in equilibrium, **P**, **Q** and **S**, must be concurrent and in the same plane.

Hence

> Three coplanar non-parallel forces which are in equilibrium can be represented in magnitude and direction by the sides of a triangle, taken in order; the three forces must also be concurrent.

Three forces which can be represented in magnitude and direction by the sides of a triangle, but which are *not* concurrent, form a couple.

Example 1

A force of magnitude 80 N acts along the positive *x*-axis, and another force of magnitude 50 N is inclined at an angle of 120° to this axis. Find their resultant.

The given forces are shown in the parallelogram OABC (Figure 14.3). The diagonal OB of the parallelogram represents the resultant.

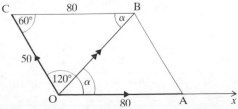

Figure 14.3

Applying the cosine rule to triangle OBC, we have

$$OB^2 = OC^2 + CB^2 - 2 \times OC \times CB \times \cos C$$

In this triangle, angle C equals 60°. Substituting this and the rest of the data gives

$$
\begin{aligned}
OB^2 &= 50^2 + 80^2 - (2 \times 50 \times 80 \times 0.5) \\
&= 2500 + 6400 - 4000 \\
&= 4900 \\
\therefore \quad OB &= 70
\end{aligned}
$$

Hence the magnitude of the resultant is 70 N.

The direction of the resultant is given by angle α in Figure 14.3; using the sine rule, we have

$$\frac{\sin \alpha}{50} = \frac{\sin 60°}{70}$$

$$\therefore \quad \sin \alpha = \left(\frac{\sin 60°}{70}\right) \times 50$$

$$\therefore \qquad \alpha = 38.2° \quad \text{(correct to one decimal place)}$$

The resultant makes an angle of 38.2° with the positive x-axis.

Qu. 1 Solve Example 1 by dropping the perpendicular from B onto OA, and applying elementary trigonometry (i.e. not the cosine and sine rules) to the right-angled triangles so formed.

Example 2

A load of 100 N is suspended by two light inextensible cables AC and BC whose lengths are 4 m and 3 m respectively; the cables are joined at C and the load is attached to this point. The ends A and B are fixed to two points, on the same horizontal level, 5 m apart. Calculate the tension in each cable.

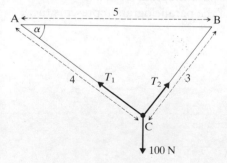

Figure 14.4

Let the tensions in AC and BC be T_1 and T_2, respectively. Let angle BAC be α; since this is a (3,4,5) triangle, it is right-angled at C. From this it follows that angle ABC = $(90° - \alpha)$, and hence that T_2 is inclined at an angle α to the vertical. The three forces, T_1, T_2 and the weight are in equilibrium, and so they can be represented by the sides of a triangle (see triangle PQR in Figure 14.5).

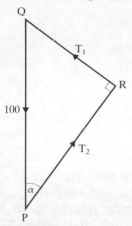

Figure 14.5

Note that the angle between the tensions is a right-angle, and that T_2 is inclined at angle α to the vertical. The problem can now be solved using elementary trigonometry.

$T_1 = 100 \sin \alpha$

From triangle ABC we can see that $\sin \alpha = \frac{3}{5} = 0.6$, so

$T_1 = 60$

Also $T_2 = 100 \cos \alpha$
$= 100 \times 0.8$
$= 80$

Hence the tensions in AC and BC are 60 N and 80 N, respectively.

In the next example an object, namely a sphere, will be described as 'uniform'. This implies that its weight may be considered to act at its geometrical centre. There will also be contact between the object and a 'smooth' wall. The word 'smooth' implies that there is no friction force, and hence the only contact force is the reaction perpendicular to the wall.

Example 3

A large uniform sphere of radius r and weight W rests against a smooth vertical wall. It is supported by a light inextensible cable AB of length r, one end of which is attached to the surface of the sphere at B, and the other end is fixed to a point A on the wall. Find the tension in the cable in terms of W.

The point of contact between the sphere and the wall is point C in Figure 14.6, and O is the centre of the sphere.

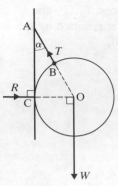

Figure 14.6

The weight of the sphere, W, is vertical and it acts at O. The reaction, R, of the wall on the sphere is horizontal, and this also passes through point O. The third force acting on the sphere is the tension T in the cable, and since the sphere is in equilibrium under the action of these three forces, T must also pass through point O; hence ABO is a straight line.

Let angle CAO be α, then

$$\sin \alpha = \frac{CO}{OA}$$

$$= \frac{r}{2r}$$

$$= \tfrac{1}{2}$$

Hence $\alpha = 30°$

We can now draw the triangle of forces (see Figure 14.7).

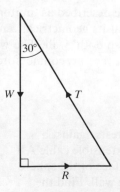

Figure 14.7

From this triangle we see that

$$T \cos 30° = W$$
$$\therefore \qquad T = W \div \cos 30°$$
$$= \frac{2W}{\sqrt{3}}$$

Hence the tension in the cable is $\dfrac{2W}{\sqrt{3}}$.

In the next example a heavy rod will be described as 'uniform'. This means that its weight is evenly distributed along its length. We may then treat the weight as if it were that of a particle situated at the geometrical centre of the object, in this instance the midpoint of the rod. We shall discuss this in more detail in Chapter 16.

The rod will also be described as 'freely jointed'. Readers who at some time have played with a child's building kit, such as 'Meccano', should think of a strip of the kit joined at one end to the rest of the structure by a nut and bolt which is not done up tightly, leaving the strip free to rotate about the joint. If the nut and bolt were fully tightened up, the strip would in effect become a rigidly attached part of the rest of the structure, and could be arranged at any required angle. This strip could then support a load without any further means of maintaining equilibrium, whereas a 'freely jointed' rod would require some other force or forces to prevent it rotating about the joint.

In the following example one end of the uniform rod is 'freely jointed' and the other is supported by a cable.

Example 4

A heavy uniform rod AB, of weight 500 N and length 4 m, is freely jointed at A to a point on a vertical wall. It is supported by a light inextensible cable joining the end B to a point C on the wall. The cable is inclined at an angle of 30° to the vertical. Find the magnitudes of the tension in the cable and the reaction at A.

In questions like this, it is very important to analyse the geometry of the diagram carefully. To this end, the reader is advised to draw two diagrams, one showing the forces (Figure 14.8a on the next page), and the other (Figure 14.8b) showing the geometrical features of the problem.

In these diagrams G, M and N are the midpoints of AB, BC and CA, respectively.

$$BC = 4 \div \sin 30° = 8$$
$$AN = NC = 4 \cos 30° = 2\sqrt{3}$$
and $NM = 2$

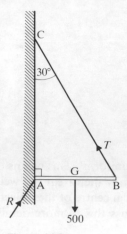

Figure 14.8a

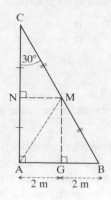

Figure 14.8b

It follows that

$$\tan \text{NAM} = \frac{2}{(2\sqrt{3})}$$

$$= \frac{1}{\sqrt{3}}$$

$$\therefore \quad \angle \text{NAM} = 30°$$

There are three forces acting on the rod: its weight, the tension T N
in the cable, and the reaction R N acting at point A. These three
forces must form a vector triangle, and they must be concurrent.
The weight of the rod is a vertical force passing through G and M;
the tension in the cable acts along BC and so this also passes
through M. Hence the remaining force, R, must also pass through
this point. Consequently R acts along AM, i.e. it is inclined at 30° to
the vertical.

We can now draw the triangle of forces:

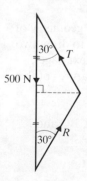

Figure 14.9

Since this is an isosceles triangle, T and R are equal in magnitude, and, by drawing the mediator of this triangle, they can be found as follows:

$$T - R = 250 \div \cos 30°$$
$$= 289 \quad \text{(correct to three significant figures)}$$

The tension in the cable and the reaction at A are both 289 newtons, correct to three significant figures.

Exercise 14a

1 A force of magnitude 80 N acts along the positive x-axis and a force of magnitude 60 N acts along the positive y-axis. Calculate the magnitude and direction of their resultant.

2 Two forces, whose magnitudes are 50 N and 100 N respectively, have a resultant of magnitude 120 N. Find the angle between them.

3 To remove a spike from a vertical wall requires a horizontal force of 400 N. A strong smooth light cord is attached to a fixed point A, which is 4 m away from the spike horizontally, and 2 m above it (see Figure 14.10).

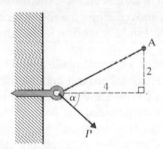

Figure 14.10

The cord passes through an eye at the end of the spike. A force P N is applied to the free end of the cord at an angle α below the horizontal; P is just sufficient to remove the spike. Find α and P.

4 Three coplanar forces $\mathbf{P}$, $\mathbf{Q}$ and $\mathbf{R}$, acting at O, the origin of a set of axes, are in equilibrium. The force $\mathbf{P}$ has a magnitude of 50 N and makes an angle of 60° with the positive x-axis. The force $\mathbf{Q}$ has a magnitude of 80 N and is at an angle α *below* the x-axis. The force $\mathbf{R}$ acts along the negative x-axis. Find (a) the angle α, and (b) the magnitude of $\mathbf{R}$.

5 Three forces of magnitudes 60 N, 80 N and 100 N, act along the sides AB, BC and CA, respectively, in the directions indicated by the letters, of a triangle ABC, in which $AB = 3$ m, $BC = 4$ m and $CA = 5$ m. Show that the resultant force is zero. Explain why these forces do not produce equilibrium, and show that they could form a couple, with the 60 N force acting along AB, as one of its members. State the magnitude, direction and line of action of the other member of the couple.

6 A heavy bob, which can be regarded as a particle of weight 50 N, is suspended by a light inelastic cord, of length 4 m, from a fixed point A. The bob is held in equilibrium by a horizontal force of 20 N, with the cord inclined at an angle α to the vertical. Find α and the magnitude of the tension in the cord.

7 A picture which weighs 100 N is suspended by means of two smooth rings attached to two points on its frame, 1 m apart, and on the same horizontal level. A smooth light inextensible cord passes through the rings and forms a closed loop, of total length 3 m. The picture is then hung by placing the loop over a hook fixed to a vertical wall. Forces such as the friction between the picture and the wall may be neglected. Find
  (a) the force exerted by the hook;
  (b) the tension in the cord;
  (c) the resultant force acting on each ring.

8 A load of 100 N is suspended by two light inextensible cables AC and BC whose lengths are 4 m and 2 m respectively; the load is attached to point C. The ends A and B are fixed to two points, on the same horizontal level, 5 m apart. Calculate the tensions in the cables.

9 A light rigid rod AB, of length 8 m, is freely pivoted at A. The other end, B, is attached by means of a light inelastic cord, to a fixed point C which is 6 m vertically above A; the rod is horizontal. A load of weight 200 N is suspended from a point P on AB. The point P is gradually moved along the rod from A towards B, and when $AP = 6$ m, the cord snaps. Find the tension in the cord just before it snaps.

10 A uniform ladder, whose weight is 400 N, rests in a vertical plane against a smooth vertical wall. Its foot rests on rough horizontal ground and the friction force between the ground and the ladder is 100 N. Show that the resultant of the normal reaction and the friction force between the foot of the ladder and the ground, is inclined to the horizontal at an angle $\tan^{-1} 4$. Hence show that, with the ladder in this state, the ladder makes an angle $\tan^{-1} 2$ with the horizontal.

14.4 Combining forces by vector methods

As we have seen in the last section, two forces may be combined by adding them by the parallelogram law. This means we can use the methods of vector algebra.

Some readers may find it helpful to revise Chapter 5 at this point. One particular technique will be used very frequently: if **F** represents a vector which is inclined at an angle α to the x-axis, (see Figure 14.11) then it may be written

$$\mathbf{F} = (F\cos\alpha)\mathbf{i} + (F\sin\alpha)\mathbf{j}$$

where F represents the magnitude of vector **F**, and **i** and **j** are the unit vectors parallel to the x-axis and the y-axis respectively.

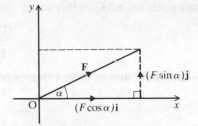

Figure 14.11

Example 1 (p 269) could be solved in vector notation, as follows:

$$\overrightarrow{OA} = 80\mathbf{i} \quad \text{and}$$
$$\overrightarrow{OC} = (50\cos 120°)\mathbf{i} + (50\sin 120°)\mathbf{j}$$
$$= (-50\cos 60°)\mathbf{i} + (50\sin 60°)\mathbf{j}$$
$$= -25\mathbf{i} + (25\sqrt{3})\mathbf{j}$$

The resultant, $\overrightarrow{OB}$, is the vector sum of $\overrightarrow{OA}$ and $\overrightarrow{OC}$, i.e.

$$\overrightarrow{OB} = \overrightarrow{OA} + \overrightarrow{OC}$$
$$= (80 - 25)\mathbf{i} + (25\sqrt{3})\mathbf{j}$$
$$= 55\mathbf{i} + (25\sqrt{3})\mathbf{j}$$

The magnitude of $\overrightarrow{OB}$ can then be found, using Pythagoras' theorem:

$$OB^2 = 55^2 + (25\sqrt{3})^2$$
$$= 3025 + 1875$$
$$= 4900$$
$$\therefore \quad OB = 70$$

Hence the magnitude of the resultant (which is represented by OB) is 70 N.

If the resultant $\overrightarrow{OB}$ makes an angle α with the x-axis, then

$$\tan \alpha = \frac{(25\sqrt{3})}{55}$$

$$\therefore \quad \alpha = 38.2°$$

Hence the resultant is inclined at 38.2° to the direction of the 80 N force.

Example 2 (p 270) could be expressed as follows:

the weight $= -100\mathbf{j}$
the tension in AC $= (-T_1 \cos \alpha)\mathbf{i} + (T_1 \sin \alpha)\mathbf{j}$
the tension in BC $= (T_2 \sin \alpha)\mathbf{i} + (T_2 \cos \alpha)\mathbf{j}$

Since the system is in equilibrium, the sum of these three vectors is zero, i.e.

$$-100\mathbf{j} + (-T_1 \cos \alpha)\mathbf{i} + (T_1 \sin \alpha)\mathbf{j} + (T_2 \sin \alpha)\mathbf{i} + (T_2 \cos \alpha)\mathbf{j} = 0$$

Hence

$$(-T_1 \cos \alpha + T_2 \sin \alpha)\mathbf{i} + (-100 + T_1 \sin \alpha + T_2 \cos \alpha)\mathbf{j} = 0$$

But a vector can only be zero if both its components are zero; in this case:

$$-T_1 \cos \alpha + T_2 \sin \alpha = 0 \qquad \qquad \cdots \quad (1)$$

and $\quad -100 + T_1 \sin \alpha + T_2 \cos \alpha = 0 \qquad \qquad \cdots \quad (2)$

Now, in this example, $\cos \alpha = 0.8$ and $\sin \alpha = 0.6$. Substituting these values, and rearranging equations (1) and (2), gives

$$0.8 T_1 = 0.6 T_2$$
$$\therefore \qquad T_1 = 0.75 T_2$$

and $0.6 T_1 + 0.8 T_2 = 100$

Hence

$$0.6 \times 0.75 T_2 + 0.8 T_2 = 100$$
$$0.45 T_2 + 0.8 T_2 = 100$$
$$1.25 T_2 = 100$$
$$T_2 = 80$$

and hence $\qquad \qquad T_1 = 60$

The tension in AC is 60 N, and the tension in BC is 80 N.

The working, in the example above, can be shortened by writing down equations (1) and (2) immediately. They represent the horizontal and vertical components of the resultant force, and we say that we have **resolved horizontally** to obtain equation (1), and **vertically** for equation (2).

Vector methods become especially convenient when more than two forces are involved, as the next example shows.

Example 5

Four forces, which act at the origin, have the following magnitudes and directions (all angles are measured from the positive x-axis):

(i)   magnitude 10 N, direction 40°
(ii) magnitude 20 N, direction 120°
(iii) magnitude 15 N, direction 210°
(iv) magnitude 25 N, direction 310°

Find their resultant, expressing it in the form $X\mathbf{i} + Y\mathbf{j}$.

Find also the extra force required to maintain equilibrium, expressing it in the same form.

Let the resultant of the four forces be $X\mathbf{i} + Y\mathbf{j}$.
Resolving parallel to the x-axis:

$$X = 10 \cos 40° + 20 \cos 120° + 15 \cos 210° + 25 \cos 310°$$
$$= 0.739\,75 \quad \text{(correct to five decimal places)}$$

Resolving parallel to the y-axis:

$$Y = 10 \sin 40° + 20 \sin 120° + 15 \sin 210° + 25 \sin 310°$$
$$= -2.902\,73 \quad \text{(correct to five decimal places)}$$

Hence the resultant force is $(0.740i - 2.90j)$ newtons, correct to three significant figures; it acts at the origin.

In order to maintain equilibrium, the extra force must be equal in magnitude to the resultant; it must have the same line of action, but the opposite direction. Consequently the required force is $(-0.740i + 2.90j)$ newtons, acting at the origin.

In the problems we have considered so far in this section, the forces have been concurrent, that is, they have all passed through a common point; in such cases the resultant also passes through this point. When the forces are *not* concurrent, we may still use vector algebra to find the magnitude and the direction of the resultant, but it may be difficult to identify its line of action. We shall see how to cope with this difficulty in the next chapter.

Qu. 2 Two forces, **P** and **Q**, each have a magnitude of 10 N. Force **P** acts along the positive x-axis, and **Q** along the positive y-axis. Find the magnitude of their resultant and the equation of its line of action. A third force **R**, also of magnitude 10 N, is now added, acting along the line $y = 2$. Find the point which is common to the lines of action of this force and the resultant of **P** and **Q**. Find the magnitude of the resultant of **P**, **Q** and **R**, and the equation of its line of action.

Exercise 14b

1 Two horizontal forces F and G act at a point O. The force F has a magnitude of 35 N and acts due North. The force G has a magnitude $20\sqrt{2}$ N, and it acts to the South East. Calculate the magnitude and bearing of the resultant of F and G.

2 Find the resultant of each of the following sets of forces, which are given in newtons, expressing your answers in the form $xi + yj$:
   (a) $16i$ and $(4i - 3j)$ and $(-2i + 8j)$;
   (b) 50 N at 40° to the positive x-axis and 30 N at $-20°$ to the same axis;
   (c) a force of 80 N acting at 30° to the positive x-axis, and a force of 50 N, making an angle of 120° with the same axis;
   (d) $2i + 3j$, $8i - 4j$ and a force of 5 N at 60° to the positive x-axis.

3 A force of magnitude 10 N acts along the positive x-axis, and another force, whose magnitude is 20 N, makes an angle θ with the same axis. Given that their resultant acts along the positive y-axis, find θ. Find also the magnitude of the resultant.

4 Find the magnitude of the resultant of two forces $(2i + 4j + 5k)$ newtons and $(i + 7k)$ newtons, and find the angles which this resultant makes with the coordinate axes.

5 Three cables are attached to an eye-bolt, as shown in Figure 14.12.

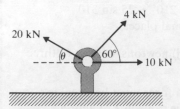

Figure 14.12

The 10 kN force is horizontal, the 4 kN force is inclined to it at 60°, and the 20 kN force acts at an angle θ to the horizontal, as shown; the forces are coplanar. Find the value of θ for which the resultant force on the eye-bolt is vertical, and, for this value of θ, find the resultant.

6 A structure made from scaffolding includes three light poles which meet at a point O, as in Figure 14.13. The forces exerted on O are: 2 kN along OB and 5 kN along OC.

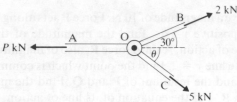

Figure 14.13

Assuming the poles are freely jointed at O, calculate the angle θ, and the force, P kN, acting along OA.

7 A jet aircraft, whose mass is 25 tonnes, is climbing at constant speed at an angle of 20° to the horizontal. Its engines exert a total thrust of T kN and there is a drag of 10 kN; both of these forces act along the line of the fuselage. Calculate the value of T, and the aerodynamic lift (which is perpendicular to the line of the fuselage) required, given that the resultant force on the aircraft is zero. (Take g to be 9.81 m s^{-2}.)

8 A heavy smooth ring R, whose weight is 8 N, is threaded on a light inextensible string of total length 160 cm, the ends of which are attached to two points A and B on the same horizontal level, 80 cm apart. A horizontal force is applied to the ring and as a result it rests in equilibrium vertically below B. Calculate the length BR, the tension in the string, and the magnitude of the horizontal force.

9 A photographer uses a tripod to support a camera which has a mass of 1.5 kg. The legs of the tripod are each 1 m long, and the feet of the tripod are the vertices of an equilateral triangle, each side of which is 0.5 m long. Taking g to be 9.81 m s^{-2}, calculate the force exerted by each leg of the tripod.

10 A radio mast consists of a light vertical pole, supported by three cables OA, OB and OC, each 5 m long, attached to a point O, which is 3 m from F, the foot of the pole. The ends A, B and C are fixed to three points on horizontal ground, each 4 m from F. The points A, B and C form an isosceles triangle, with $AB = AC$; the remaining side, BC, is 6 m long. The tension in each of the cables OB and OC is 50 N.

(a) Show that the resultant of the tensions in OB and OC has a magnitude of 80 N.
(b) Find the tension in OA.
(c) Find the thrust exerted on the ground by the foot of the pole.

14.5 Friction

We have already discussed friction in §9.2 and §9.3. For convenience, Coulomb's laws of friction are repeated here.

(1) The friction force between two surfaces opposes the relative motion.
(2) The magnitude of the friction force is independent of the area of the surfaces in contact.
(3) The magnitude of the friction force does not depend on the relative velocity of the surfaces.
(4) The magnitude of the friction force has a limiting value which is proportional to the normal reaction (in the usual notation $F \leqslant \mu R$).

In Chapter 9, we were chiefly concerned with objects which were sliding, consequently the *limiting* value of the friction force was normally required. In this chapter we are concerned with forces in equilibrium, so the limiting friction will only be required when the equilibrium is about to be broken; in all other cases, the friction force will be *less* than its limiting value. In other words, using the usual notation, $F < \mu R$, unless slipping is about to occur, in which case $F = \mu R$.

Example 6

A particle, of weight 5 N, is placed on a rough inclined plane. The coefficient of friction between the particle and the plane is 0.4, and the plane is inclined to the horizontal at an angle of 20°. Show that the particle does not slip and find the magnitude of the friction force.

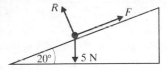

Figure 14.14

The diagram shows the particle and the forces acting on it. The weight is 5 N (given) and the normal reaction is R newtons. The friction force, F newtons, is shown acting up the slope (i.e. opposing the tendency to slip down the slope, as required by the first of the

laws of friction.) In this example it is convenient to resolve parallel
to the slope and perpendicular to it, since this gives us R and F
directly.

Resolving perpendicular to the slope:

$$R - 5\cos 20° = 0$$
$$\therefore \qquad\qquad R = 5\cos 20° \quad (\approx 4.70)$$

Resolving parallel to the slope:

$$F - 5\sin 20° = 0$$
$$\therefore \qquad\qquad F = 5\sin 20° \quad (\approx 1.71)$$

The limiting friction $= \mu R$
$$= 0.4 \times 5\cos 20°$$
$$= 1.88$$

Since the friction force required to maintain the equilibrium (1.71 N)
is less than the limiting friction (1.88 N), it is possible for the
equilibrium to be maintained by the friction, without assistance
from any further external forces. The magnitude of the friction force
is 1.71 N. (All numerical values, above, have been corrected to three
significant figures.)

Example 7

A particle, of weight W, is placed on a rough inclined plane. The
particle does not slip. The plane is inclined at an angle α to the
horizontal, and the coefficient of friction between the particle and
the plane is μ. Find, in terms of W, μ and α, the magnitude of the
force, P, acting parallel to the plane, which is just large enough to
make the particle slide up the plane.

Since the particle slides, the friction force is equal to μR, where R is
the normal reaction, and since the particle only *just* moves, we may
take the acceleration to be zero.

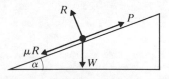

Figure 14.15

Resolving perpendicular to the plane:

$$R - W\cos \alpha = 0$$
$$\therefore \qquad\qquad R = W\cos \alpha$$

Resolving parallel to the plane:

$$P - W \sin \alpha - \mu R = 0$$
$$\therefore \qquad P = W \sin \alpha + \mu R$$

Substituting for R, we have:

$$P = W \sin \alpha + \mu W \cos \alpha$$

Hence

$$P = W(\sin \alpha + \mu \cos \alpha)$$

Example 8

A particle of weight W is placed on a rough inclined plane, and maintained in equilibrium by a horizontal force P, acting towards the inclined plane. The coefficient of friction between the particle and the plane is μ, and the angle of inclination of the plane is α. Given that $\mu < \tan \alpha < 1/\mu$, show that

$$\frac{W(\sin \alpha - \mu \cos \alpha)}{(\cos \alpha + \mu \sin \alpha)} \leqslant P \leqslant \frac{W(\sin \alpha + \mu \cos \alpha)}{(\cos \alpha - \mu \sin \alpha)}$$

Let P_1 be the horizontal force when the particle is on the point of slipping up the slope, and let R be the corresponding normal reaction. The friction force is limiting, i.e. equal to μR, and it acts down the slope (see Figure 14.16a).

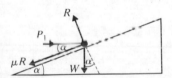

Figure 14.16a

Resolving perpendicular to the slope:

$$R - (P_1 \sin \alpha + W \cos \alpha) = 0$$
$$\therefore \quad R = P_1 \sin \alpha + W \cos \alpha$$

Resolving parallel to the slope:

$$P_1 \cos \alpha - (\mu R + W \sin \alpha) = 0$$
$$\therefore \quad P_1 \cos \alpha = \mu R + W \sin \alpha$$

Substituting for R, we have

$$P_1 \cos \alpha = \mu(P_1 \sin \alpha + W \cos \alpha) + W \sin \alpha$$

Hence

$$P_1(\cos \alpha - \mu \sin \alpha) = W(\mu \cos \alpha + \sin \alpha)$$

[At this point, note that the right-hand side of this equation is certainly positive, and so is P_1. Consequently the factor $(\cos\alpha - \mu\sin\alpha)$ must also be positive, so $\cos\alpha > \mu\sin\alpha$, i.e. $\tan\alpha < 1/\mu$ (which was given). If this condition is not satisfied, it is impossible to move the particle up the slope by applying a horizontal force.]

Dividing both sides of the above equation by $(\cos\alpha - \mu\sin\alpha)$, we have

$$P_1 = \frac{W(\mu\cos\alpha + \sin\alpha)}{(\cos\alpha - \mu\sin\alpha)}$$

Since P_1 is the force when the particle is about to move, the horizontal force P must be less than or equal to P_1.

Let P_2 be the horizontal force which just prevents the particle from slipping down the slope, and let S be the corresponding normal reaction. The friction force is limiting so it is equal to μS; it acts up the slope, i.e. opposing the particle's tendency to slip down the slope (see Figure 14.16b).

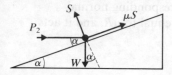

Figure 14.16b

Resolving perpendicular to the slope:

$$S - (P_2\sin\alpha + W\cos\alpha) = 0$$
$$\therefore \quad S = P_2\sin\alpha + W\cos\alpha$$

Resolving parallel to the slope:

$$P_2\cos\alpha + \mu S - W\sin\alpha = 0$$
$$\therefore \quad P_2\cos\alpha = W\sin\alpha - \mu S$$

Substituting for S, we have

$$P_2\cos\alpha = W\sin\alpha - \mu(P_2\sin\alpha + W\cos\alpha)$$

Hence

$$P_2(\cos\alpha + \mu\sin\alpha) = W(\sin\alpha - \mu\cos\alpha)$$

[Here note that $\sin\alpha > \mu\cos\alpha$, i.e. $\tan\alpha > \mu$ (which was given), otherwise the particle will remain in equilibrium, even without a horizontal force supporting it.]

Dividing both sides of the above equation by $(\cos\alpha - \mu\sin\alpha)$, we have

$$P_2 = \frac{W(\sin\alpha - \mu\cos\alpha)}{(\cos\alpha + \mu\sin\alpha)}$$

[It is worth noting that this is the expression for P_1, with the sign of μ reversed; this is because the only difference between the two situations is the direction of the friction force.]

Since P_2 is the force which only just maintains the equilibrium, the horizontal force P must be greater than or equal to P_2.
Hence, in equilibrium

$$\frac{W(\sin\alpha - \mu\cos\alpha)}{(\cos\alpha + \mu\sin\alpha)} \leqslant P \leqslant \frac{W(\sin\alpha + \mu\cos\alpha)}{(\cos\alpha - \mu\sin\alpha)}$$

[However in the real world, it would be impossible to achieve exact equality, so this possibility is of academic interest only.]

Exercise 14c

1 A particle, whose weight is 5 N, lies on a rough plane inclined at an angle 30° to the horizontal. A light inelastic string attached to the particle, lies in the line of greatest slope of the incline. The string passes over a smooth light pulley at the top of the slope, and the remaining portion of the string hangs vertically, with a particle, whose weight is 3 N, attached to the free end. Given that the particle on the slope is about to slip up the slope, find the coefficient of friction between the 5 N particle and the plane.

2 A particle, whose weight is 10 N, lies on a rough plane inclined at an angle 20° to the horizontal. A light inelastic string, attached to the particle, lies in the line of greatest slope of the incline. The string passes over a smooth light pulley at the top of the slope, and the remaining portion of the string hangs vertically, with a particle, whose weight is 2 N attached to the free end. Given that the particle on the slope is about to slip down the slope, find the coefficient of friction between the 10 N particle and the plane.

3 A box, whose weight is 500 N, rests on rough horizontal ground. An attempt is made to move it by pulling it along with a force of P newtons acting at 30° to the horizontal. When $P = 100$, the box does not move, calculate the magnitude of the friction force opposing the motion. Given that the box just begins to move when $P = 200$, find the coefficient of friction.

4 A box, whose mass is 200 kg, is placed on a rough plane which is inclined at 30° to the horizontal, and a horizontal force of P newtons, acting towards the plane, is applied to the box. The coefficient of friction between the box and the plane is 0.2. Taking g to be 9.81 m s^{-2}, find the magnitude of the friction force when (a) $P = 1000$, (b) $P = 200$.

5 A rectangular wooden block of weight W is placed on rough horizontal ground; the coefficient of friction between the block and the ground is μ_1. Another block weighing $2W$ is placed on top of the first block; the coefficient

of friction between the two blocks is μ_2. A horizontal force, which starts from zero and gradually increases until the equilibrium is broken, is applied to the upper block. Show that, if $\mu_1 = \mu_2$, the equilibrium will be broken by the upper block slipping.

Find the condition for the equilibrium to be broken by the lower block slipping.

6 A particle of weight 20 N rests on rough horizontal ground and the coefficient of friction between the particle and the ground is μ. A force P N, acting upwards at 30° to the horizontal, is just sufficient to move the particle. Show that

$$P = \left(\frac{40\mu}{\sqrt{3} + \mu}\right)$$

Show that P increases with μ. It is estimated that μ is 0.6, correct to *one decimal place*. Find the limits between which P lies.

7 A particle of weight $2W$ lies on a rough plane inclined at an angle α to the horizontal, where $\sin \alpha = 0.6$. The coefficient of friction between the particle and the plane is μ. A light inelastic string attached to the particle, lies in the line of greatest slope of the incline. The string passes over a smooth light pulley and its remaining portion hangs vertically with a particle, whose weight is W, attached to the free end. Given that the particle on the slope is about to slip down the slope, find the value of μ.

The particle of weight W is now replaced by another particle of weight W', which is just sufficient to make the particle of weight $2W$ move up the slope. Find W' in terms of W.

8 A particle of weight W rests on rough horizontal ground and the coefficient of friction between the particle and the ground is $\tan \lambda$. A force P, acting upwards at an angle θ to the horizontal, is applied to the particle. Given that the particle is on the point of slipping, show that

$$P = \frac{W \sin \lambda}{\cos (\theta - \lambda)}$$

Hence show that, as θ varies, the minimum value of P is $W \sin \lambda$, and state the value of θ for which this occurs.

9 A uniform ladder rests in a vertical plane against a smooth vertical wall, with the ladder making an angle α with the vertical. The foot of the ladder rests on rough horizontal ground and the coefficient of friction between the ground and the ladder is μ. Show that the resultant of the normal reaction and the friction force between the foot of the ladder and the ground, is inclined at an angle θ to the vertical, where $\tan \alpha = 2 \tan \theta$.

10 A student stands with both arms stretched out in front of her, and one finger of each hand held out horizontally, on the same level, and less than one metre apart. A metre rule is placed across her fingers and she then gradually moves them towards each other. Explain what happens. (It is suggested that the reader should actually try out this experiment before attempting to answer the question.)

14.6 The angle of friction, λ

Definition

The angle of friction, λ, is the angle whose tangent is equal to the coefficient of friction for the two surfaces in contact, i.e. if the coefficient of friction is μ, then

$$\tan \lambda = \mu$$

Suppose the normal reaction between a certain object and the surface on which it rests is R, and that the friction force between them is F. The resultant of F and R is called the **total reaction**; it is denoted by S in Figure 14.17. In that diagram,

$$\tan \theta = F/R$$

where θ is the angle between R and S.

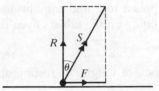

Figure 14.17

In equilibrium, $F \leqslant \mu R$, so

$$\tan \theta \leqslant \mu R/R$$
i.e. $\tan \theta \leqslant \mu$

and replacing μ by $\tan \lambda$, we have

$$\tan \theta \leqslant \tan \lambda$$
$$\therefore \qquad \theta \leqslant \lambda$$

In other words, the total reaction makes an angle with the normal reaction which is less than or equal to λ; (it equals λ, when the object is on the point of slipping). Diagrammatically this means that the total reaction must lie within the triangular region which is shaded in Figure 14.18. In three dimensions this becomes a cone, formed by rotating the triangular region about its vertical axis.

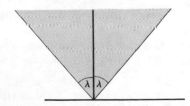

Figure 14.18

Qu. 3 A particle is placed on an inclined plane, and the angle of inclination is gradually increased until the particle begins to slip. Show that, when the particle is on the point of slipping, the angle of inclination is equal to the angle of friction.

Exercise 14d (Miscellaneous)

1 A horizontal force **R** is of magnitude 12 N and acts due East from a point O. The horizontal forces **P** and **Q** act from O in directions 030° and due South respectively. Given that $P + Q = R$, calculate the magnitudes of **P** and **Q**.
(L)

2 Two horizontal forces **P** and **Q** act at a point O. The force **P** is of magnitude 1 N and acts due North. The force **Q** is of magnitude $3\sqrt{2}$ N and acts North-East. Calculate the magnitude and the direction of the resultant of **P** and **Q**.
(L)

3 A particle P, of mass $7m$, is placed on a rough horizontal table, the coefficient of friction between P and the table being μ. A force of magnitude $2mg$ acting upwards at an angle α to the horizontal, is applied to P and equilibrium is on the point of being broken by the particle sliding on the table. Given that $\tan \alpha = \frac{5}{12}$, find the value of μ.
(L)

4 Figure 14.19 shows the forces acting on a block of weight W resting on a plane inclined at angle α to the horizontal.

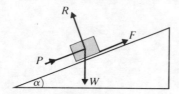

Figure 14.19

The normal reaction exerted by the plane on the block is R, and the frictional force acting on the block is F. The coefficient of friction between the block and the plane is μ. The force P is the least force which, when acting parallel to the line of greatest slope, prevents the block from sliding *down* the slope. Prove that

$$P = W \sin \alpha - \mu W \cos \alpha.$$
(L)

5 When the force acting up the slope in question 4 is $2P$, the block is in limiting equilibrium and about to slide *up* the slope.
(a) Draw a clearly labelled force diagram for this case.
(b) Find another expression for P in terms of W, μ and α.
(c) Deduce that $\mu = \frac{1}{3} \tan \alpha$.
(L)

6 Two forces, of 3 and 5 units, act in directions inclined at 40°.
(a) Calculate the components of the resultant (i) parallel to the larger force,

(ii) perpendicular to the larger force. Hence use Pythagoras' theorem to calculate the magnitude of the resultant.

(b) Verify the last answer by making use of the cosine rule to calculate the magnitude of the resultant.

*(SMP)

7 A toy truck of mass 2 kg is pulled along a level floor by a string inclined at 30° to the horizontal. When it just moves, explain why the tension in the string is not more than 40 N. Hence show that, if it does move, the resistance to motion must be certainly less than 35 N. (Take g to be 10 m s^{-2} in this question.)

*(SMP)

8 A block of weight W stands on a rough level floor, and a force P newtons, acting downwards and making an angle θ with the horizontal, is applied to it. The angle of friction for the floor and the block is λ. Show that, when P is just sufficiently large to move the block,

$$P = \frac{W \sin \lambda}{\cos (\theta + \lambda)}$$

Explain what happens if $\theta + \lambda = \frac{1}{2}\pi$.

9 Two light rings can slide on a rough horizontal rod. The rings are connected by a light inextensible string of length a, to the midpoint of which is attached a weight W. Show that the greatest distance between the two rings consistent with equilibrium is

$$\frac{\mu a}{(1 + \mu^2)^{\frac{1}{2}}}$$

where μ is the coefficient of friction between either ring and the rod.

(O & C)

10 A body of mass m kg is held in equilibrium on a rough plane, inclined at an angle θ to the horizontal, by a force of P N, acting up the line of greatest slope. When $P = 15.6$, the body is about to slide up the plane, and when $P = 8.4$, the body is about to slide down the plane.

Given that $\tan \theta = 0.75$, find the values of m and μ, the coefficient of friction between the body and the plane.

(Take g to be 10 m s^{-2} in this question.)

*(C)

11 A block of mass 1.5 kg is placed on a rough plane inclined at an angle α to the horizontal, where $\sin \alpha = 0.6$. The coefficient of friction between the block and the plane is 0.25.

The block is acted on by a force of P N *perpendicular to the plane* and is on the point of slipping downwards.

Draw a diagram showing all the forces acting on the block and calculate the value of P.

(Take g to be 10 m s^{-2} in this question.)

*(C)

12 With reference to Cartesian axes Ox, Oy, the vertices of a square are A (a,a), B $(-a,a)$, C $(-a,-a)$, D $(a,-a)$. A particle P, whose coordinates are (x,y), is subject to forces acting along the lines joining P to the four vertices.

(a) In the case when the forces are $k\overrightarrow{PA}$, $k\overrightarrow{BP}$, $k\overrightarrow{PC}$, $k\overrightarrow{DP}$, show that P is in equilibrium.

(b) In the case when the forces are $k\overrightarrow{PA}$, $k\overrightarrow{PB}$, $k\overrightarrow{PC}$, $k\overrightarrow{PD}$ show, by using components or otherwise, that the resultant force is $4k\overrightarrow{PO}$.

(JMB)

13 Two particles of masses M and m are attached to the ends A and B respectively of a light inextensible string, which passes over a smooth pulley at the top of a rough plane inclined at an angle α to the horizontal. The masses are at rest with M in contact with the plane, BC vertical and AC parallel to the line of greatest slope of the plane.

Determine the components normal and parallel to the plane, of the force exerted by the plane on the mass M. Deduce that

$$M(\sin \alpha + \mu \cos \alpha) \geqslant m \geqslant M (\sin \alpha - \mu \cos \alpha)$$

where μ is the coefficient of friction between the particle and the plane.

Experiments are conducted with a set of inclined planes made of identical material. Find μ, given that the mass M is on the point of moving down a plane for which $\alpha = 60°$, and on the point of moving up the plane for which $\alpha = 30°$.

(MEI)

Chapter 15

The moment of a force

15.1 Introduction

Most readers will be able to visualise the motion of a seesaw from memories of childhood days, and will be able to recall that it is possible for a small child, sitting a long way from the pivot, to balance a large child sitting near to it. We say that the small child has the same 'turning effect' or **moment** about the pivot as the large one.

It is quite easy to verify this impression experimentally, and we find that, with the seesaw horizontal, the product of the weight and the distance from the pivot, O, is the same in each case. That is, in Figure 15.1, $aw = bW$.

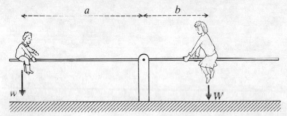

Figure 15.1

This leads us to define the moment, about a point O, of a force in a situation like that in Figure 15.1, as the product of the force and the distance of its line of action from the point O.

However, anyone who has ridden a bicycle will realise that this is not the full story. If we try to ride the cycle by applying our weight vertically to a pedal, we get a greater turning effect with the pedal in the 'three o'clock' position (Figure 15.2a), than we would with the pedal at 'one o'clock' (Figure 15.2b).

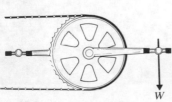

Figure 15.2a

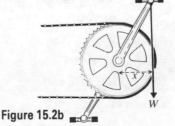

Figure 15.2b

The moment of a force

In order to take account of this, we define the moment, about a point O, of a force, as the product of the force and the *perpendicular* distance from O to its line of action. In Figure 15.2b, the moment of the weight is $W \times x$.

15.2   The moment of a force

The moment of the force, F, in Figure 15.3, is aF. This definition includes the simpler version, in §15.1, as a special case. (The moment of a force is often called its **torque**.) The SI unit for the moment of a force is the newton metre (abbreviation N m).

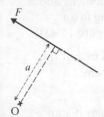

Figure 15.3

Qu. 1 A girl, whose weight is 300 N, sits on a seesaw at a distance of 2 m from the pivot. Find the distance from the pivot where her brother, whose weight is 400 N, should sit in order to balance his sister.

Qu. 2 A spanner is used to turn a nut. The magnitude of the force exerted is 100 N, and the distance from the centre of the nut to the point where the force acts is 40 cm. Find the moment of the force about the centre of the nut, when it is applied

(a) at right angles to the spanner;

(b) along a line making 60° with the spanner.

There is a further complication: the turning effect may be either clockwise or anticlockwise. To allow for this, we follow the normal convention in mathematics: *anticlockwise* rotations, or in the present context, moments, are taken to be *positive* quantities, and *clockwise* ones are taken to be *negative*. The direction of a rotation, clockwise or anticlockwise, is usually called its **sense**.

Example 1

Find the moment about the origin of the following forces. (Take the lengths to be in metres.)

(a) A force of 10 N parallel to the positive y-axis, and acting at the point whose coordinates are $(4, 0)$.

(b) A force of 10 N parallel to the positive y-axis, and acting at $(-4, 0)$.

(c) A force of 10 N parallel to the positive x-axis, acting at $(3, 5)$.

(d) A force represented by the vector $10\mathbf{j}$ newtons acting at the point whose position vector is $(2\mathbf{i} - 3\mathbf{j})$.

(e) A force represented by $(10\mathbf{i} + 10\mathbf{j})$ newtons acting at $(2\mathbf{i} + 2\mathbf{j})$.

(f) A force of 10 N, acting upwards along the line $y = 2x - 6$.

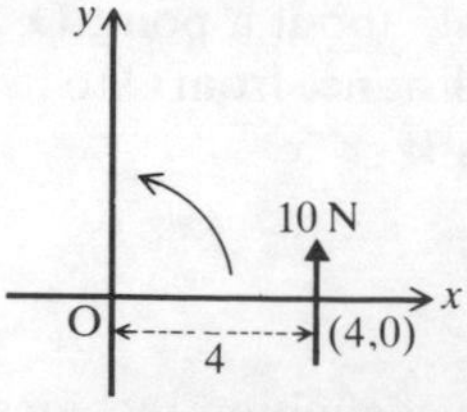

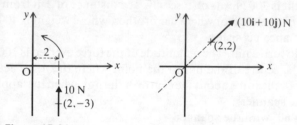

Figure 15.4a 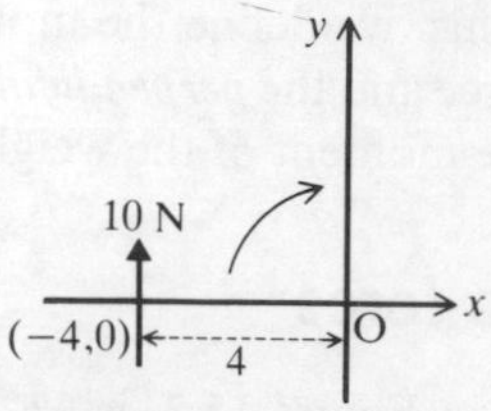   **Figure 15.4b** 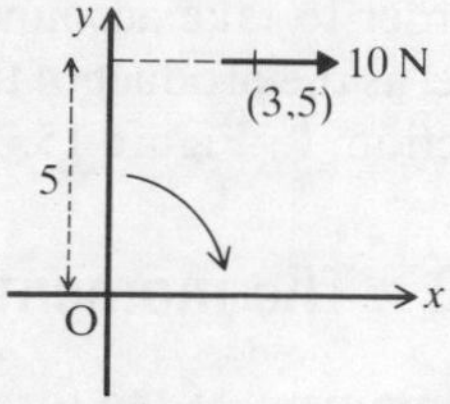   **Figure 15.4c**

(a) The magnitude of the force is 10 N, and the perpendicular distance from O to its line of action is 4 m, so the moment about O of the force has a magnitude of 40 N m. It is acting in an anticlockwise sense, so the moment is + 40 N m (Figure 15.4a).

(b) As in part (a), the magnitude of the moment is 40 N m, but this time the sense is clockwise, so the moment is − 40 N m (Figure 15.4b).

(c) The magnitude of the force is 10 N. The perpendicular distance from O to its line of action is 5 m, and its sense is clockwise, so the moment is − 50 N m (Figure 15.4c).

Figure 15.4d   **Figure 15.4e**

(d) This force also has a magnitude of 10 N, it acts parallel to the positive y-axis, and the perpendicular distance from O to the line of action is 2 m. The sense is anticlockwise, so the moment is + 20 N m (Figure 15.4d).

(e) The direction of this force makes an angle of 45° with the positive x-axis, and it passes through the point (2,2). Consequently its line of action passes through O, so it has no turning effect about O, i.e. its moment is zero (Figure 15.4e).

(f) The moment in this part is $10p$ N m, where p is the perpendicular distance from the origin to the line $y = 2x - 6$, (see Figure 15.4f). The line crosses the x-axis at the point A (3,0), and $p = OA \sin \alpha$, where α is the angle which the line makes with the positive x-axis, i.e.

$$p = 3 \sin \alpha$$

Hence the moment is $30 \sin \alpha$.

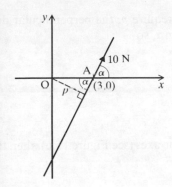

Figure 15.4f

Now $\tan \alpha$ is the gradient of the line $y = 2x - 6$, i.e. $\tan \alpha = 2$, so

$$\text{moment} = 30 \sin (\tan^{-1} 2)\, \text{N m}$$
$$= 26.8\, \text{N m}$$

Part (f) of the above example illustrates a common problem in this topic, namely, finding the perpendicular distance required may be quite difficult. In the next section we shall see that this problem can be bypassed, by resolving the force parallel to the axes.

15.3 The moment of a force, using components

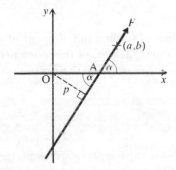

Figure 15.5

Suppose the force F, indicated in Figure 15.5, passes through the point (a,b) and that it makes an angle α with the positive x-axis. Then the equation of its line of action is

$$y - b = \tan \alpha (x - a)$$

and this intersects the x-axis when $y = 0$, i.e. when

$$-b = \tan \alpha (x - a)$$
$$\therefore \quad x - a = -b \cot \alpha$$
$$\therefore \qquad x = a - b \cot \alpha$$

Hence, in the diagram, $OA = a - b \cot \alpha$. We require p, the perpendicular distance from O to the line, and this equals $OA \sin \alpha$, so

$$p = (a - b \cot \alpha) \sin \alpha$$
$$= a \sin \alpha - b \cos \alpha$$

Hence the moment of the force F is

$$(a \sin \alpha - b \cos \alpha)F \qquad \qquad \ldots \ (1)$$

If we resolve F into its components parallel to the axes (see Figure 15.6), then the sum of their moments about O is

$$a(F \sin \alpha) - b(F \cos \alpha)$$

Factorising this gives

$$(a \sin \alpha - b \cos \alpha)F$$

and this is the same as (1) above.

Hence, the moment of F about O, is equal to the sum of the moments of its components.

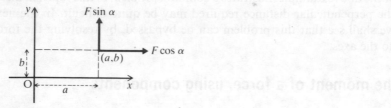

Figure 15.6

This result is very useful since it simplifies the task of finding the moment of a force about a given point. We can use it to prove that the sum of the moments, about a given point, of a pair of forces is equal to the moment of their resultant.

Let the two forces be

$$\mathbf{P} = X_1\mathbf{i} + Y_1\mathbf{j} \text{ and } \mathbf{Q} = X_2\mathbf{i} + Y_2\mathbf{j}$$
then $\mathbf{P} + \mathbf{Q} = (X_1 + X_2)\mathbf{i} + (Y_1 + Y_2)\mathbf{j}$

and suppose that they pass through the point (a,b). Using the previous result, the moment of $\mathbf{P}$ is $(aY_1 - bX_1)$ and the moment of $\mathbf{Q}$ is $(aY_2 - bX_2)$, so the sum of the moments of $\mathbf{P}$ and $\mathbf{Q}$ about the origin

$$= (aY_1 - bX_1) + (aY_2 - bX_2)$$
$$= a(Y_1 + Y_2) - b(X_1 + X_2)$$
$$= \text{the moment about the origin of } (\mathbf{P} + \mathbf{Q})$$

This may be extended to the sum of any number of coplanar forces, i.e.

> The sum of the moments about a given point of a set of coplanar forces is equal to the moment of their resultant.

This is called 'Varignon's Theorem'.

15.4 The line of action of the resultant of a set of coplanar forces

We saw in the last chapter that we can find the magnitude and direction of a set of coplanar forces by resolving them in two perpendicular directions and summing the components.

Qu. 3 Find the magnitude and direction of the resultant of the following set of coplanar forces:

10 N along the x-axis;

20 N through the origin, inclined at 30° to the positive x-axis;

30 N, through (1,0), parallel to the positive y-axis.

We noted earlier, (see §14.4, Qu. 2), that finding the actual line of action of the resultant can be quite difficult; but now that we have developed the techniques for taking moments of forces, we can tackle this problem more easily. The next example illustrates how it can be done.

Example 2

OABC is a square with its vertices at O (0,0), A (2,0), B (2,2) and C (0,2), with the following forces acting along its sides: 10 N along AO, 15 N along AB, 20 N along CB and 5 N along OC, (in the directions indicated by the letters). Find the resultant in the form $X\mathbf{i} + Y\mathbf{j}$, and find the equation of its line of action.

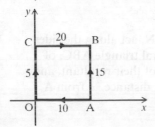

Figure 15.7

The forces can be expressed as $-10\mathbf{i}$, $15\mathbf{j}$, $20\mathbf{i}$ and $5\mathbf{j}$ newtons, respectively, and their resultant, **R** newtons, is given by

$$\mathbf{R} = -10\mathbf{i} + 15\mathbf{j} + 20\mathbf{i} + 5\mathbf{j}$$
$$= 10\mathbf{i} + 20\mathbf{j}$$

This vector is inclined at an angle α to the positive x-axis, where $\tan \alpha = 2$, so it acts along a line whose gradient is 2; such a line has an equation of the form $y = 2x + c$. In this equation, c is the intercept on the y-axis, that is, the line intersects the y-axis at $(0, c)$. Figure 15.8 illustrates this and shows the force $\mathbf{R} = 10\mathbf{i} + 20\mathbf{j}$, acting at $(0,c)$.

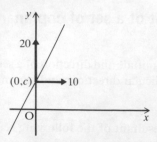

Figure 15.8

The moment of **R** about O is $-10c$. On the other hand, the sum of the moments about O of the separate forces in Figure 15.7 is

$$(15 \times 2 - 20 \times 2) = -10$$

Using Varignon's theorem, we have

$$-10c = -10$$
$$\therefore \qquad c = 1$$

Hence the equation of the line of action of the resultant is $y = 2x + 1$.

Qu. 4 Find the point where the resultant of the forces in Qu. 3 intersects the x-axis.

Example 3

Three forces, of magnitudes 5 N, 10 N and 8 N, act along the sides AB, BC and AC, respectively, of an equilateral triangle ABC, of side a (see Figure 15.9). Find the magnitude of their resultant, and show that its line of action intersects AB at a distance $\frac{5}{9}a$ from A.

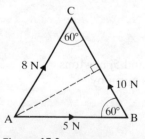

Figure 15.9

Let the component of the resultant along AB be X newtons, then

$$X = 5 - 10 \cos 60° + 8 \cos 60°$$
$$= 4$$

and let the component of the resultant perpendicular to AB be
Y newtons, then

$$Y = 10 \sin 60° + 8 \sin 60°$$
$$= 18 \sin 60°$$
$$= 9\sqrt{3}$$

Let the magnitude of the resultant be R newtons, then using
Pythagoras' theorem,

$$R^2 = X^2 + Y^2$$
$$= 4^2 + (9\sqrt{3})^2$$
$$= 16 + 243$$
$$= 259$$

Hence the magnitude of the resultant is $\sqrt{259}$ newtons,
i.e. 16.1 newtons, correct to three significant figures.

Suppose the line of action of the resultant intersects AB at a distance
x from A. Then the moment of the resultant about A is xY, i.e.
$(9\sqrt{3})x$.

The forces which act along AB and AC have no moment about A;
the moment about A of the force acting along BC is

$$10 \times a \sin 60° = (5\sqrt{3})a$$

Hence

$$(9\sqrt{3})x = (5\sqrt{3})a$$
$$\therefore \quad x = \tfrac{5}{9}a$$

The line of action of the resultant intersects AB at a distance $\tfrac{5}{9}a$
from A.

15.5 The moment of a couple

It is possible for a set of coplanar forces to have a resultant whose magnitude is
zero, but a torque which is not zero. The simplest example of this would be two
forces of equal magnitude, acting along parallel lines in opposite directions, as
illustrated in Figure 15.10.

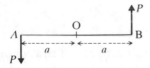

Figure 15.10

A motorist using a wheel brace to undo a wheel nut may well be exerting a couple;
another example would be the pair of forces we might exert to undo the top of a
pot of jam, by gripping the lid at opposite ends of a diameter and rotating it. It is

typical of the action of a couple that it causes the object on which it acts to rotate, without a translation.

If, in Figure 15.10, $AO = OB = a$, i.e. $AB = 2a$, then the moment of the couple about point O is $2aP$. The turning effect of a couple has several interesting properties, one of which is that its moment about *any* point in its plane is the same. If X is the point in AB such that $AX = x$, then in Figure 15.11a, $XB = (2a - x)$, and the moment of the couple about X is

$$(2a - x)P + xP$$
$$= 2aP - xP + xP$$
$$= 2aP$$

The moment of the couple about X equals its moment about O.

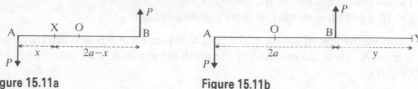

Figure 15.11a   **Figure 15.11b**

If we take moments about a point Y which is not between A and B, as in Figure 15.11b, then the moment of the couple about Y is

$$-yP + (2a + y)P$$

which is also equal to $2aP$.

Another interesting property of a couple is that if it is combined with a force, the force is translated (that is, the force is moved without any change in its magnitude or direction). Suppose a force F acts at right angles to the x-axis and that it passes through the point $(x,0)$. Its moment about O is then xF. If we now introduce a couple of moment T, the magnitude and direction is unaltered (because the resultant of the forces making up the couple is zero.) However the total torque about the origin is now $T + xF$. So the whole system is equivalent to a force of magnitude F parallel to the original force, but acting through a point $(d,0)$, whose x-coordinate, d, we can find by taking moments about O, i.e.

$$dF = T + xF$$

$$d = \frac{(T + xF)}{F}$$

Qu. 5 A force represented by the vector 10**i** acts at the point $(0,5)$. When a couple, whose moment is 30 units is introduced, the resultant is a force represented by 10**i**, passing through the point $(0,c)$. Find c.

In Chapter 14 we saw that the resultant of a set of *concurrent* forces which can be represented in magnitude and direction by the sides of a triangle, taken in order, is zero. That is, such a set of forces would be in equilibrium. If they are not concurrent, the resultant force is still zero, but the torque is not. If the forces actually act along the sides of the triangle, as in Figure 15.12, a couple is formed: the forces represented by $\overrightarrow{CA}$ and $\overrightarrow{AB}$ would have a resultant whose magnitude and direction is represented by $\overrightarrow{CB}$, but this resultant would have a line of action passing through point A. This combined with the force $\overrightarrow{BC}$ produces the couple.

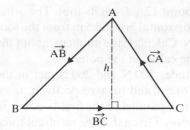

Figure 15.12

The moment of this couple can be found by taking moments about A, i.e.

the moment of the couple $= h \times BC$,

where h is the altitude of the triangle, as shown in Figure 15.12. It is interesting to note that the area of triangle ABC is $\frac{1}{2}h \times BC$, so the moment of the couple is equal in magnitude to twice the area of the triangle. (Note that the moment of the couple about B or C would have the same magnitude. This is not surprising, since we have already shown that the moment of a couple about any point in its plane is the same.)

We normally say that any system of coplanar forces which has a zero resultant, but a non-zero torque can be 'reduced to a couple'.

15.6 The work done by a couple

If the arm AOB in Figure 15.10 rotates about O through an angle θ radians, the points A and B each rotate through an arc of length $a\theta$, so the work done by each of the two forces is $a\theta P$, and hence the total work done is $2a\theta P$. Now the moment of the couple is $2aP$, so we can write

the work done by a couple $= \theta T$

where T is the moment, or torque, of the couple.

Exercise 15a

Take all the forces in this exercise to be in newtons, and all lengths to be in metres.

1 Find the moment of each of the following forces about the origin;
   (a) 10 N parallel to the positive x-axis, through the point $(2, -3)$;
   (b) a force represented by the vector $10\mathbf{j}$, acting at the point with position vector $(2\mathbf{i} + 5\mathbf{j})$;

    (c) a force of 20 N acting upwards along a line whose equation is $4y = 3x + 8$;

    (d) a force represented by the vector $(4\mathbf{i} - 3\mathbf{j})$ passing through the point (2,5);

    (e) a force represented by the vector $\overrightarrow{AB}$, where A is the point (2.5) and B is (8,13).

2 A cable is attached to a vertical pole at a point 12 m from its foot. The other end of the cable is attached to a point on horizontal ground 5 m from the foot of the pole. The tension in the cable is 260 N. Calculate the moment, about the foot of the pole, of the force exerted by the cable on the pole.

3 Two parallel horizontal forces, of magnitudes 300 N and 200 N, act in the same direction on a vertical pole at heights of 5 m and 10 m, respectively, from the foot of the pole. A third horizontal force, parallel to the first two, but in the opposite direction, acts at a height h metres. Given that the resultant force acting on the pole and the resultant torque are both zero, find h.

4 Forces represented by $15\mathbf{j}$, $10\mathbf{j}$, $20\mathbf{j}$ and $5\mathbf{j}$ newtons act at the points (0,0), (10,0), (20,0) and (30,0). Find their resultant and find the point on the x-axis through which it passes.

5 The vertices of a triangle ABC are A (1,1), B (5,1) and C (5,4). Forces represented by the vectors $\overrightarrow{AB}$, $\overrightarrow{BC}$ and $\overrightarrow{CA}$ act along the sides of the triangle Find the resultant torque about O. Verify that this value of the torque can also be found by taking moments about any of the vertices of the triangle.

6 A force represented by the vector $20\mathbf{j}$ newtons acts at the point (5,0). A couple of moment 40 N m is then combined with it. Find the line of action of their resultant.

7 A force represented by the vector $(4\mathbf{i} + 3\mathbf{j})$ newtons passes through the point (0,10). A couple of moment 20 N m is then combined with it. Find the point on the y-axis through which their resultant passes.

8 A force $P\mathbf{j}$ acts at the point $(a,0)$. A couple, which can be regarded as a pair of forces $-P\mathbf{j}$ acting at $(a,0)$ and $+P\mathbf{j}$ acting at $(a + b, 0)$ is now combined with the original force. Describe the effect it has.

9 The vertices of a square OABC are at the points (0,0), (5,0), (5,5) and (0,5) respectively, and the following forces act along the sides:

    (a) 10 N along OA   (b) 20 N along AB

    (c) 10 N along CB   (d) 5 N along CO

Find the magnitude of the resultant and the equation of its line of action.

10 The coordinates of the vertices of a triangle ABC are A (1,1), B (5,1) and C (5,5). Forces of magnitudes 5 N, 10 N and $10\sqrt{2}$ N act along the sides AB, BC and CA, respectively. Find the magnitude of the resultant and the equation of its line of action.

15.7 The centre of gravity of a rigid body

The term **rigid body** is frequently used in this subject to describe an object in which the various parts do not move relative to one another, or at least an object in which the effect of such movement is negligible. We can regard a plank of wood as a rigid body; a pail of water, on the other hand is not, but if the water were frozen into a solid block of ice, we could then regard the pail and its contents as a rigid body.

The centre of gravity of a rigid body is the point where the total weight may be considered to act. We shall examine this concept in more detail in the next chapter. For the moment, the position of the centre of gravity will be given. In the case of a *uniform* rigid body, we assume that it is at the object's geometrical centre. For a uniform rod, this would be at the midpoint; for a uniform circular disc, it would be at the centre of the circle and for a uniform rectangular plate it would be at the intersection of its diagonals.

15.8 Coplanar forces in equilibrium

Suppose we have a set of coplanar forces (that is, forces which are all in the same plane) acting on a rigid body; what conditions are necessary in order to maintain the rigid body in equilibrium? Firstly the resultant force must be zero, otherwise the object will be translated; we saw in the last chapter that if the forces are concurrent then this condition is sufficient. If, however, the forces are not concurrent, it is possible for the resultant force to be zero, but the resultant torque non-zero. Such a set of forces would cause the rigid body to rotate. Consequently, as well as the resultant force being zero, we must be sure that the resultant torque is also zero. In short, the conditions for the rigid body to be in equilibrium are

(a) the resultant force must be zero;
(b) the resultant torque must be zero.

We can ensure that the first condition is satisfied by resolving the forces in two distinct directions, usually at right-angles, and equating the sum of the resolved parts in each direction to zero.

For condition (b) the sum of the moments of the forces about a point must be zero. We have seen that the moment of a couple about *any* point is always the same, so any point may be chosen; it is only a matter of convenience.

The following examples show how these principles are used.

Example 4

A heavy non-uniform plank AB, whose weight is 100 N rests in a horizontal position on vertical supports at A and B. The length of AB is 4 m and the centre of gravity of the plank is 1.5 m from A. Find the force exerted by each support.

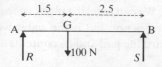

Figure 15.13

Let the forces at A and B be R newtons and S newtons respectively, as shown in Figure 15.13. In this example there are no horizontal forces, so we can ensure the resultant force is zero by resolving vertically, i.e.

$$R + S - 100 = 0$$
$$\therefore \qquad R + S = 100 \qquad\qquad\qquad . . . (1)$$

To ensure that the resultant torque is zero we take moments about A, i.e.

$$4 \times S - 1.5 \times 100 = 0$$
$$\therefore \qquad\qquad 4S = 150$$
$$\therefore \qquad\qquad S = 37.5$$

Substituting this into equation (1) gives

$$R + 37.5 = 100$$
$$\therefore \qquad\qquad R = 62.5$$

Hence the upthrusts at A and B are 62.5 N and 37.5 N, respectively.

Example 5

A uniform ladder, of weight 200 N and length 4 m, rests with its foot on rough horizontal ground. The coefficient of friction between the ladder and the ground is 0.3. The top of the ladder rests against a smooth vertical wall. The plane containing the ladder is perpendicular to the wall. Find the angle the ladder makes with the horizontal when it is on the point of slipping.

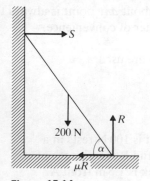

Figure 15.14

Let the normal reaction at the ground be R N, then since the ladder is on the point of slipping, the friction at this point is limiting, i.e. its magnitude is μR and it is directed towards the wall, as shown in Figure 15.14.

Let the normal reaction between the wall and the top of the ladder be S N. The wall is smooth, so there is no friction force at this point.

The resultant force must be zero, and we ensure this condition is satisfied by resolving horizontally and vertically.

Resolving horizontally:

$$S - \mu R = 0 \qquad\qquad\qquad \cdots \quad (1)$$

Resolving vertically:

$$R - 200 = 0$$
$$R = 200$$

Hence from equation (1)

$$S = \mu \times 200$$
$$= 0.3 \times 200$$
$$= 60$$

Now we ensure that the resultant torque is zero by taking moments about the foot of the ladder. (This point is chosen because the force R and μR both pass through it and hence have no moment about it, making the subsequent equation simpler.) Since the ladder is 'uniform', we may assume that the weight, 200 N, acts at its midpoint

Taking moments about the foot of the ladder:

$$200 \times 2 \cos \alpha - S \times 4 \sin \alpha = 0$$

where α is the angle the ladder makes with the horizontal (see Figure 15.14). Hence

$$4S \sin \alpha = 400 \cos \alpha$$

but $S = 60$, so

$$240 \sin \alpha = 400 \cos \alpha$$

$$\therefore \qquad \tan \alpha = \frac{400}{240}$$

$$\therefore \qquad \alpha = 59.0°$$

The ladder makes an angle of 59.0° with the horizontal.

Example 6

A uniform rod AB has weight W and length $2a$. The end A is
attached to a fixed point on a vertical wall by a smooth hinge. The
rod is maintained in a horizontal position by a light inextensible
cable which is attached to point B; the other end of the cable is
attached to a point C on the wall at a distance a vertically above A,
as shown in Figure 15.15a. A load of weight w is suspended from B.
Show that the tension T in the cable is given by

$$T = \frac{\sqrt{5}}{2}(W + 2w)$$

Find also the vertical and horizontal components of the reaction
at A.

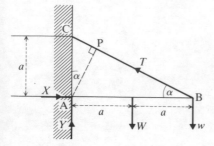

Figure 15.15a   **Figure 15.15b**

(Note that because the hinge is 'smooth' it does not exert a turning
effect on the rod.)

Let the angle which the cable makes with the horizontal be α, then
$\tan \alpha = \frac{1}{2}$. This angle is shown in the triangle in Figure 15.15b and,
by applying Pythagoras' theorem, we can deduce that the length of
the hypotenuse is $\sqrt{5}$, and hence that $\sin \alpha = \frac{1}{\sqrt{5}}$ and $\cos \alpha = \frac{2}{\sqrt{5}}$.
The perpendicular distance from A to the line of action of the cable
(AP in the diagram) is $a \cos \alpha$, which equals $\frac{2a}{\sqrt{5}}$.

Taking moments about A:

$$T \times \frac{2a}{\sqrt{5}} - aW - 2aw = 0$$

$$\therefore \qquad T \times \frac{2a}{\sqrt{5}} = aW + 2aw$$

$$\therefore \qquad T = \frac{\sqrt{5}}{2}(W + 2w)$$

Let the reaction at A have a vertical component Y and a horizontal
component X, as shown in Figure 15.15a.

Resolving vertically:

$$Y + T \sin \alpha - W - w = 0$$
$$\therefore \quad Y = W + w - T \sin \alpha$$

$$= W + w - \frac{\sqrt{5}}{2}(W + 2w) \times \frac{1}{\sqrt{5}}$$

$$= W + w - \tfrac{1}{2}(W + 2w)$$

$$= \tfrac{1}{2}W$$

Resolving horizontally:

$$X - T \cos \alpha = 0$$
$$\therefore \quad X = T \cos \alpha$$

$$= \frac{\sqrt{5}}{2}(W + 2w) \times \frac{2}{\sqrt{5}}$$

$$= W + 2w$$

Hence the vertical component of the reaction at A is $\tfrac{1}{2}W$, and the horizontal component is $(W + 2w)$.

Qu. 6 Find X and Y in Example 6 by taking moments about point C and point B respectively.

Exercise 15b

1 A seesaw consists of a uniform plank, of length 4 m and weight 500 N, pivoted at its midpoint. A child whose weight is 200 N sits at one end and is balanced by her mother, whose weight is 750 N, sitting on the other side. How far from the pivot should the mother sit? What is the upthrust of the pivot on the seesaw?

2 Two children, whose weights are 200 N and 250 N, sit on one side of the seesaw in question 1, at distances from the pivot of 1.8 m and 1.5 m, respectively. They are balanced by their father, who weighs 800 N, sitting on the other side. How far from the pivot is he?

3 A light rod AB, whose length is L, rests on vertical supports at its ends A and B. A load W is suspended from a point P on the rod. Given that $AP = x$, where $x < L$, find the reaction at B in terms of W, x and L.
Find also the reaction at A
   (a) by resolving vertically;
   (b) by taking moments about B.

4 A light rod AB of length 8 m is suspended in a horizontal position from two vertical strings, as shown in Figure 15.16. Weights are suspended from it in positions which are shown in the diagram.

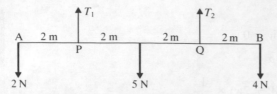

Figure 15.16

Calculate the tensions in the strings.

5 In question 4, the weight 4 N at end B is removed and replaced by a weight W newtons. Find the value of W for which the tension at P becomes zero.

6 Given that the strings in Figure 15.16 break if the tension in them exceeds 20 N, find the greatest weight which could replace the 2 N load at end A.

7 The diagram represents a shopping trolley. The total weight of the trolley and its contents is 200 N. The shopper exerts a vertical force F newtons, as shown in Figure 15.17.

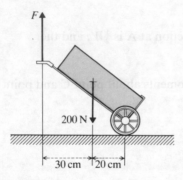

Figure 15.17

Calculate the magnitude of F.

Assuming that the weight is evenly distributed between the two wheels, find the force exerted on the floor by each of them.

8 Figure 15.18 shows a tray, of weight 4 N and length 40 cm, placed overlapping the edge of a table. A cup, whose weight is 2.5 N, is placed with its centre of gravity 5 cm from the unsupported end of the tray.

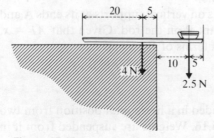

Figure 15.18

Show that the tray will topple over. Find the least weight which should be placed at the far end of the tray to maintain the equilibrium.

9 The diagram in Figure 15.19 illustrates a mobile crane whose weight is 300 kN; all lengths are in metres. Its centre of gravity G is shown.
Find the greatest load, W kN, it can carry without tilting about the front wheels, when $x = 1.5$.

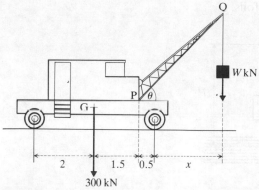

Figure 15.19

10 The jib PQ of the crane in question 9 is 12 m long and it makes an angle θ with the horizontal. Find the value of θ, given that the crane begins to tilt about the front wheels when $W = 100$.

11 In Figure 15.20, a mobile crane is operating from a platform CD, which weighs 200 kN, and which is 12 m long. The platform is supported by two vertical piles placed at A and B, 1.5 m from its ends, as shown.
The crane's weight is 400 kN and its centre of gravity is at G; it carries a load of 100 kN. Show that the crane does not tilt about its front wheels. Find the upthrust exerted by the piles.

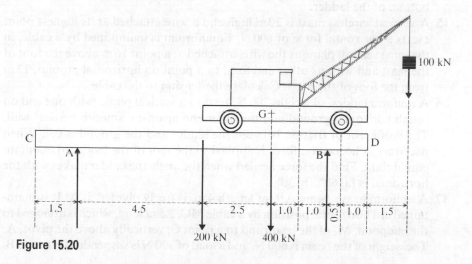

Figure 15.20

12 A rectangular door is 2 m high and 1 m wide, and its weight, which is uniformly distributed, is 200 N. It turns on two hinges fixed to an edge, with the upper one 20 cm from the top of the door and the lower one 20 cm from the bottom. Given that the reaction at the upper hinge is horizontal, find its magnitude. Find also the magnitude and direction of the reaction at the lower hinge.

13 The diagram (Figure 15.21) represents a boat, of weight W, travelling at constant velocity on hydrofoils.

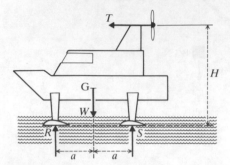

Figure 15.21

There are two forward hydrofoils and two more towards the stern. The centre of gravity of the boat is at G. The boat is driven forward, at constant speed, by a propellor which exerts a thrust T. Given that the lift exerted by the foils is vertical, calculate the magnitude of the lift, R, provided by each of the forward foils in terms of W, T and the dimensions shown in the diagram. Find also the lift, S, exerted by each of the stern foils.

14 A uniform ladder, of weight 200 N and length 10 m, rests in a vertical plane with its upper end against a smooth vertical wall and its lower end on rough horizontal ground. Given that the ladder makes an angle of 60° with the horizontal, calculate the magnitude of the normal reactions at the top and bottom of the ladder.

15 A vertical wireless mast is 20 m high and a wire attached at its highest point exerts a horizontal force of 600 N. Equilibrium is maintained by a cable, in the same vertical plane as the wire, attached at a point 16 m above the foot of the mast and with its other end fixed to a point on horizontal ground, 12 m from the foot of the mast. Calculate the tension in the cable.

16 A uniform ladder, of weight 200 N, rests in a vertical plane, with one end on rough horizontal ground and the other end against a smooth vertical wall. The coefficient of friction between the ladder and the ground is 0.4. When necessary, a horizontal force is applied to the foot of the ladder to maintain equilibrium. Find the force needed when the angle the ladder makes with the horizontal is (a) 60°, (b) 30°.

17 A uniform heavy beam AB, of length 4 m, is freely pivoted at A. It is maintained in a horizontal position by a cable MC, 2.5 m long, which is attached to the midpoint, M, of the beam and to a point C, vertically above the pivot, A. The weight of the beam is 600 N, and a load of 300 N is suspended from end B.

Find the tension in the cable. Find also the magnitude and direction of the force acting on the beam at A.

18 A uniform ladder, of weight w and length $2a$, rests in a vertical plane with one end on rough horizontal ground and the other end against a smooth vertical wall. The coefficient of friction between the ladder and the ground is $\frac{5}{12}$, and the ladder makes an angle $\tan^{-1} 2$ with the horizontal. A man of weight W begins to climb the ladder; when he has climbed a distance x, it begins to slip. Show that

$$x = \frac{a(2w + 5W)}{3W}$$

Hence show that the man can reach the top of the ladder provided $W \leqslant 2w$.

19 A uniform ladder, of weight W, rests in a vertical plane with one end on rough horizontal ground and the other end against a vertical wall. The coefficient of friction between the ladder and the ground is μ; between the ladder and the wall it is μ'. The ladder makes an angle α with the horizontal. Using the methods of this chapter show that, when the ladder is on the point of slipping,

$$\tan \alpha = \frac{1 - \mu\mu'}{2\mu}$$

[The reader should compare this solution with that obtained for Exercise 14d (Miscellaneous), questions 14, (see p. 290) where the method of taking moments was not used.]

20 A uniform rod, of length $2a$, rests in a fixed smooth hemispherical bowl, of radius a, whose rim is in a horizontal plane. The rod makes an angle α with the horizontal. One end of the rod is in contact with the inner surface of the bowl, and its upper end projects beyond the bowl's rim. Find α, correct to one decimal place.

15.9 Jointed rods

When applying the methods of this chapter to jointed rods, and other similar composite bodies, it is important to remember that, at a point of contact, Newton's third law tells us that the action and reaction are equal in magnitude, and opposite in direction. Readers are advised to draw several diagrams when tackling such problems. The following two examples show how to proceed.

Example 7

A uniform rod AB, of weight W and length $2a$, is freely hinged to a fixed point at A. It is freely hinged at B to an identical rod BC, and a horizontal force P acts at C. In equilibrium AB and BC make angles α and β respectively with the downward vertical. Find, in terms of P and W, the horizontal and vertical components of the force acting on rod AB at A. Find also the tangents of α and β in terms of P and W.

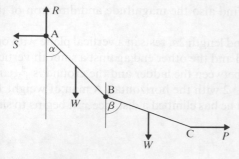

Figure 15.22a

The diagram (Figure 15.22a) shows the rods and the vertical and horizontal components, R and S respectively, of the reaction at the hinge A.

CONSIDER THE WHOLE SYSTEM

Resolve vertically:

$$R - 2W = 0$$
$$\therefore \quad R = 2W \qquad\qquad \cdots \quad (1)$$

Resolve horizontally:

$$P - S = 0$$
$$\therefore \quad S = P \qquad\qquad \cdots \quad (2)$$

Hence the horizontal component of the reaction at A is P, and the vertical component is $2W$.

Figures 15.22b and 15.22c show the forces acting on AB and BC respectively.

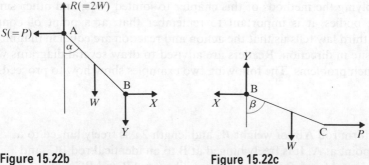

Figure 15.22b   **Figure 15.22c**

(Note that the horizontal and vertical forces, X and Y, acting on AB at B, are equal and opposite to those acting on BC at this point, as required by Newton's third law of motion.)

CONSIDER THE ROD AB

Resolving horizontally:

$X = S$

(and hence, from equation (2), $X = S = P$)

Resolving vertically:

$R = W + Y$

But, from equation (1), $R = 2W$, so

$2W = W + Y$

and hence

$Y = W$

Taking moments about A (using $X = P$, and $Y = W$):

$Wa \sin \alpha + W \times 2a \sin \alpha = P \times 2a \cos \alpha$

$\therefore \quad 3W \sin \alpha = 2P \cos \alpha$

Hence

$$\tan \alpha = \frac{2P}{3W}$$

CONSIDER THE ROD BC

(Resolving horizontally and vertically gives $X = P$ and $Y = W$, which we already know.)

Take moments about B:

$Wa \sin \beta = P \times 2a \cos \beta$

$\therefore \quad W \sin \beta = 2P \cos \beta$

$\therefore \quad \tan \beta = \dfrac{2P}{W}$

Example 8

A pair of steps can be regarded as a pair of uniform rods, AB and BC, each of weight W and length $2a$, freely jointed at B. They stand in a vertical plane, with their feet, A and C, on a smooth horizontal floor and joined by a light inextensible string of length $4b$. A man of weight $3W$ climbs half way up side AB. Find, in terms of W, the reactions at A and C, and the horizontal and vertical components of the reaction at B.

Find also the tension T in the string in terms of α, the angle which AB makes with the vertical.

CONSIDER THE WHOLE SYSTEM

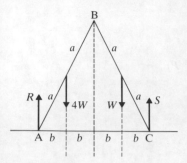

Figure 15.23a

Taking moments about A:

$$(4b \times S) - (b \times 4W) - (3b \times W) = 0$$

$$\therefore \quad 4bS = 7bW$$
$$\therefore \quad\quad S = \tfrac{7}{4}W$$

Resolving vertically:

$$R + S - 5W = 0$$
$$\therefore \quad\quad\quad R = 5W - S$$
$$= 5W - \tfrac{7}{4}W$$
$$= \tfrac{13}{4}W$$

The reactions at A and C are $\tfrac{13}{4}W$ and $\tfrac{7}{4}W$, respectively.

CONSIDER AB

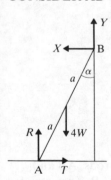

Figure 15.23b

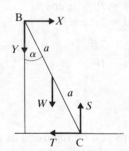

Figure 15.23c

(By Newton's third law, the forces X and Y, acting on AB at B, are equal and opposite to those acting on BC.)

Resolving vertically for AB (see Figure 15.23b):

$$Y + \tfrac{13}{4}W = 4W$$
$$\therefore \quad\quad\quad Y = \tfrac{3}{4}W$$

Resolving horizontally for BC (see Figure 15.23c):

$$T - X = 0$$
$$\therefore \qquad T = X$$

(The reader should note that these results could also be found by considering rod BC.)

Take moments about B for AB (see Figure 15.23b):

$$4W \times a\sin\alpha + T \times 2a\cos\alpha - \tfrac{13}{4}W \times 2a\sin\alpha = 0$$
$$\therefore \quad 2T\cos\alpha = (\tfrac{13}{2}W - 4W)\sin\alpha$$
$$\therefore \quad 2T\cos\alpha = \tfrac{5}{2}W\sin\alpha$$
$$\therefore \qquad T = \tfrac{5}{4}W\tan\alpha$$

The tension in the string is $\tfrac{5}{4}W\tan\alpha$.

Qu. 7 Find T in Example 8 by considering BC.

Exercise 15c (Miscellaneous)

1 A uniform plank AB, of length 4 m and weight 70 N, rests horizontally on supports at P and Q, where $PQ = 2$ m and P lies between A and Q. A block, of weight 30 N, hangs from the plank at A. Given that the forces on the supports at P and Q are equal in magnitude, find the distance AP.

*(L)

2 A uniform straight plank AB, of mass 12 kg and length 2 m, rests horizontally on two supports, one at C and the other at D, where $AC = CD = 0.6$ m. A particle P of mass X kg is hung from B and the plank is on the point of tilting.
(a) Find the value of X.

The particle is removed from B and hung from A.
(b) Find, in newtons, the magnitude of the force exerted on the plank at each support. (Take g to be $10\,\text{m s}^{-2}$.)

*(L)

3 The rectangle ABCD has $AB = 3a$ and $AD = 2a$. Forces of magnitudes P, Q, R and 10 newtons act along $\overrightarrow{BC}$, $\overrightarrow{BA}$, $\overrightarrow{DA}$ and $\overrightarrow{DC}$ respectively. Given that the forces are in equilibrium, find the values of P, Q and R.

*(L)

4 The centre of gravity of a non-uniform rod AB, of length $3a$ and weight $2W$, is at the point G, where $AG = a$. The rod is placed in limiting equilibrium with the end A on rough horizontal ground and the end B against a smooth vertical wall. The vertical plane containing AB is perpendicular to the wall. The coefficient of friction between the rod and the ground is $\tfrac{1}{2}$.
(a) Sketch a diagram showing all the forces acting on the rod.
(b) Calculate, to the nearest degree, the angle made by AB with the horizontal.

Find, in terms of W, the magnitude of the greatest horizontal force, acting directly towards the wall, which could be applied to the rod at A, without disturbing the equilibrium.

*(L)

5 The end A of a uniform rod AB, shown in Figure 15.24, is smoothly hinged to a fixed point.

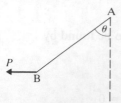

Figure 15.24

The rod has length 2 m and weight 80 N. The rod is maintained in equilibrium at an angle θ to the downward vertical by a horizontal force P, of magnitude 50 N, acting at B in the vertical plane containing AB.
(a) Find an expression, in terms of θ, for the magnitude of the moment, in N m, of P about A.
(b) Show that $\tan \theta = \frac{5}{4}$.

Given that the magnitude of the force exerted by the hinge on the rod is Q newtons and acts at an angle α to the upward vertical, calculate
(c) the value of Q to three significant figures;
(d) the value of α to the nearest degree.

*(L)

6 Forces of magnitude 7, 5, P and Q newtons act along the sides $\overrightarrow{AB}$, $\overrightarrow{BC}$, $\overrightarrow{CD}$, $\overrightarrow{DA}$ respectively of a square ABCD of side 3 m.
(a) Given that this system of forces reduces to a couple, find P, Q and the moment of this couple.
(b) Given instead that $P = 5$ and $Q = 7$,
(i) show that the system of forces reduces to a single force R parallel to DB, and calculate the magnitude of R;
(ii) show that R passes through the point E on BA produced, where $AE = 15$ m.

*(L)

7 A uniform ladder, of length 4 m and mass 60 kg, stands on rough horizontal ground. It is leaning against a smooth vertical wall and makes an angle θ with the ground, where $\tan \theta = 1.5$.

A man of mass 80 kg climbs up the ladder. When he reaches a point x m from the lowest point of the ladder (measured along the ladder), the ladder slips. Given that the coefficient of friction between the ladder and the ground is 0.4, calculate the value of x.

*(C)

8 A uniform rod AGB, of weight w newtons, rests horizontally on two supports C and D. $AG = GB = 6$ cm. The support C can be placed in any position from A to G, and the support D can be placed in any position from G to B. The reactions at the supports C and D are P newtons and Q newtons, respectively.

Denoting AC by x cm and DB by y cm, express P and Q in terms of x, y and w.

Given that $P = 2Q$,
(a) show that $2x - y = 6$;
(b) find the value of x when the support D is placed at B.

*(C)

9 Figure 15.25 shows four forces acting along the sides of a square ABCD.

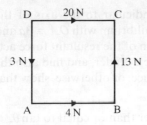

Figure 15.25

(a) Calculate the magnitude of their resultant and the angle which this result-
     ant makes with the direction of AB.
(b) Prove that the resultant passes through the centre of the square.

*(C)

10 Figure 15.26 shows a uniform bar AD of weight 30 N and length 1 m.

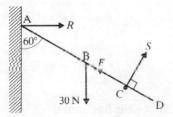

Figure 15.26

It rests on a rough peg at C, where $CD = 20$ cm and at an angle of 60° to the
vertical, against a smooth vertical wall at A. The figure also shows the forces,
in newtons, acting on the bar. Calculate
(a) the values of R, S and F;
(b) the least possible value of the coefficient of friction between the bar and
     the peg for equilibrium to be possible in this position.

*(C)

11 One end A of a uniform straight rod AB, of weight W and length $4a$, rests on
a smooth horizontal plane. The rod also rests on a rough semicircular
cylinder whose plane face is fixed to the horizontal plane, as shown in Figure
15.27 overleaf.

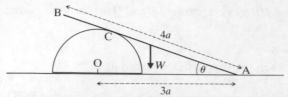

Figure 15.27

The vertical plane containing the rod is perpendicular to the axis of the cylinder and meets this axis at O. The rod is in equilibrium with $OA = 3a$ and angle $OAB = \theta$. Find the magnitude and direction of the resultant force acting on the rod at the point of contact, C, with the cylinder, and find also the magnitude and direction of the reaction at A. Hence, or otherwise, show that $\cos \theta \geqslant \frac{2}{3}$.

Show that the coefficient of friction at C is greater than or equal to $\tan \theta$.

(C)

12 A uniform rod AB, of weight $2W$ and length $2a$, is freely hinged to a fixed point at A and is freely hinged at B to a uniform rod BC of weight W and length $2b$. A horizontal force acts at C, as shown in Figure 15.28.

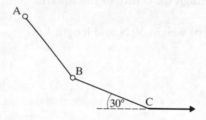

Figure 15.28

In equilibrium, BC is inclined at an angle of 30° to the horizontal.

Find the magnitude of the force applied at C.

Find also
(a) the magnitude of the force on AB at A,
(b) the inclination to the horizontal of the force on AB at A, to the nearest 0.1 of a degree.

Prove that the inclination of AB to the horizontal is θ, where $\tan \theta = \frac{4}{3}\sqrt{3}$.

(C)

13 Figure 15.29 shows a uniform rectangular plate ABCD, of weight W, resting on horizontal ground at A and against a vertical wall at D.

The lengths of AB and AD are a and $2a$ respectively. The coefficient of friction at A is $\frac{1}{2}$. The plate is in equilibrium in a vertical plane perpendicular to the wall, with AD inclined at an angle θ to the horizontal.

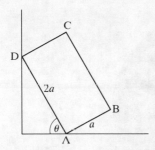

Figure 15.29

(a) Given that the contact at D is smooth and that equilibrium is limiting, show that $\tan \theta = \frac{2}{3}$.

(b) Given that the coefficient of friction at D is $\frac{1}{3}$ and that the equilibrium is limiting, find the value of $\tan \theta$.

(C)

14 A uniform rod AB of weight W and length $2a$ has the end A freely hinged to a fixed point. The other end B is freely hinged to a uniform rod BC of weight W and length $6a$. The ends A and C are joined by a light inelastic string whose length is such that AB is perpendicular to BC, and the system hangs in equilibrium. Show that AB is inclined at 45° to the vertical.

Find the tension in AC and the horizontal and vertical components of the reaction on BC at B.

(C)

15 A uniform ladder, of length 7 m, rests against a vertical wall with which it makes an angle of 45°. The coefficients of friction between the ladder and the wall and the horizontal ground are $\frac{1}{3}$ and $\frac{1}{2}$ respectively. A boy whose weight is one half that of the ladder slowly ascends the ladder. How far will he be from the foot of the ladder when it slips?

(O & C)

Centre of gravity

Chapter 16

Centre of gravity

16.1 Finding the centre of gravity of a rigid body experimentally

We saw in the last chapter that it is frequently necessary to know the position of the centre of gravity of a rigid body. In that chapter, the rigid bodies were either uniform, (that is their weight was uniformly distributed and consequently the centre of gravity was at the geometrical centre), or the position of the centre of gravity was given. In this chapter we shall consider methods for finding the position of the centre of gravity of a rigid body.

Remember that the position of the centre of gravity of a rigid body is the point where we may consider the total weight to act. This point is not necessarily a point within the substance of the body, for instance, the centre of gravity of a uniform circular ring is at its centre, i.e. in the middle of the hole!

If a rigid body rests in equilibrium when it is freely suspended from a fixed point, then there are just two forces acting on it: the force supporting it at the point of suspension, and its own weight acting vertically downwards at the centre of gravity. As we saw in Chapter 14, when *two* forces are in equilibrium they have equal magnitudes and they act along the same straight line, but in opposite directions. Hence the force at the point of suspension is equal in magnitude to the weight of the rigid body and it acts vertically, so the centre of gravity lies vertically below the point of suspension. This gives us a very simple method for determining the position of the centre of gravity of a rigid body by suspending it from a fixed point: if we draw a vertical line through the point of suspension, this line will also pass through the centre of gravity. If we now repeat this operation for a different point of suspension, (not on the line already drawn), the centre of gravity will lie at the point where the two lines intersect. (Plainly it would be a sensible precaution to use more than two lines.)

Exercise 16a

Cut out the following shapes (next page) on stiff card and find the centre of gravity of them by the method described above.

1 Great Britain (i.e. England, Scotland and Wales); name the nearest large town to the centre of gravity.
2 England; name the nearest large town to the centre of gravity.
3 A right-angled triangle ABC, with sides $AB = 9$ cm, $BC = 12$ cm and $CA = 15$ cm; give the distances of the centre of gravity from sides AB and BC.
4 An isosceles triangle ABC, with $AB = BC = 15$ cm and angle ABC = 120°; give the distance of the centre of gravity from the vertex B.
5 From a square ABCD, centre O, with each side 30 cm long, cut out a square AMON, where M and N are the midpoints of AB and AD respectively; give the length of OG, where G is the centre of gravity of the L-shaped figure OMBCDN.

16.2 Finding the centre of gravity by calculation

Finding the position of the centre of gravity of a rigid body experimentally may be sufficient for some purposes, but an engineer is more likely to require the precision which can only be achieved by calculation. We shall now consider how this can be done, starting with a very simple situation, namely by considering a number of small weights attached to a light rigid rod. Although this may not be a very realistic situation, the calculation is very simple because the object is, in effect, one-dimensional. When we have done this we shall proceed to two dimensional shapes and then we shall be ready to consider solid objects.

Suppose we have a light rigid rod OX, of length 4 m, with weights 2 N, 4 N, 3 N and 1 N suspended from points A, B, C and X respectively, where

$$OA = AB = BC = CX = 1 \text{ m} \quad \text{(see Figure 16.1)}$$

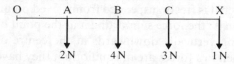

Figure 16.1

The centre of gravity, G, is the point where we may consider the total weight to act. As we saw in the last chapter, for a single force to be equivalent to a set of individual forces, it must have the same resultant (in this case, the total weight, 10 N, acting vertically downwards) and it must have the same turning effect. So we can find the position of G by taking moments about a suitable point, say O. It is the normal practice in this work to denote the distance of the centre of gravity from the origin by $\bar{x}$ (read 'x bar'), so, in this case, we write $OG = \bar{x}$.
Taking moments about O:

$$10\bar{x} = (2 \times 1) + (4 \times 2) + (3 \times 3) + (1 \times 4)$$
$$= 23$$
$$\therefore \quad \bar{x} = 2.3$$

The centre of gravity G is 2.3 metres from O.

The rod, with the weights attached, could be supported in equilibrium by a force of 10 N, acting vertically upwards at G.

Qu. 1 Find the weight which should be attached to point O of the rod OX (see figure 16.1), so that the centre of gravity of the five weights is at point B.

We can generalise this procedure as follows: suppose weights of magnitude $w_1, w_2, w_3, \ldots, w_n$, are attached to the rod OX, at points whose distances from O are $x_1, x_2, x_3, \ldots, x_n$, respectively, the position of the centre of gravity, G, is found, as above, by taking moments about O, i.e.

$$(w_1 + w_2 + w_3 + \ldots + w_n)\bar{x} = (w_1 x_1 + w_2 x_2 + w_3 x_3 + \ldots + w_n x_n)$$

$$\bar{x} = \frac{(w_1 x_1 + w_2 x_2 + w_3 x_3 + \ldots + w_n x_n)}{(w_1 + w_2 + w_3 + \ldots + w_n)}$$

Using the **sigma notation**† this can be written:

$$\bar{x} = \frac{\Sigma (w_i x_i)}{\Sigma w_i} \qquad \ldots \ (1)$$

16.3 Centre of gravity and centre of mass

We could replace each w_i in (1), by $m_i g$, where m_i is the mass of the ith weight. The factor g would appear in every term in the numerator, and every term in the denominator of equation (1), and consequently it could be cancelled giving

$$\bar{x} = \frac{\Sigma (m_i x_i)}{\Sigma m_i} \qquad \ldots \ (2)$$

Strictly speaking, we say that (1) determines the centre of *gravity* and (2) determines the centre of *mass*. For all practical purposes these are the same, but in theory it is possible to imagine a situation with a very large rigid body in which the value of g is not constant; in these circumstances we would not be able to cancel it in (1), and consequently the value of $\bar{x}$ given by (1) would be different from that given by (2).

In this context, the reader will also encounter the term **centroid**. This is the geometrical centre of the rigid body; for example, for a rectangle it is the point where the diagonals meet, and for a triangle it is the intersection of the medians. For a body whose weight is uniformly distributed, the centroid will also be the centre of gravity, but for an object whose density (mass per unit volume) is not constant, they could be in different places.

In this chapter we shall only consider uniform objects, and we shall regard g as constant. Consequently the centre of gravity, the centre of mass and the centroid will all be located at the same point.

† See *Pure Mathematics Book 1*, by J. K. Backhouse and S. P. T. Houldsworth, revised by P. J. F. Horril, published by Longman, Chapter 13.

16.4 Centre of gravity in two dimensions

We shall now extend the techniques we met in the previous section to cover two-dimensional objects.

We shall frequently meet the word **lamina** in this context: it comes from Latin, where it means 'a flat sheet'. We meet the same concept in everyday speech when we refer to 'laminated wood', which is a sheet of wood formed by bonding together several thin layers of wood, and a similar idea is involved when we say a car has a 'laminated windscreen'. We use it in mathematics to describe a shape, with a continuous closed boundary, drawn on a two-dimensional plane of negligible thickness. (The reader should think of a shape like a map of an island, drawn on a thin sheet of paper.)

In our earlier discussion, we examined the case of a number of small weights attached to a thin light rod; the purpose of the rod was simply to keep the weights in place, and we neglected its own weight. In this section, we shall use a lamina for a similar purpose, namely to keep a number of small weights rigidly in position. In a later section we shall consider the effect of the weight of the lamina itself.

Imagine a lamina, lying in a horizontal plane, with Cartesian axes drawn on it, and suppose the following small weights are fixed to it:

2 N at A (1,5)   4 N at B (2,7)   3 N at C (3,2)   and   1 N at D (4,3)

These are shown in Figure 16.2.

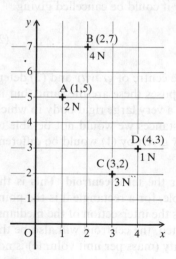

Figure 16.2

Suppose the centre of gravity is at G $(\bar{x}, \bar{y})$. To find these coordinates we shall take moments about the axes (both of which are horizontal). When we take moments about the y-axis, the perpendicular distance to a particular weight is given by its x-coordinate.

Taking moments about the y-axis:

$$10\bar{x} = (2 \times 1) + (4 \times 2) + (3 \times 3) + (1 \times 4)$$
$$= 23$$
$$\therefore \qquad \bar{x} = 2.3$$

When we take moments about the x-axis, the perpendicular distance to a particular weight is given by its y-coordinate.

Taking moments about the x-axis:

$$10\bar{y} = (2 \times 5) + (4 \times 7) + (3 \times 2) + (1 \times 3)$$
$$= 47$$
$$\therefore \qquad \bar{y} = 4.7$$

Hence the coordinates of G, the centre of gravity, are (2.3, 4.7).

As in §16.2, we may generalise this result for a set of n particles. If a typical particle has w_i and position (x_i,y_i), then the position of the centre of gravity is $(\bar{x},\bar{y})$, where

$$\bar{x} = \frac{(w_1x_1 + w_2x_2 + w_3x_3 + \ldots + w_nx_n)}{(w_1 + w_2 + w_3 + \ldots + w_n)}$$

and

$$\bar{y} = \frac{(w_1y_1 + w_2y_2 + w_3y_3 + \ldots + w_ny_n)}{(w_1 + w_2 + w_3 + \ldots + w_n)}$$

If we are using vectors, and the particle w_i has position vector

$$\mathbf{r}_i = x_i\mathbf{i} + y_i\mathbf{j}$$

then the position vector of the centre of gravity, $\bar{\mathbf{r}}$, is given by

$$\bar{\mathbf{r}} = \frac{(w_1\mathbf{r}_1 + w_2\mathbf{r}_2 + w_3\mathbf{r}_3 + \ldots + w_n\mathbf{r}_n)}{(w_1 + w_2 + w_3 + \ldots + w_n)}$$

Exercise 16b

1 A light rigid rod AB, is 1 m long. It has small particles, whose masses are 10 g, 20 g, 40 g and 30 g, attached to it at points A, X, Y and B respectively, where $AX = YB = 20$ cm. Find the distance of the centre of mass of the particles from end A.

2 Four particles, whose weights are 5 N, 7 N, 6 N and 2 N, replace those attached to the points A, X, Y and B, respectively, of the rod in question 1. Given that the rod can be supported by a single vertical force acting at a point P of the rod, find the length AP.

3 Four particles, whose masses are m, $2m$, $3m$ and $4m$, are fixed at points O (0,0), A ($2a$,0), B ($2a$,a) and C (0,a), respectively, in a Cartesian plane. Find the coordinates of G, their centre of mass.

4 Three particles, whose masses are m, m and M, are fixed to points O $(0,0)$, A $(2a,0)$ and B $(a,4h)$, respectively. Given that their centre of mass is at the point (a,h), show that $2m = 3M$.

5 Identical particles, each of mass m, are attached to points O $(0,0)$, A $(a,0)$ and B (c,h). Find the coordinates of G, their centre of mass. Show that G lies on the line OM, where M is the midpoint of AB, and that $OG = \frac{2}{3}OM$. Show also that $AG = \frac{2}{3}AN$, where N is the midpoint of OB.

6 Identical particles, each of mass m, are fixed to points A, B, C and D, whose position vectors are

$$\mathbf{a} = 2\mathbf{i} + 3\mathbf{j} \qquad \mathbf{b} = 5\mathbf{i} + 2\mathbf{j} \qquad \mathbf{c} = 4\mathbf{i} + 7\mathbf{j} \qquad \mathbf{d} = 5\mathbf{i}$$

Find the position vector of their centre of mass.

7 Find the position vector of the centre of mass, when particles of mass, m, $2m$, $3m$ and $4m$, respectively, replace those attached to points A, B, C and D in question 6.

8 Identical particles, each of mass m, are attached to points A, B and C, whose position vectors are $\mathbf{a}$, $\mathbf{b}$ and $\mathbf{c}$. Find the position vector of G, their centre of mass in terms of $\mathbf{a}$, $\mathbf{b}$ and $\mathbf{c}$. Show that $\overrightarrow{AG} = \frac{2}{3}\overrightarrow{AM}$, where M is the midpoint of BC. (AM is a *median* of the triangle ABC.) Hence show that G is the point of intersection of the three medians of the triangle.

16.5 The volume of a solid of revolution

So far, in this chapter, we have been concerned with finding the centre of mass of a set of particles. Now we shall turn our attention to solid objects. Finding the centre of mass of an irregular solid could be very difficult, and an experimental method might be the only possible approach; however, we shall restrict ourselves to a special type of solid, namely a **solid of revolution**. This is a solid which is formed by rotating a region bounded by a known curve (called the **generator**) through 360° about a fixed axis. The axis will usually be the x-axis, but other lines can be used. For example, a cone is a solid of revolution which can be formed by rotating a line which is inclined to the x-axis, through 360°. If the equation of the line is written $y = rx/h$, and $0 \leqslant x \leqslant h$, we obtain a cone with its axis of symmetry along the x-axis, of base radius r and height h.

It is assumed that the reader has already met this topic in pure mathematics, and will already have found the following results:

Name	Dimensions	Generator	Volume
Cone	base radius r, height h	$y = rx/h$	$\frac{1}{3}\pi r^2 h$
Hemisphere	radius r (and $x \geqslant 0$)	$y = \sqrt{(r^2 - x^2)}$	$\frac{2}{3}\pi r^3$
Cylinder	radius r, length l	$y = r$	$\pi r^2 l$
Paraboloid	base $x = b$	$y = \sqrt{(4ax)}$	$2\pi ab^2$

(Notice that it is characteristic of these formulae for volumes that they comprise a numerical factor π, and three lengths, multiplied together.)

Obtaining these formulae requires the fundamental theorem:

$$\lim_{\delta x \to 0} \sum_{x=a}^{x=b} f(x)\delta x = \int_a^b f(x)\,dx$$

The proof of the formula for the volume of a paraboloid is given below, as a reminder.

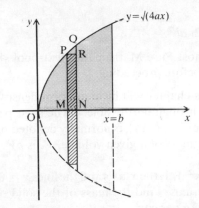

Figure 16.3

This solid is generated by rotating the region bounded by the upper half of the curve $y^2 = 4ax$, [i.e. $y = \sqrt{(4ax)}$], and the lines $y = 0$ and $x = b$, through 360°, about the x axis. Its cross section perpendicular to this axis is circular. In Figure 16.3 let $OM = x$ and $MP = y$; note that MP is the radius of this circular cross-section at this point, so the area of the cross-section is πy^2. Let $MN = \delta x$, where δx is very small compared with x, then the cylinder with base radius MP and thickness δx, formed by rotating the rectangle MPRN about the x-axis, can be regarded as a typical element of the volume. (At this stage in the argument, the reader should regard this element of volume as a very thin cylinder, such as one could form by cutting out a circle from a sheet of paper.)

The volume of this typical element is written δV and

$$\delta V = \pi y^2 \delta x$$

The total volume of the paraboloid, V, is found by summing the volumes of all such elements, for $x = 0$ to $x = b$, and letting $\delta x \to 0$, i.e.

$$V = \lim_{\delta x \to 0} \sum_{x=0}^{x=b} (\pi y^2)\delta x$$

This sum is evaluated by finding the corresponding definite integral, i.e.

$$V = \int_0^b (\pi y^2)\,dx$$

So far this argument could be applied to *any* solid of revolution; for this particular solid $y^2 = 4ax$, so

$$V = \int_0^b (\pi 4ax)\,dx$$

$$= \left[2\pi ax^2 \right]_0^b$$

$$= 2\pi ab^2$$

Therefore the volume of the paraboloid is $2\pi ab^2$.

This process is covered in more detail in most Pure Mathematics text-books†; readers who need to revise it should do so before proceeding.

The objects we shall be dealing with in this chapter will have uniform density, that is, the mass will be evenly distributed throughout the object. The mass per unit of volume, for example grams per cm^3 or kg per m^3, is normally denoted by the Greek letter ρ (pronounced 'ro'), so the mass of a given volume, V, is ρV.

Qu. 2 The region bounded by the curve $y = x^2 + 1$, the x-axis and the lines $x = 0$, $x = 1$, is rotated through 360° about the x-axis. Find the mass of the solid of revolution formed, given that its density is 3 g per cm^3.

16.6 The centre of mass of a solid of revolution

We now consider how we can adapt the formula for finding centre of mass, which we first met in §16.3, namely

$$\left(\sum m_i \right)\bar{x} = \sum (m_i x_i) \qquad \qquad . \ . \ . \ (1)$$

The term $\sum m_i$, on the left-hand side of equation (1), is the total mass, so this will be found by multiplying the total volume by ρ, the density.

The solid of revolution is divided into cylindrical elements, and the volume of a typical element can then be found in the usual way, and multiplied by ρ to give the mass of the element; this will take the place of the m_i on the right-hand side of (1). The term x_i will be the distance of the centre of the element from O, the origin. We then sum all such products, and find the limit of this sum as δx, the thickness of the typical element, tends to zero; this will require the formation of a definite integral. We shall use the phrase 'take moments about O', but strictly speaking we take moments about an axis through O, perpendicular to the plane of the diagram.

Let us consider this in more detail. Figure 16.4 represents a curve, given by $y = f(x)$, which is to be rotated through 360° about the x-axis, to form a solid of revolution, bounded by $x = a$ and $x = b$.

† See *Pure Mathematics Book 1*, Chapter 8, by J. K. Backhouse and S. P. T. Houldsworth, revised by P. J. F. Horril, published by Longman.

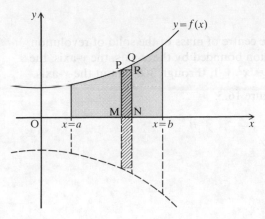

Figure 16.4

The volume, V, of the solid is given by the definite integral

$$\int_a^b (\pi y^2)\,dx$$

and the total mass is ρV.

Consider a typical element, formed by rotating the rectangle MPRN about the x-axis, where $OM = x$. This typical element is a very thin cylinder whose cross-section is a circle of radius $MP\,(=y)$, and area πy^2. The volume of the element is $(\pi y^2)\delta x$, where $\delta x\,(=MN)$ is the thickness of the element. Hence the mass of the element is $\rho(\pi y^2)\delta x$, and its moment about O is $x\rho(\pi y^2)\delta x$. The sum of all such terms for the whole solid is then

$$\sum_{x=a}^{x=b} x\rho(\pi y^2)\,\delta x$$

We evaluate this, when δx tends to zero, by calculating the definite integral

$$\int_a^b x\rho(\pi y^2)\,dx$$

Let the coordinates of the centre of mass be $(\bar{x},0)$. Equating the moment of the total mass, and the sum of the moments of the elements making up the solid, we have

$$\bar{x}\int_a^b \rho(\pi y^2)\,dx = \int_a^b x\rho(\pi y^2)\,dx \qquad \qquad \ldots \quad (2)$$

It is suggested that the reader should work through this procedure whenever it is needed, rather than try to commit (2) to memory.

Example 1

Find the coordinates of the centre of mass of the solid of revolution formed by rotating the region bounded by the x-axis, the y-axis, the line $x = 1$ and the curve $y = x^2 + 1$, through 360° about the x-axis.

The region is shown in Figure 16.5.

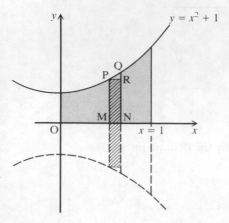

Figure 16.5

The first step is to find the volume of the solid. The typical element, formed by rotating MPRN about the x-axis has volume δV, where

$$\delta V = (\pi y^2)\delta x$$

Summing all such quantities, for $x = 0$ to $x = 1$, and letting $\delta x \to 0$, gives

$$V = \int_0^1 (\pi y^2)\,dx$$

For this curve, $y = x^2 + 1$, so

$$V = \int_0^1 \pi(x^2 + 1)^2\,dx$$

$$= \int_0^1 \pi(x^4 + 2x^2 + 1)\,dx$$

$$= \pi\left[\tfrac{1}{5}x^5 + \tfrac{2}{3}x^3 + x\right]_0^1$$

$$= \pi[\tfrac{1}{5} + \tfrac{2}{3} + 1]$$

$$= \tfrac{28}{15}\pi$$

The volume of the solid is $\tfrac{28}{15}\pi$, and hence its mass is $\tfrac{28}{15}\rho\pi$, where ρ is the density.

Now we consider the moment of the small element. Its mass is $\rho(\pi y^2)\,\delta x$, and so its moment about O is

$$x\rho(\pi y^2)\,\delta x$$

Summing all such quantities, for $x = 0$ to $x = 1$, and letting $\delta x \to 0$, gives

$$\int_0^1 x\rho(\pi y^2)\,dx$$

and since, in this case $y = x^2 + 1$, we have

$$
\begin{aligned}
\int_0^1 x\rho\pi(x^2 + 1)^2\,dx &= \rho\pi \int_0^1 x(x^4 + 2x^2 + 1)\,dx \\
&= \rho\pi \int_0^1 (x^5 + 2x^3 + x)\,dx \\
&= \rho\pi \left[\tfrac{1}{6}x^6 + \tfrac{1}{2}x^4 + \tfrac{1}{2}x^2 \right]_0^1 \\
&= \rho\pi[\tfrac{1}{6} + \tfrac{1}{2} + \tfrac{1}{2}] \\
&= \tfrac{7}{6}\rho\pi
\end{aligned}
$$

Let the coordinates of the centre of mass be $(\bar{x},0)$. Equating the moment of the total mass and the sum of the moments of the elements, we have

$$
\begin{aligned}
\bar{x}\left(\tfrac{28}{15}\rho\pi\right) &= \tfrac{7}{6}\rho\pi \\
\therefore \quad \bar{x} &= \tfrac{7}{6} \times \tfrac{15}{28} \\
&= \tfrac{5}{8}
\end{aligned}
$$

The centre of mass is at $(\tfrac{5}{8}, 0)$.

(It is a wise precaution to consider whether the answer places the centre of mass in a 'sensible' position: in this case it is between 0 and 1, which is plainly essential, and it is nearer the 1 than the 0, and since the solid gets wider towards $x = 1$, this makes sense.)

Example 2

Find the coordinates of the centre of mass of the solid of revolution (called a paraboloid) formed by rotating the region in the first quadrant, bounded by the x-axis, the line $x = b$ and the curve $y^2 = 4ax$.

It was shown in §16.5 that the volume is $2\pi ab^2$.

The mass of a typical element is $\rho(\pi y^2)\delta x$, which in this case becomes $(4\rho\pi ax)\delta x$. Its moment about O is $x(4\rho\pi ax)\delta x$. Summing for $x = 0$ to $x = b$, and letting $\delta x \to 0$, gives

the sum of the moments of the elements

$$= \int_0^b (4\rho\pi ax^2)\,dx$$

$$= 4\rho\pi a\left[\tfrac{1}{3}x^3\right]_0^b$$

$$= \tfrac{4}{3}\rho\pi ab^3$$

Let the coordinates of the centre of mass be $(\bar{x},0)$. Equating the moment of the total mass and the sum of the moments of the elements, we have

$$\bar{x}(2\rho\pi ab^2) = \tfrac{4}{3}\rho\pi ab^3$$

$$\therefore \qquad 2\bar{x} = \tfrac{4}{3}b$$

$$\therefore \qquad \bar{x} = \tfrac{2}{3}b$$

Hence the coordinates of the centre of mass of the paraboloid are $(\tfrac{2}{3}b,0)$.

(The reader is reminded that one should check that the answer is 'sensible'.)

Exercise 16c

Find the x-coordinate of the centre of mass of each of the solids of revolution, below. In each question, the region *in the first quadrant* bounded by the curve and lines given, is to be rotated through 360° about the x-axis. The reader is advised to sketch the region in each case. Leave the answers as fractions where appropriate.

1 $y = \tfrac{1}{2}x$, $y = 0$ and $x = 8$
2 $y = x + 1$, $y = 0$ and $x = 6$
3 $y = \tfrac{1}{2}x^2$, $y = 0$ and $x = 10$
4 $y = 2\sqrt{x}$, $y = 0$ and $x = 4$
5 $y = \sqrt{(16 - x^2)}$, $y = 0$ and $x = 0$
6 $y = 1/x^2$, $y = 0$, $x = 1$ and $x = 2$
7 $y = 6/x$, $y = 0$, $x = 1$ and $x = 2$
8 $y = e^x$, $y = 0$, $x = 0$ and $x = 1$
   (Hint: the 'integration by parts' rule is required.)
9 $y = (rx)/h$, $y = 0$ and $x = h$
   (Note that this is a cone, base radius r and height h.)
10 $y = \sqrt{(r^2 - x^2)}$, $y = 0$ and $x = 0$
   (Note that this is a hemisphere, radius r.)

16.7 The centre of mass of a triangular lamina

The reader is reminded that a lamina is a flat, two-dimensional, surface, bounded by a closed curve, or a polygon, such as one might form by cutting out a shape from a thin sheet of card. In this section we shall be considering a triangular lamina ABC (see Figure 16.6). Imagine the triangle divided into narrow strips parallel to BC, as shown.

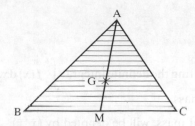

Figure 16.6

Since the strips are very narrow, we can regard each one as a uniform bar, and consequently, its centre of mass is at its midpoint. The centre of mass of the whole triangle must lie somewhere on the line passing through these points, in other words it lies on the median AM of the triangle, M being the midpoint of BC. We could repeat the argument, dividing the triangle into strips parallel to the other two sides, so G, the centre of mass of the whole triangle, lies on the intersection of the medians.

This point G is called the **centroid** of the triangle; it lies two-thirds of the way along a median, measured from a vertex. In Figure 16.6, $AG = \frac{2}{3}AM$, where M is the midpoint of BC. When using Cartesian axes, the coordinates of G can be calculated by finding the average of the coordinates of the vertices of the triangle †

Qu. 3 Find the centre of mass of a triangle whose vertices are $(1,5)$ $(-2,7)$ and $(7,-3)$.

16.8 The centre of mass of a lamina

In this section we shall apply the methods of calculus to find the centre of mass of a lamina bounded by the x-axis, a curve given by $y = f(x)$, and where necessary, lines given by $x = a$ and $x = b$, (see Figure 16.7). We shall take the plane to be placed horizontally, so both of the coordinate axes will be horizontal, and, as in §16.4, we shall take moments about these axes. We shall assume that the mass is uniformly distributed, and write the mass per unit of area as σ (the Greek lower case letter 'sigma'). [The reader is reminded that, for practical purposes, 'centre of gravity' and 'centre of mass' may be regarded as synonymous, (see §16.3).]

† See *Pure Mathematics Book 1*, by J. K. Backhouse and S. P. T. Houldsworth, revised by P. J. F. Horril, published by Longman, Chapter 15, section 15.9.

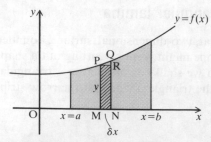

Figure 16.7

The area of the lamina can be found by evaluating the definite integral $\int_a^b f(x)\,dx$, and this is multiplied by σ to give the total mass.

As before, the coordinates of G, the centre of mass, will be denoted by $(\bar{x}, \bar{y})$.

To find $\bar{x}$, we divide the lamina into narrow strips parallel to the y-axis. We consider a typical element MPRN, with $OM = x$ and $MN = \delta x$, and take moments about the y-axis, as follows:

height of element $\qquad\qquad = y$
area of element $\qquad\qquad\quad = y\delta x$
mass of element $\qquad\qquad\quad = \sigma y \delta x$
moment of the mass of the element about the y-axis $= x(\sigma y \delta x)$

Summing, for all such elements, from $x = a$ to $x = b$, and letting $\delta x \to 0$, gives the definite integral

$$\int_a^b \sigma xy\,dx = \sigma \int_a^b x f(x)\,dx$$

(Note that, since σ is constant, it may be taken out of the integral, but if it were variable, it would have to remain as a factor of the integrand.)

We now equate the moment about y-axis of the total mass, regarded as a point mass situated at G, the centre of mass, and the sum of the moments of the elements:

$$\bar{x}\left(\sigma \int_a^b f(x)\,dx\right) = \sigma \int_a^b x f(x)\,dx$$

$$\bar{x} = \frac{\displaystyle\int_a^b x f(x)\,dx}{\displaystyle\int_a^b f(x)\,dx}$$

(It is not suggested that the reader should memorise this formula, it is more important to try to remember the method for finding it.)

To find $\bar{y}$, we take moments about the x-axis. There is however an important difference from the previous argument. The moment of the element MPRN, about the x-axis is its mass $(\sigma y \, \delta x)$ multiplied by the distance of its own centre of mass from the x-axis. Since the density is uniform, this point is half way up the element, i.e. its distance from the x-axis is $\frac{1}{2}y$. Consequently

the moment of the element about the x-axis $= \frac{1}{2}y(\sigma y \, \delta x)$
$$= \tfrac{1}{2}\sigma y^2 \, \delta x$$

Summing, for all such elements, from $x = a$ to $x = b$, and letting $\delta x \to 0$, gives the

definite integral $\displaystyle\int_a^b \tfrac{1}{2}\sigma y^2 \, dx$.

Hence, taking moments about the x-axis gives

$$\bar{y}\left(\sigma \int_a^b y \, dx\right) - \tfrac{1}{2}\sigma \int_a^b y^2 \, dx$$

$$\therefore \qquad \bar{y} = \frac{\dfrac{1}{2}\displaystyle\int_a^b y^2 \, dx}{\displaystyle\int_a^b y \, dx}$$

where $y = f(x)$.

Example 3

Find the coordinates of the centre of mass of a uniform lamina bounded by the x-axis, the line $x = 12$, and the curve $y = x^2$.

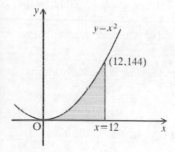

$y-x^2$

$(12,144)$

$x=12$

Figure 16.8

Let the mass per unit of area be σ and let the position of the centre of mass be $(\bar{x}, \bar{y})$.

Taking moments about the y-axis:

$$\bar{x}\sigma \int_0^{12} y \, dx = \sigma \int_0^{12} xy \, dx$$

In this example $y = x^2$, so, using this and dividing both sides by σ, we have

$$\bar{x} \int_0^{12} x^2 \, \mathrm{d}x = \int_0^{12} x^3 \, \mathrm{d}x$$

$$\bar{x}\left[\tfrac{1}{3}x^3\right]_0^{12} = \left[\tfrac{1}{4}x^4\right]_0^{12}$$

$$\bar{x}[\tfrac{1}{3} \times 12^3] = \tfrac{1}{4} \times 12^4$$

$$\therefore \qquad\qquad \bar{x} = \tfrac{3}{4} \times 12$$

$$\therefore \qquad\qquad \bar{x} = 9$$

The x-coordinate of the centre of mass is 9.

Taking moments about the x-axis:

$$\bar{y}\sigma \int_0^{12} y \, \mathrm{d}x = \sigma \int_0^{12} \tfrac{1}{2}y^2 \, \mathrm{d}x$$

Putting $y = x^2$, and dividing both sides by σ, gives

$$\bar{y} \int_0^{12} x^2 \, \mathrm{d}x = \tfrac{1}{2} \int_0^{12} x^4 \, \mathrm{d}x$$

As before, the term representing the area on the left-hand side equals $\tfrac{1}{3} \times 12^3$, so we have

$$[\tfrac{1}{3} \times 12^3]\bar{y} = \tfrac{1}{2}\left[\tfrac{1}{5}x^5\right]_0^{12}$$

$$= \tfrac{1}{10} \times 12^5$$

$$\therefore \qquad\qquad \bar{y} = \tfrac{3}{10} \times 12^2$$

$$\therefore \qquad\qquad \bar{y} = 43.2$$

The y-coordinate of the centre of mass is 43.2.

The position of the centre of mass is (9, 43.2).

Example 4

Find the centre of mass of the lamina bounded by the curve $y^2 = 4ax$, and the line $x = b$, where $b > 0$.

In one respect this example is simpler than the last one: since the lamina is symmetrical about the x-axis, the centre of mass lies on the x-axis, so let its coordinates be $(\bar{x},0)$.

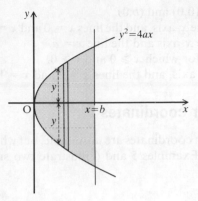

Figure 16.9

The length of a typical strip in this case is $2y$; otherwise, we proceed as before.

Taking moments about the y-axis:

$$\bar{x}\sigma \int_0^b 2y \, dx = \int_0^b x(2y) \, dx$$

Putting $y^2 = 4ax$, and dividing both sides by σ, gives

$$2\bar{x}\int_0^b \sqrt{(4ax)} = 2\int_0^b x\sqrt{(4ax)} \, dx$$

$$4\bar{x}\int_0^b a^{\frac{1}{2}} x^{\frac{1}{2}} \, dx = 4\int_0^b a^{\frac{1}{2}} x^{\frac{3}{2}} \, dx$$

$$\therefore \qquad \bar{x}\left[\tfrac{2}{3}x^{\frac{3}{2}}\right]_0^b = \left[\tfrac{2}{5}x^{\frac{5}{2}}\right]_0^b$$

$$\therefore \qquad \bar{x}\left[\tfrac{2}{3}b^{\frac{3}{2}}\right] = \left[\tfrac{2}{5}b^{\frac{5}{2}}\right]$$

$$\therefore \qquad \bar{x} = \tfrac{3}{5}b$$

The position of the centre of mass is $(\tfrac{3}{5}b, 0)$.

Exercise 16d

Find, by integration, the position of the centre of mass of each of the uniform laminae described below. Note that questions 1–5 have a line of symmetry.

1 The triangle whose vertices are $(0,0)$, (h,b) and $(h,-b)$.
2 The lamina bounded by $y^2 = 18x$ and the line $x = 2$.
3 The semicircle bounded by $y^2 = 100 - x^2$ and the y-axis $(x \geqslant 0)$.
4 The lamina bounded by $y = x^2$ and the line $y = 4$.
5 The lamina bounded by $y = \cos x$ and the x-axis, between $x = -\tfrac{1}{2}\pi$ and $x = +\tfrac{1}{2}\pi$.

6 The triangle whose vertices are $(0,h)$, $(0,0)$ and $(b,0)$.
7 The lamina bounded by $y = x^2 + 1$, the x-axis, and the lines $x = 0$ and $x = 1$.
8 The lamina bounded by $a^2 y = x^3$, the x-axis and the line $x = a$.
9 The quadrant of a circle, of radius r, for which $x \geq 0$ and $y \geq 0$.
10 The lamina bounded by $y = e^x$, the x-axis, and the lines $x = 0$ and $x = 1$.

16.9 Centre of mass, using polar coordinates

There are some shapes for which Cartesian coordinates are unsuitable, but which can be tackled using polar coordinates. Examples 5 and 6 illustrate two such cases.

Example 5

Find the centre of mass of a uniform sector of a circle, radius r, with an angle 2α between the radii.

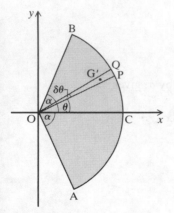

Figure 16.10

In the diagram (Figure 16.10) a sector of a circle is shown, with angle AOB equal to 2α, and its axis of symmetry along the x-axis, i.e. the line whose polar equation is $\theta = 0$. The area of the sector is $\frac{1}{2}(2\alpha)r^2 = \alpha r^2$. So the total mass of the sector is $\sigma\alpha r^2$, where σ represents the mass per unit of area. By symmetry, the centre of mass lies on Ox, so let its coordinates be $(\bar{r},0)$. Consequently, the moment about O of the total mass is

$$\bar{r}(\sigma\alpha r^2) \qquad \qquad \cdots \quad (1)$$

Using polar coordinates, we divide the sector into small sectors, with POQ, in Figure 16.10, representing a typical element: let $\angle$ COP $= \theta$, and $\angle$ POQ $= \delta\theta$.

The area of POQ is $\frac{1}{2}r^2\,\delta\theta$, and so its mass is $\frac{1}{2}\sigma r^2\,\delta\theta$. Since $\delta\theta$ is small, we may regard the sector POQ as a triangle, with its centre of mass (the point G' in Figure 16.10) at a distance $\frac{2}{3}r$ from O. We may take the distance of G′ from the y-axis to be $\frac{2}{3}r\cos\theta$. Hence the moment of the mass of the sector POQ about the y-axis is

$$(\tfrac{1}{2}\sigma r^2\,\delta\theta)(\tfrac{2}{3}r\cos\theta) = \tfrac{1}{3}\sigma r^3\cos\theta\,\delta\theta$$

Summing for all values of θ from $\theta = -\alpha$ to $\theta = +\alpha$, and letting $\delta\theta \to 0$, we have

$$\int_{-\alpha}^{+\alpha} \tfrac{1}{3}\sigma r^3\cos\theta\,d\theta = \tfrac{1}{3}\sigma r^3\Big[\sin\theta\Big]_{-\alpha}^{+\alpha}$$

$$= \tfrac{1}{3}\sigma r^3\left[\sin(+\alpha) - \sin(-\alpha)\right]$$

$$= \tfrac{1}{3}\sigma r^3\left[\sin\alpha + \sin\alpha\right]$$

$$= \tfrac{2}{3}\sigma r^3\sin\alpha$$

Equating this (the sum of the moments of the elements which make up the sector) and the moment of the total mass of the sector (see (1) above), we have

$$\bar{r}(\sigma\alpha r^2) = \tfrac{2}{3}\sigma r^3\sin\alpha$$

$$\therefore \qquad \bar{r} = \frac{2r\sin\alpha}{3\alpha}$$

The centre of mass of the sector lies on the axis of symmetry, at a distance $\left(\dfrac{2r\sin\alpha}{3\alpha}\right)$ from O.

Note that a semicircular lamina (i.e. a shape rather like a protractor) can be regarded as a sector of a circle, with $\alpha = \tfrac{1}{2}\pi$, in which case the above result becomes $\dfrac{4r}{3\pi}$.

Example 6

Find the centre of mass of a thin uniform hemispherical shell, of radius r.

(The reader should try to imagine an object like a ping-pong ball cut in half.)

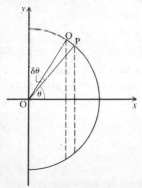

Figure 16.11

By symmetry, the centre of gravity lies on the x-axis; let its coordinates be $(\bar{r},0)$.

The surface area of the shell is $2\pi r^2$, and so we write its mass as $2\sigma\pi r^2$, where σ represents the mass per unit area. The moment of the total mass about a line through O, perpendicular to the coordinate axes, is then

$$\bar{r}(2\sigma\pi r^2) \qquad \qquad \text{. . . (1)}$$

Consider the small element formed by rotating the small arc PQ in Figure 16.11, through 360° about the x-axis, where radius OP makes an angle θ with the x-axis and angle POQ $= \delta\theta$. (The reader should imagine this element to be like one of those thin circular metal bands seen on old-fashioned wooden beer barrels.) For such an element

width of band	$= r\,\delta\theta$
radius	$= r\sin\theta$
circumference	$= 2\pi r\sin\theta$
area of band	$= (2\pi r\sin\theta)\,r\,\delta\theta$
	$= (2\pi r^2\sin\theta)\,\delta\theta$
mass of band	$= (2\sigma\pi r^2\sin\theta)\,\delta\theta$
distance from y-axis	$= r\cos\theta$
moment about O	$= (r\cos\theta)(2\sigma\pi r^2\sin\theta)\,\delta\theta$
	$= (2\sigma\pi r^3\sin\theta\cos\theta)\,\delta\theta$

Summing for all such elements, from $\theta = 0$ to $\theta = \frac{1}{2}\pi$, and letting $\delta\theta \to 0$, we have

$$\int_0^{\frac{1}{2}\pi} (2\sigma\pi r^3 \sin\theta\cos\theta)\,d\theta = \sigma\pi r^3 \int_0^{\frac{1}{2}\pi} (2\sin\theta\cos\theta)\,d\theta$$

$$= \sigma\pi r^3 \left[\sin^2\theta\right]_0^{\frac{1}{2}\pi}$$

$$= \sigma\pi r^3 \sin^2\left(\tfrac{1}{2}\pi\right)$$

$$= \sigma\pi r^3$$

Equating this, the sum of the moments of all the elements making up the hemisphere, and the moment of its total mass (see (1) above), we have

$$\bar{r}(2\sigma\pi r^2) = \sigma\pi r^3$$
$$\therefore \qquad \bar{r} = \tfrac{1}{2}r$$

The centre of mass of the hemispherical shell is on its axis of symmetry, at a distance $\frac{1}{2}r$ from O.

Summary of the results in the preceding section

Name	Dimensions	Position of centre of mass
solid cone	height h, radius r	$\frac{3}{4}h$ from vertex
solid hemisphere	radius r	$\frac{3}{8}r$ from centre of base
triangle		at intersection of medians
sector of circle	angle at centre 2α, radius r	$\dfrac{2r\sin\alpha}{3\alpha}$
semicircular lamina	radius r	$\dfrac{4r}{3\pi}$
hemispherical shell	radius r	$\frac{1}{2}r$

Exercise 16e (Miscellaneous)

1 Find the position vector of four particles whose weights are m, $2m$, $3m$ and $4m$ with position vectors $(2\mathbf{i} + 3\mathbf{j} + 3\mathbf{k})$, $(\mathbf{i} - 6\mathbf{j} + 7\mathbf{k})$, $(\mathbf{j} + \mathbf{k})$ and $(4\mathbf{i} + 4\mathbf{j})$, respectively.

2 Four identical particles, of mass m, are placed at points A, B, C and D, not necessarily coplanar, with position vectors $\mathbf{a}$, $\mathbf{b}$, $\mathbf{c}$ and $\mathbf{d}$. Find the position vector of G, their centre of mass.
Show that G lies at the intersection of the midpoints of AB and CD.

3 Find the coordinates of the centre of gravity of the solid of revolution formed by rotating the region bounded by $y = x^2$ and $y = h$ about the y-axis.

4 A hollow cone of height h, has no base and its thickness is negligible. Prove that its centre of mass is at a distance $\frac{2}{3}h$ from the vertex.

5 Find the centre of mass of a uniform lamina enclosed by the x-axis and the curve $hy = (a^2 - x^2)$.

6 A solid of revolution is formed by rotating a region, R, about the x-axis. R is bounded by $x = 0$, $x = 4$, $y = 0$, and a curve which passes through the points $(0,1)$, $(1,1)$, $(2,2)$, $(3,2.5)$ and $(4,3)$. Use the trapezium rule to estimate (a) its volume, (b) the x-coordinate of its centre of gravity.

7 A thin uniform wire AB is bent into a circular arc which is part of a circle centre O, radius r. The angle AOB is 2α, where α is in radians. Show that the centre of mass of the wire is at a distance $(r\sin\alpha)/\alpha$ from O.
A uniform wire, of length 20 cm, is formed into such an arc with radius 10 cm. Calculate the distance of its centre of mass from O.

8 Using the result of question 7, prove that the centre of mass of a uniform sector of a circle, centre O, radius r, which subtends an angle 2α at O, is at a distance $\dfrac{2r\sin\alpha}{3\alpha}$ from O. (See Example 5 for an alternative method.)

9 Using the result of Example 6 (page 339), prove that the centre of mass of a uniform solid hemisphere is at a distance $\frac{3}{8}r$ from the centre of its plane surface. (See Exercise 16c, question 10, for an alternative method.)

10 Using the result of Example 6 (page 339), prove that the centre of gravity of a uniform *thick* hemispherical shell, with inner radius r and outer radius R, is

$$\frac{3(R^4 - r^4)}{8(R^3 - r^3)}$$

By factorising this expression and letting $r \to R$, verify that this formula gives the formula for the centre of mass of a *thin* hemispherical shell.

Note that this topic normally appears in examinations as the introduction to a longer question; see examples in Exercise 17c (Miscellaneous) and Revision Exercise 4.

Chapter 17

Miscellaneous statics

17.1 Introduction

In this chapter we shall draw together the ideas and techniques developed in the previous three chapters.

Firstly, we shall find the centre of mass of a 'composite body'; that is, a rigid body formed by combining two or more basic shapes, such as those we examined in the last chapter. When doing this we shall frequently quote the standard results we have already found. As in the earlier chapters, all the objects we shall meet will have uniform density, unless stated otherwise.

We shall then go on to use these results to examine the stability of rigid bodies.

17.2 The centre of mass of a composite body

Example 1

A simple model of a man, of total mass 70 kg and height 1.80 m, is made (see Figure 17.1) from the following components:

Part	Shape	Mass
Head	sphere, radius 10 cm	5 kg
Trunk	cylinder, length 80 cm	35 kg
Arms	cylinders, length 80 cm	3 kg (each)
Legs	cylinders, length 80 cm	12 kg (each)

Find the distance of the centre of mass from the feet of the model (point O in Figure 17.1).

Static equilibrium

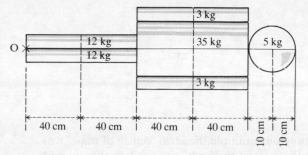

Figure 17.1

The centre of mass of the whole body will be on the axis of symmetry; let its distance from O be $\bar{x}$ cm (note that we may use cm for all the lengths).

Part	Mass (in kg)	Distance of centre of mass from O (in cm)
Legs	24	40
Arms	6	120
Trunk	35	120
Head	5	170
Whole body	70	$\bar{x}$

The moment of the mass of the whole body, about an axis through O, is $70\bar{x}$; we equate this to the sum of the moments of the component parts:

$$70\bar{x} = (24 \times 40) + (6 \times 120) + (35 \times 120) + (5 \times 170)$$
$$= 960 + 720 + 4200 + 850$$
$$= 6730$$
$$\therefore \quad \bar{x} = 96.1 \quad \text{(correct to three significant figures)}$$

The centre of mass of the model is 96.1 cm from its feet.

Qu. 1 The arms of the model in Example 1 are now raised above its head, by rotating them through 180° about a pivot which may be assumed to pass through the tops of the cylinders representing the arms and the trunk. Find the distance of the centre of mass of the model from its feet when it is in this position.

Example 2

A rigid body consists of a solid cylinder, radius r and height r, and a solid hemisphere, of the same density and radius r, the plane surface of which is fixed to one end of the cylinder (see Figure 17.2). Show that the centre of gravity of the composite body is at a distance $\frac{17}{20}r$ from the other end of the cylinder.

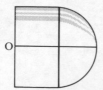

Figure 17.2

(Note that in the previous example the term 'centre of mass' was employed because the *mass* of the body was used; in this example 'centre of gravity' is the correct term, because the *weight* is considered. In practice this makes no difference to the answer. (See §16.3.)

Let the distance of the centre of gravity of the composite body, from the base of the cylinder be $\bar{x}$. Let the density be ρ, then the weight of the cylinder is $(\pi r^2 \times r)\rho = \pi r^3 \rho$, and that of the hemisphere is $(\frac{2}{3}\pi r^3)\rho$; it will be convenient to denote these weights by W and $\frac{2}{3}W$, respectively. (Remember that the centre of gravity of a solid hemisphere is at a distance $\frac{3}{8}r$ from its plane face.) The information is tabulated below:

Part	Weight	Distance of centre of gravity from base
Cylinder	W	$\frac{1}{2}r$
Hemisphere	$\frac{2}{3}W$	$(r + \frac{3}{8}r) = \frac{11}{8}r$
Whole solid	$\frac{5}{3}W$	$\bar{x}$

Taking moments about an axis through point O in Figure 17.2, we have

$$(\tfrac{5}{3}W)\bar{x} = \tfrac{1}{2}Wr + \tfrac{2}{3}W \times \tfrac{11}{8}r$$

$$\therefore \qquad = (\tfrac{1}{2} + \tfrac{11}{12})Wr$$

$$= \tfrac{17}{12}Wr$$

$$\therefore \qquad 5\bar{x} = \tfrac{17}{4}r$$

Hence

$$\bar{x} = \tfrac{17}{20}r$$

The centre of gravity of the composite body is at a distance $\frac{17}{20}r$ from its plane face.

Qu. 2 Solve Example 2 by taking moments about the centre of the plane face of the hemisphere.

Example 3

A right solid cone has base radius $2r$ and height $2h$. The top part of it, consisting of a small cone, radius r and height h, is removed. Find the distance from the base of the large cone to the centre of gravity of the portion that remains.

(Note that the remaining portion is called a **frustum**. Also, the word 'right', which appears in the first sentence, means that the cone is symmetrical, i.e. its axis is perpendicular to its base. A cone which does not have this property is called a 'skew' cone.)

This example differs from the first two, in that the part we are interested in, the frustum, is formed by removing part of the original object. The method, however, is the same as before; the frustum and the small cone can be combined to form the big cone.

The small cone and the big cone are similar, and their lengths are in the ratio $1:2$. Consequently their volumes, and hence their weights, are in the ratio $1:8$, so we shall take them to be W and $8W$, respectively. The weight of the frustum is then $7W$. (Some readers may be unfamiliar with this argument; even so they will find that it is a very convenient method in problems like this. Alternatively, the weights can be written down, as in earlier examples, by first finding the volumes of the two cones.)

Figure 17.3 shows the cone with its axis horizontal. Let the height of the centre of gravity of the frustum (OG in the diagram) above the base, be $\bar{x}$. Remember that the centre of gravity of a solid cone of height h, is at a distance $h/4$ from its base; for the large cone this becomes $2h/4$ i.e. $\frac{1}{2}h$.

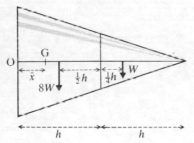

Figure 17.3

The information is tabulated below.

Part	Weight	Distance of centre of gravity from base
Large cone	$8W$	$\frac{1}{2}h$
Small cone	W	$(h + \frac{1}{4}h) = \frac{5}{4}h$
Frustum	$7W$	$\bar{x}$

The moment of the weight of the large cone, about an axis through O, is

$$(8W) \times (\tfrac{1}{2}h) = 4Wh$$

We now equate this to the sum of the moments of the small cone and the frustum.

$$4Wh = W \times \tfrac{5}{4}h + 7W \times \bar{x}$$
$$= \tfrac{5}{4}Wh + 7W\bar{x}$$
$$\therefore \quad 7\bar{x} = 4h - \tfrac{5}{4}h$$
$$= \tfrac{11}{4}h$$

Hence

$$\bar{x} = \tfrac{11}{28}h$$

The centre of gravity of the frustum is at a distance $\tfrac{11}{28}h$ from the base.

(It is always wise to check that the answer is 'sensible': in this case, cutting off the top of the large cone has moved the centre of gravity towards the base, which is what one would expect.)

Example 4

A circular disc, of radius $4a$ and centre O, has a circular hole, of radius a, punched in it. Relative to axes in the plane of the disc, with O as the origin, the centre of the hole is at $(2a,0)$. Find the coordinates of the centre of gravity of the remainder.

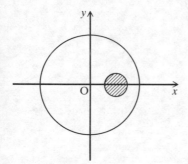

Figure 17.4

Since the radii of the circles are in the ratio 1:4, their areas, and hence their weights, are in the ratio 1:16, so we shall write them as W and $16W$. The weight of the remainder is therefore $15W$.

Cutting the hole on the disc will move the centre of gravity to the left. Suppose the distance it moves is $\bar{x}$, then the coordinates of the centre of gravity of the remainder are $(-\bar{x}, 0)$. The information is tabulated below.

Part	Weight	Distance of the centre of gravity from O
Big circle	$16W$	0
Small circle	W	$2a$
Remainder	$15W$	$\bar{x}$

The moment of the big circle about O is zero, so the sum of the moments of the other two parts is also zero (note that they have opposite signs.) Hence

$$2aW + (-15W\bar{x}) = 0$$
$$\therefore \qquad 15W\bar{x} = 2aW$$
$$\therefore \qquad \bar{x} = \tfrac{2}{15}a$$

The position of the centre of gravity of the remainder is $(-\tfrac{2}{15}a, 0)$.

Example 5

Find the coordinates relative to the axes shown, of the centre of gravity of a uniform lamina in the shape of a letter F, shown in Figure 17.5.

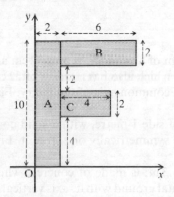

Figure 17.5

This object, unlike those in the previous examples, has no line of symmetry, so it is necessary to take moments about both axes in order to find its centre of gravity; let the coordinates of this point be $(\bar{x}, \bar{y})$.

We can divide the object into three rectangles, labelled A, B and C, in Figure 17.5. Since it is a uniform lamina, the weight can be taken to be proportional to the area; so we shall write the weights of A, B and C as 20w, 12w and 8w, respectively. The total weight is then 40w. The information is tabulated below.

| Part | Weight | Distance of centre of gravity from: | |
		the x-axis	the y-axis
A	20w	5	1
B	12w	9	5
C	8w	5	4
Whole	40w	$\bar{y}$	$\bar{x}$

Taking moments about the x-axis:

$$40w\bar{y} = (20w \times 5) + (12w \times 9) + (8w \times 5)$$
$$= 100w + 108w + 40w$$
$$= 248w$$
$$\therefore \qquad \bar{y} = 6.2$$

Taking moments about the y-axis:

$$40w\bar{x} = (20w \times 1) + (12w \times 5) + (8w \times 4)$$
$$= 20w + 60w + 32w$$
$$= 112w$$
$$\therefore \qquad \bar{x} = 2.8$$

The centre of weight is at the point (2.8,6.2).

Exercise 17a

1 A solid concrete bollard is made in the form of a cylinder, radius 32 cm and height 64 cm, surmounted by a hemisphere which also has a radius of 32 cm, so that the cylinder and hemisphere have a common axis of symmetry. Find the height of its centre of gravity.

2 Another bollard consists of a solid cube of side 1 metre, with a solid cone, base radius 50 cm and height 30 cm placed symmetrically on top of it. Find the height of the centre of gravity.

3 A solid cylinder of radius 2 m and length 3 m, is made of concrete whose density is 2400 kg m^{-3}. It stands on horizontal ground with its axis vertical. A solid cone, of base radius 2 m and height 1 m, is made of cast iron and is placed symmetrically on top of the cylinder. The density of the cast iron is 7200 kg m^{-3}. Find the height of the centre of gravity.

4 A thin hemispherical shell, of radius r, is fitted with a circular base of radius r, made from the same material. Find the distance of the centre of gravity from the centre of the plane base.

5 A solid cylinder of length 2l and radius 2r, has a cylindrical hole of radius r and depth l, drilled in it. The axis of the hole coincides with the axis of the cylinder. Find the position of the centre of gravity of the part which remains, measured from the centre of the end where the hole is drilled.

6 Find the position of the centre of gravity of the lamina in the form of a letter T, shown in Figure 17.6.

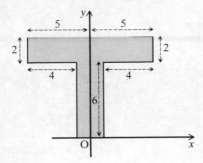

Figure 17.6

7 Find the position of the centre of gravity of the lamina in the form of a letter L, shown in Figure 17.7.

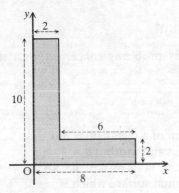

Figure 17.7

8 Find the position of the centre of gravity of the lamina in the form of a letter b, formed from a rectangle and a semicircle, shown in Figure 17.8.

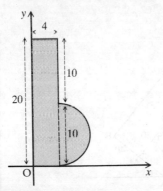

Figure 17.8

9 Find the position of the centre of gravity of the lamina in the form of a square of side 2*a* with an equilateral triangle of side 2*a* fixed to one edge, shown in Figure 17.9.

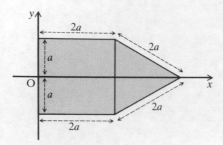

Figure 17.9

10 A semicircular lamina, centre O, has radius 2*r*. A semicircle, centre O and radius *r* is removed from it. Find the distance of the centre of gravity from O.

17.3 Centre of gravity and equilibrium

The examples which follow illustrate various problems concerned with the equilibrium of a rigid body.

Example 6

The front face of a heavy chest, in the form of a rectangular box, is a rectangle ABCD, with height $AD = 150$ cm and width $AB = 80$ cm. The chest is tilted so that the rectangle turns, in its own plane, about A. The point A is in contact with a horizontal surface which is sufficiently rough to prevent it slipping. The centre of gravity of the chest is on the line of symmetry parallel to AD, at a distance of 50 cm from AB. Initially AB is horizontal.

Find the largest angle through which the chest can be tilted before it topples over.

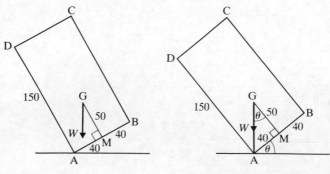

Figure 17.10a   **Figure 17.10b**

In the diagram, G is the centre of gravity of the chest, and M is the midpoint of AB. When the chest is in a position such as that shown in Figure 17.10a, its weight W has a clockwise moment about A; consequently if it were released from this position the chest would tend to move back to its upright position, with AD vertical.

If it is tilted until AG is vertical, as in Figure 17.10b, the line of action of the weight would pass through point A, so, in theory, the chest would balance in this position, (in practice it would be virtually impossible to achieve this). Let the angle which AB makes with the horizontal be θ, as shown in Figure 17.10b. Then the angle which GM makes with the vertical is also θ.

$$\tan \theta = \tfrac{40}{50} = 0.8$$

The largest angle through which the chest can be tilted is $\tan^{-1}(0.8)$.

(Note that $\tan^{-1}(0.8) = 38.7°$, correct to three significant figures, so if the chest is tilted through 38° and released it will return to the upright position, but if it is tilted through 39° it will topple over.)

Example 7

A composite rigid body consists of a uniform solid cylinder radius r, height $2r$ and weight $3W$, surmounted by a uniform solid hemisphere radius r, weight W, the cylinder and hemisphere having a common axis of symmetry. Find the distance of its centre of gravity from the centre of its plane face.

The object stands with its plane face in contact with an inclined plane, which is sufficiently rough to prevent sliding. Given that the object is on the point of tilting, find the angle θ which the plane makes with the horizontal.

Remember that the centre of gravity of a solid hemisphere is at a distance $\tfrac{3}{8}r$ from its plane face. (See summary on page 341, chapter 16.)

The information is tabulated below.

Part	Weight	Distance of the centre of gravity from the base
Cylinder	$3W$	r
Hemisphere	W	$2r + \tfrac{3}{8}r = \tfrac{19}{8}r$
Whole	$4W$	$\bar{x}$

Taking moments about M, the centre of the plane face, we have

$$4W\bar{x} = 3Wr + W(\tfrac{19}{8}r)$$

$$\therefore \quad 4\bar{x} = \tfrac{43}{8}r$$

$$\therefore \quad \bar{x} = \tfrac{43}{32}r$$

The centre of gravity is at a distance $\tfrac{43}{32}r$ from the centre of the plane face.

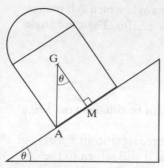

Figure 17.11

Figure 17.11 shows the object on the inclined plane when it is about to tilt about point A, i.e. with G, the centre of gravity, vertically above A. The plane makes an angle θ with the horizontal, and this is also the angle which MG makes with the vertical.

$$\tan \theta = \frac{AM}{MG}$$

$$= r \div (\tfrac{43}{32}r)$$

$$= \tfrac{32}{43}$$

Hence the object is on the point of tilting when the plane makes an angle $\tan^{-1}(\tfrac{32}{43})$ with the horizontal.

Example 8

A plane uniform lamina ABCDE consists of a square ABCD, of side $2a$, joined to an isosceles triangle ADE, in which $EA = ED$; M is the midpoint of AD, and $ME = 3h$. Find the distance of the centre of gravity of the lamina from M.

The lamina is freely suspended from A and it hangs in equilibrium in a vertical plane. Given that the side AB is inclined at an angle α to the horizontal, show that

$$\tan \alpha = \frac{4a^2 - 3h^2}{4a^2 + 3ah}$$

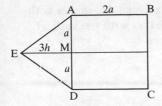

Figure 17.12

The centre of gravity of the triangle is at a distance $\frac{1}{3}ME\,(=h)$ from M. Let the weight per unit of area be σ. Since the area of the square is $4a^2$, its weight is $4a^2\sigma$, and the area of the triangle is $\frac{1}{2} \times 2a \times 3h$, so its weight is $3ah\sigma$. The information is tabulated below.

Part	Weight	Displacement of the centre of gravity to the right of M
Square	$4a^2\sigma$	$+a$
Triangle	$3ah\sigma$	$-h$
Lamina	$(4a^2 + 3ah)\sigma$	$\bar{x}$

Taking moments about M:

$$(4a^2 + 3ah)\sigma\bar{x} = 4a^3\sigma - 3ah^2\sigma$$
$$= a\sigma(4a^2 - 3h^2)$$
$$\bar{x} = \frac{4a^2 - 3h^2}{4a + 3h}$$

Hence the centre of gravity of the lamina is at a distance
$$\frac{4a^2 - 3h^2}{4a + 3h}$$ from M.

When the lamina is suspended from A, its centre of gravity, G, will be vertically below A.

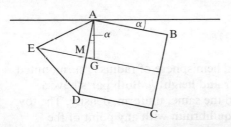

Figure 17.13

Since angle DAB is a right-angle, the angle α which AB makes with the horizontal equals the angle which AM makes with the vertical, (see Figure 17.13), i.e. angle MAG = α. Hence

$$\tan \alpha = \frac{GM}{MA}$$

$$= \left(\frac{4a^2 - 3h^2}{4a + 3h}\right) \div a$$

$$\therefore \quad \tan \alpha = \frac{4a^2 - 3h^2}{4a^2 + 3ah}$$

17.4 Stable, unstable and neutral equilibrium

The state of equilibrium can take three possible forms: stable, unstable and neutral. All three forms can be demonstrated with a solid cone. When a cone stands with its base on a horizontal table, the equilibrium is stable. If it is slightly displaced from such a position and released, it topples back towards the upright position. With a great deal of patience, we might just be able to balance the cone on its vertex; it would be extremely difficult, but in theory it could be done. The slightest displacement from this position, however, would cause the cone to fall over; we say it is in unstable equilibrium. When a cone is placed with its curved surface in contact with a horizontal table it will remain in that state. If it is slightly displaced, it simply stays resting in its new position; we say it is in neutral equilibrium.

Objects with uniform density and a circular cross-section, such as cylinders and spheres, can always be placed in neutral equilibrium on a horizontal surface, because of an important geometrical property of a circle: the tangent at any point is perpendicular to the radius at that point. So the centre of gravity, G, of the object will always be vertically above P, the point of contact with the surface (see Figure 17.14).

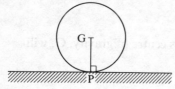

Figure 17.14

Example 9

A child's toy consists of a solid hemisphere of radius r, surmounted by a solid cone of base radius r and height h. Both parts have a common axis of symmetry and the same, uniform, density. The toy is designed to rest in neutral equilibrium with any point of the curved surface of the hemisphere in contact with a horizontal plane. Show that $h = r\sqrt{3}$.

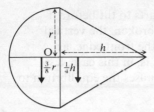

Figure 17.15

For neutral equilibrium, the centre of gravity of the toy must be at the centre of the plane surface of the hemisphere (point O in Figure 17.15).

Let the density be ρ. The information is tabulated below.

Part	Weight	Displacement of the centre of gravity to the right of O
Cone	$\frac{1}{3}\pi r^2 h\rho$	$+\frac{1}{4}h$
Hemisphere	$\frac{2}{3}\pi r^3 \rho$	$-\frac{3}{8}r$
Toy	$(\frac{2}{3}\pi r^3\rho + \frac{1}{3}\pi r^2 h\rho)$	0

Taking moments about O gives

$$(\tfrac{1}{3}\pi r^2 h\rho)(\tfrac{1}{4}h) - (\tfrac{2}{3}\pi r^3\rho)(\tfrac{3}{8}r) = 0$$

Hence

$$(\tfrac{1}{3}\pi r^2 h\rho)(\tfrac{1}{4}h) = (\tfrac{2}{3}\pi r^3\rho)(\tfrac{3}{8}r)$$

$$\therefore \qquad \tfrac{1}{12}r^2 h^2 = \tfrac{1}{4}r^4$$

$$\therefore \qquad h^2 = 3r^2$$

Hence $h = r\sqrt{3}$.

17.5 Sliding and tilting

Anyone who has tried to push or pull a heavy piece of furniture, say an upright piano, along a horizontal floor will know that one has to be careful, or one might tip it over. In this section we shall look at the conditions for such an object to slide or tilt.

Example 10

A heavy uniform chest, in the shape of a cube of side $2a$, has weight W. It stands on a rough horizontal floor and the coefficient of friction between the chest and the floor is μ. A horizontal force is applied to the midpoint of one of the edges of the upper face, at right-angles to this edge. The force is gradually increased, starting from zero. Given that the equilibrium is broken by the chest tilting, prove that $\mu > \frac{1}{2}$.

Consider first the possibility that the chest starts to tilt before it slides: at the instant when the equilibrium is broken, the vertical reaction, R, of the floor on the chest acts on the front edge, as shown in Figure 17.16a. (The friction force F is, in this case, less than its limiting value.) Let the force which causes the equilibrium to be broken in this way be P.

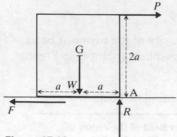

Figure 17.16a

Taking moments about A (or more accurately, about the axis through A perpendicular to the plane of the diagram), we have

$$2aP = aW$$
$$\therefore \quad P = \tfrac{1}{2}W$$

Now let us consider the possibility that the equilibrium is broken by sliding rather than tilting. Let the value of the force in this case be Q. The vertical reaction, R, is still equal to the weight W, but its line of action does not pass through point A. When the chest is about to slip, the friction is limiting, i.e. $F = \mu R = \mu W$, (see Figure 17.16b).

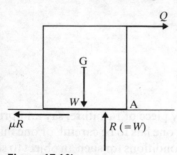

Figure 17.16b

Resolving horizontally:

$$Q = \mu W$$

The force gradually increases from zero, so the chest will tilt rather than slide if $P < Q$, i.e.

$$\tfrac{1}{2}W < \mu W$$
i.e. $\tfrac{1}{2} < \mu$

The chest will tilt before it slides if $\mu > \tfrac{1}{2}$.

If, in Example 10, a force T less than P and less than Q, is applied then the chest will remain in equilibrium. The friction force will be less than limiting; the normal reaction, R, will be vertically upwards and equal in magnitude to W, but its line of action will be at some distance to the left of A. Let this distance be x. (See Figure 17.16c.)

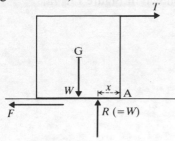

Figure 17.16c

Taking moments about point A gives:

$$xR + 2aT - aW = 0$$

But $R = W$, so

$$xW = a(W - 2T)$$

i.e. $x = \dfrac{a(W - 2T)}{W}$

Notice that x is positive if T is less than $\frac{1}{2}W$, (that is, less than P) and becomes zero when $T = \frac{1}{2}W$.

Example 11

A solid cone, weight W, base radius r and height h, is placed on an inclined plane whose angle of inclination can be varied. The coefficient of friction between the cone and the plane is μ. The angle of inclination is gradually increased, starting from zero. Show that the cone will slide before it tilts if $\mu < 4r/h$.

Let us first consider the possibility that the equilibrium is broken by sliding, (see Figure 17.17a). Let the angle of inclination in this case be α.

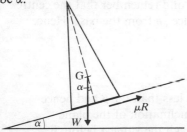

Figure 17.17a

In this case the friction is limiting.

Resolving perpendicular to the plane:

$$R = W \cos \alpha$$

(Note that the normal reaction, R, is not shown in Figure 17.17a. See Qu. 3 on page 361).

Resolving parallel to the plane:

$$W \sin \alpha = \mu R$$
$$= \mu W \cos \alpha$$

Hence

$$\tan \alpha = \mu$$

On the other hand, suppose the equilibrium is broken by tilting. The friction force F will be less than limiting, but when the cone is about to tilt, the normal reaction, S, passes through A, the lowest point of contact between the cone and the plane (see Figure 17.17b). At this instant the centre of gravity, G, of the cone is vertically above A; let the angle of inclination of the plane in this case be β.

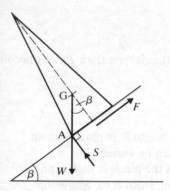

Figure 17.17b

Notice that the angle between the line of action of the weight and the axis of the cone is also equal to β, and remember that the centre of gravity of a solid cone is at a distance $\frac{1}{4}h$ from the base. Hence

$$\tan \beta = \frac{r}{(\frac{1}{4}h)} = \frac{4r}{h}$$

We are given that $\mu < 4r/h$, so $\tan \alpha$ is less than $\tan \beta$ and hence $\alpha < \beta$. Consequently, as the angle of inclination of the plane increases, starting from zero, it will reach the value α before it reaches β, so the equilibrium will be broken by the cone sliding.

Qu. 3 Through which point in Figure 17.17a does the normal reaction pass?

Exercise 17b

1 A uniform wire is bent to form three sides AB, BC and CD of a square. Show that when the wire is freely suspended from A, the side AB makes an angle $\tan^{-1}(\frac{4}{3})$ with the horizontal.

2 A uniform wire, of mass $3m$, is bent into the form of an equilateral triangle PQR and a particle of mass m is attached to vertex Q. The triangle is freely suspended from vertex P. Show that the side PQ makes an angle $\tan^{-1}(\frac{1}{5}\sqrt{3})$ with the vertical.

3 A uniform wire is bent into a semicircular arc, together with its diameter AB. The wire is freely suspended from A. Calculate, correct to the nearest degree, the angle that AB makes with the vertical.

4 A lamina in the shape of a trapezium ABCD has right-angles at B and C, and $AB = 6a$, $BC = 4a$, $CD = 3a$. The lamina is freely suspended from A. Show that AB makes an angle $\tan^{-1}(\frac{16}{33})$ with the vertical.

5 A solid cone, of height $3h$ and base radius $3r$, has the top part, consisting of a cone of height h and base radius r, removed. Find, in terms of h, the distance of the centre of gravity of the remaining frustum, from the smaller plane face. Show that the frustum can stand with its curved surface in contact with a horizontal plane provided $17h^2 > 13r^2$.

6 A solid uniform cylinder of radius r and length r, has a solid cone of base radius r and height r fixed symmetrically to it. It is placed with its circular face on a rough plane which is inclined at an angle α to the horizontal. The plane is sufficiently rough to prevent sliding. Show that the object will topple over if $\tan\alpha > \frac{16}{11}$.

7 A child's toy consists of a solid hemisphere, radius r and uniform density d, surmounted by a uniform cylinder of radius $\frac{1}{2}r$ and length r, with uniform density D. The hemisphere and the cylinder have a common axis. Show that the toy can stand with the hemisphere downwards and its axis vertical in stable equilibrium, provided $d > \frac{1}{2}D$.

8 A solid uniform cone, of base radius r and height $4r$, stands on a rough horizontal plane. A horizontal force, P, which is not sufficiently large to disturb the equilibrium, is applied to its vertex. Prove that the line of action of the normal reaction of the plane on the cone is at a distance $4Pr/W$ from the axis of the cone. Hence show that $P < \frac{1}{4}W$.

9 A uniform chest has a rectangular cross-section, with base b and height h. It is placed on a plane, which can be tilted, with this cross-section in a vertical plane and parallel to the line of greatest slope. The coefficient of friction between the plane and the chest is μ. The plane is initially horizontal, but it is then gradually raised until the equilibrium is broken. Show that the chest will slide rather than topple if $\mu < h/b$.

10 A uniform solid cone, of base radius r and height h, stands on a rough plane which is inclined at an angle α to the horizontal. The conditions are such that the cone does not slide down the plane, nor does it topple over. A gradually

increasing force, parallel to the line of greatest slope up the plane, is then applied to the vertex of the cone. Show that the equilibrium will be broken by sliding, rather than toppling, if

$$\mu < \frac{r}{h} - \frac{3\tan\alpha}{4}$$

Exercise 17c (Miscellaneous)

1 A uniform beam AB of mass 30 kg and length 10 m rests horizontally on two supports C and D, where $AC = 2$ m and $DB = 1$ m. An object of mass M kg is hung from B. Calculate
(a) the reactions at C and D when $M = 50$;
(b) the value of M for which the reactions are equal;
(c) the largest value of M for which equilibrium is possible.

*(C)

2 A horizontal non-uniform rod APQB is of length 6 m. It is supported at P and Q, where $AP = 1.2$ m, $QB = 2$ m. The rod may just be turned about P by an upward force of 3 N at B; it may just be turned about Q by a downward force of 18 N at B. Calculate the mass of the rod and the distance of its centre of gravity from A. (Take g to be $10\,\mathrm{m\,s^{-2}}$.)

*(C)

3 The centre of mass of a double-decker bus is at a height of 1.5 m above ground level, and is in the plane which bisects each axle at right-angles. The length of each axle is 2.5 m. The bus stands on an inclined test platform so that the axles are parallel to the line of greatest slope. The angle of inclination of the platform to the horizontal is slowly increased. Find the angle at which the bus would topple over, assuming that it is prevented from slipping.
The mass of the bus is 10 000 kg. When the test platform is horizontal 20 passengers get on the bus, and all sit upstairs on the same side. The average mass of a passenger is 70 kg. The centre of mass of the passengers is 2.5 m above ground level, and at a distance of 1 m from the plane bisecting the axles of the bus at right-angles. The platform is slowly tilted towards the side on which the passengers are sitting. Find the new angle at which the bus would topple.

(C)

4 A uniform solid right circular cone has its top removed by cutting the cone by a plane parallel to its base, leaving a truncated cone of height h, the radii of its ends being r and $4r$. Show that the distance of the centre of gravity of the truncated cone from its broader end is $\frac{9}{28}h$.

(O & C)

5 ABCD is a uniform square lamina whose centre is O. The square with diagonal OB is cut away, and the remaining body is suspended with its plane vertical, from a fixed smooth pivot through A. Find the acute angle (correct to the nearest degree) which AD makes with the vertical in the equilibrium position in which D is below the level of A.

(O & C)

6 A solid hemisphere of radius a and a solid right circular cone of height h and base radius a are made from the same uniform material and are joined together with their plane faces completely in contact. O is the centre of the common base. Find the distance from O of the centre of gravity of the whole body.

(a) The body is freely suspended from a point A on the edge of the common base and hangs in equilibrium under gravity. If $h = 3a$, find, correct to the nearest degree, the acute angle made by AO with the vertical.

(b) The body is found to rest in equilibrium when it is placed on a horizontal table with *any* point of its hemispherical surface in contact with the table. Find h in terms of a.

(O & C)

7 Figure 17.18 shows a square OABC of side a. The midpoint of BC is D. Show that, with respect to OA and OC as axes, the coordinates of the centroid F of the triangular region ABD are $(\frac{5}{6}a, \frac{2}{3}a)$.

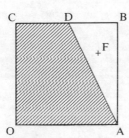

Figure 17.18

Find the coordinates of the centre of mass of a uniform lamina in the form of the figure OADC.

(JMB)

8 Prove, by integration, that the distance of centre of mass of a uniform solid right circular cone, of height h, from its plane base is $\frac{1}{4}h$.

The cone is freely hinged at its vertex and is kept in equilibrium by a light rigid rod of length h joining the centre of the base to a point $h\sqrt{3}$ directly above the vertex. Show that the tension in the rod is $\frac{1}{4}W\sqrt{3}$, where W is the weight of the cone.

Find the magnitude of the reaction at the hinge.

(L)

9 Show, by integration, that the centroid of a uniform hemisphere of radius a is at a distance $\frac{3}{8}a$ from its plane base.

A solid uniform top is made from a right circular cone of radius a and height h, and a hemisphere of radius a. The bases of the cone and hemisphere are coincident. Show that the centroid of the top is at a distance

$$\frac{3h^2 + 8ah + 3a^2}{4(2a + h)}$$

from its vertex.

The top is freely suspended from a point P on the rim of the common base of the hemisphere and cone. Show that, when $h = a$ and the top hangs in equilibrium, the angle between the downward vertical and the line joining P to the vertex is approximately $54\frac{1}{2}°$.

(L)

10 A uniform lamina is bounded by that part of the parabola $y^2 = 4ax$ (where $a > 0$) which lies in the first quadrant, by the axis $y = 0$ of the parabola and by the line $x = a$. Find the coordinates of the centroid of the lamina.

This lamina is suspended freely from the vertex of the parabola. Find the tangent of the angle of inclination to the vertical of the axis of the parabola.

(L)

Revision Exercise 4

1 Three forces F_1, F_2 and F_3 act at the top of a yacht's mast, which is in equilibrium. Given that

$$F_1 = 5i - 2j - 12k, \text{ and } F_2 = -5i - 2j - 12k$$

and that the resultant of all three forces acts vertically downwards with a magnitude of 32 units, find F_3 in terms of its component vectors.　　(O)

2 A fixed horizontal hemisphere has centre O and is fixed so that the plane of its rim is horizontal. A particle A of mass m can move on the inside surface of the sphere. The particle is acted on by a horizontal force of magnitude P, whose line of action is in the vertical plane through O and A. The diagram shows the situation when A is in equilibrium, the line OA making an acute angle θ with the vertical.

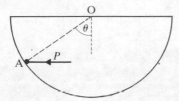

Figure 17.19

(a) Given that the inside of the hemisphere is smooth, find $\tan \theta$ in terms of P, m and g.

(b) Given instead that the inside of the hemisphere is rough, with coefficient of friction μ between the surface and A, and given that the particle is about to slip downwards, show that

$$\tan \theta = \frac{P + \mu mg}{mg - \mu P}$$

　　(C)

3 A rough fixed plane is inclined at an angle α to the horizontal. A particle of mass m rests on the plane, the coefficient of friction between the particle and the plane being μ.

(a) Given that the plane is sufficiently rough to prevent the particle sliding down, prove that $\mu \geqslant \tan \alpha$.

(b) The particle is acted on by a horizontal force whose line of action lies in a vertical plane containing the line of greatest slope. This force acts so as to push the particle up the plane. Given that the particle is about to slip up the plane when the magnitude of the horizontal force is P, find an expression for P in terms of μ, m, g and α. Deduce that $\mu < \cot \alpha$.　　(C)

4 (a) A particle is acted on by three forces with the following magnitudes and directions:

1 N in the direction due S,
2 N in the direction N 60° W,
3 N in the direction N 60° E.

Find the magnitude and direction of the resultant. (*continued*)

An additional force acts on the particle so that the particle is in equilibrium under the four forces. Find the magnitude and direction of this additional force.

(b) Find k, given that

$$k\overrightarrow{BD} = 3\overrightarrow{AC} + 3\overrightarrow{BC} - 3\overrightarrow{AB} - \overrightarrow{BD} + 6\overrightarrow{CD}$$

where ABCD is a plane quadrilateral.

(C)

5 A small rough ring P of weight W is threaded on to a fixed horizontal wire which passes through a point A, as shown in the diagram. The ring is joined by a light elastic string of natural length l and modulus λ to a fixed point B which is at a distance $2l$ vertically below A.

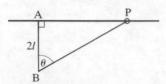

Figure 17.20

The ring is in equilibrium with the angle ABP $= \theta$. Find the tension in the string and the vertical and horizontal components of the force exerted by the wire on the ring. Show that

$$W \geqslant \lambda(2 - \cos\theta)\left(\frac{\tan\theta}{\mu} - 1\right)$$

(JMB)

6 A rectangle ABCD has $AB = 3\,\text{m}$ and $BC = 4\,\text{m}$. Forces all measured in newtons and of magnitudes 2, 4, 6, 8 and k, act along AB, BC, CD, DA and AC, respectively, the direction of each force being shown by the order of the letters. The resultant of the five forces is parallel to BD. Find k and show that the resultant has a magnitude $\frac{5}{6}$ newtons. Find the distance from A of the line of action of the resultant.

(O & C)

7 A uniform rod OA, of weight W and length $2a$, is pivoted smoothly to a fixed point at the end O. A light inextensible string, also of length $2a$, has one end attached to the end A of the rod. Its other end is attached to a small heavy ring of weight $\frac{1}{2}W$ which is threaded onto a fixed rough straight horizontal wire passing through O. The system is in equilibrium with the string taut and the rod making an angle α with the downward vertical at O.

By taking moments about O for the rod and ring together, or otherwise, find the normal component R of the reaction exerted by the wire on the ring. Find also the frictional force F on the ring, in terms of W and α.

The equilibrium is limiting when $\alpha = 60°$. Determine the coefficient of friction between the wire and the ring.

(O & C)

8 A rigid rectangular lamina ABCD, with $AB = 4a$ and $BC = 3a$, is subject to forces of magnitudes $10P$, P, $2P$, $3P$ acting along CA, AD, DC, CB respectively in the directions indicated by the order of the letters.
  (a) Find the magnitude of the resultant of the four forces.
  (b) Find the tangent of the acute angle between the line of action of the resultant and the edge AB of the lamina.
  (c) Find the distance from A of the point where the line of action of the resultant meets AB.
  (d) Indicate clearly on a diagram the line of action and the direction of the resultant.
  (e) Find the magnitude and sense of a couple G, which, if added to the system, would cause the resultant force to act through E, the midpoint of CD.
  (f) In the case when G is *not* applied, find the forces S along AB, T along AD and U along BC which, when added to the system, would produce equilibrium.

(JMB)

9 A uniform rod AB, of length $4a$, and mass m, is smoothly hinged to a vertical wall at the end A. The rod is horizontal and is kept in equilibrium at right-angles to the wall by a light inextensible string, one end of which is attached to the rod at a point C, where $AC = 3a$. The other end of the string is attached to the wall at point D, which is vertically above A and at a distance $4a$ from it.

Calculate in terms of m and g
  (a) the magnitude of the tension in the string;
  (b) the magnitude of the force exerted by the hinge at A on the rod.

Calculate also the tangent of the angle made with the horizontal by the line of action of this force at the hinge.

Given that the string breaks when the magnitude of the tension in it exceeds $2mg$, calculate the magnitude of the greatest load which can be attached to the rod at the end B without the string breaking.

(L)

10 A uniform rod AB, of length $2a$ and mass m, rests in equilibrium with its lower end A on a rough horizontal floor. Equilibrium is maintained by a horizontal elastic string, of natural length a and modulus λ. One end of the string is attached to B and the other end to a point vertically above A. Given that θ, where $\theta < \pi/3$, is the inclination of the rod to the horizontal, show that the magnitude of the tension in the string is $\frac{1}{2}mg \cot \theta$.

Prove also that

$$2\lambda = \frac{mg \cot \theta}{2 \cos \theta - 1}$$

Given that the system is in limiting equilibrium and that the coefficient of friction between the floor and the rod is $\frac{2}{3}$, find $\tan \theta$. Hence show that $\lambda = \frac{10}{9} mg$.

(L)

11 Use integration to show that the centre of mass of a uniform semicircular lamina, of radius a, is at a distance $\dfrac{4a}{3\pi}$ from O, the midpoint of its straight edge.

A semicircular lamina, of radius b and with O as the midpoint of its straight edge, is removed from the first lamina. Show that the centre of mass of the resulting lamina is at a distance $\bar{x}$ from O, where

$$\bar{x} = \frac{4(a^2 + ab + b^2)}{3\pi(a + b)}$$

Hence find the position of the centre of mass of a uniform semicircular arc of radius a.

(L)

12 Show by integration that the centre of mass of a uniform solid hemisphere of radius a is at a distance $\frac{3}{8}a$ from the centre of its plane face.

A uniform solid hemispherical bowl has internal radius $\frac{1}{2}a$ and external radius a, and the axes of the outer and inner surfaces coincide. The point O is the centre of the diameter AB of the outer rim of the bowl. Show that the centre of mass of the bowl is at a distance $\frac{45}{112}a$ from O.

The bowl is suspended by a light inextensible string, one end of which is attached at the point A, and the other end of which is attached to a fixed point C. Find the tangent of the angle which AB makes with the vertical when the bowl hangs in equilibrium.

(L)

13 A uniform triangular lamina ABC has mass m and each side is of length $2a$. The lamina rests in a fixed vertical plane with A on a rough horizontal table and the side AC is vertical. Equilibrium is maintained by a force of magnitude T which acts alongside BC. Show that $T = \frac{1}{3}mg$.

Find the magnitude of the normal reaction at A.

Show also that equilibrium is possible only when $\mu \geqslant \frac{1}{5}\sqrt{3}$, where μ is the coefficient of friction between the lamina and the table.

(L)

14 Prove by integration that the centre of mass of a uniform solid right circular cone is one quarter of the way up the axis from the base.

From a uniform solid right circular cone of height H, another cone is removed with the same base and of height h, the two axes coinciding. Show that the centre of mass of the remaining solid S is a distance $\frac{1}{4}(3H - h)$ from the vertex of the original cone.

The solid S is suspended by two vertical strings, one attached to the vertex and the other attached to a point on the bounding circular base. Given that S is in equilibrium, with its axis of symmetry horizontal, find, in terms of H and h, the ratio of the magnitude of the tension in the string attached to the vertex, to that in the other string.

(L)

15 The curved surface of a uniform solid hemisphere is in contact with a horizontal floor and a vertical wall. The plane surface of the hemisphere makes an angle α (>0) with the horizontal. Show that equilibrium is

(a) impossible if the floor is smooth;

(b) possible if the floor is rough and the wall is smooth, provided that $8\mu \geqslant 3\sin\alpha$, where μ is the coefficient of friction between the floor and the hemisphere.

Given that the floor and the wall are rough and the coefficient of friction between the hemisphere and both the wall and the floor is $\frac{1}{8}$, find the value of $\sin\alpha$ when equilibrium is limiting at both points of contact and the hemisphere is on the point of sliding down the wall.

(L)

16 A uniform rectangular lamina has vertices A, B, C, D where $AB = CD = 6a$ and $AD = BC = 10a$. The point P lies on BC at a distance $6a$ from C and the point Q lies on AD at a distance $6a$ from D. The triangular portion CDP is folded through 180° about DP so that C lies on Q. Show that the centre of mass, G, of the folded lamina is at a perpendicular distance $\frac{22}{5}a$ from AB and calculate the perpendicular distance of G from AD.

The folded lamina is suspended from P and hangs freely in equilibrium under gravity. Find the tangent of the angle which PQ makes with the vertical.

(C)

17 A uniform solid right circular cone has vertex O, height h and semi-vertical angle α. From this cone the conical portion with vertex O and height λh (where $\lambda < 1$) is removed. Find $\bar{x}$, the distance from O to the centre of mass of the remaining solid.

This solid is placed on a horizontal plane with its curved surface in contact with the plane. Show that, for equilibrium,

$$\bar{x}\cos\alpha \geqslant \lambda h \sec\alpha$$

Deduce that, for equilibrium in the case $\lambda = \frac{1}{2}$,

$$\cos^2\alpha > \tfrac{28}{45}.$$

(C)

18 A uniform straight wire AB, of length $9a$, is bent through 90° at the point O, where $AO = 4a$, and also at N, where $NB = 2a$. The four points A, O, N and B are coplanar, and A and B are on the same side of ON. Show that the distance, $\bar{x}$, of the centre of mass of the bent wire from OA is given by $\bar{x} = \frac{7}{6}a$, and find $\bar{y}$, the distance of the centre of mass from ON.

The bent wire is freely suspended from B and hangs in equilibrium. Find the tangent of the angle of inclination of BN to the vertical.

(C)

19 The diagram shows the vertical section through the centre of two identical uniform rough cubes.

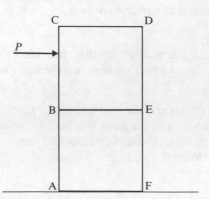

Figure 17.21

Each cube has weight W and edges of length a. The upper cube rests on the lower cube, which rests on a rough horizontal plane. The coefficient of friction at each plane of contact is μ. A horizontal force of magnitude P in the plane of the section is applied to the upper cube at a distance h above BE, where $h < a$. Obtain, in terms of a, h, μ, W, inequalities satisfied by P which will ensure

(a) that the lower cube does not slide along the plane;
(b) that the upper cube does not slide over the lower cube;
(c) that the lower cube does not topple;
(d) that the upper cube does not topple;

Determine how equilibrium would be broken if P increased steadily from zero in the case $\mu < \dfrac{a}{a+h}$, and also in the case $\mu > \dfrac{a}{a+h}$.

(C)

20 Prove by integration that the centre of mass of a uniform solid right circular cone of height h and base radius r is at a distance $\frac{3}{4}h$ from the vertex.

Such a cone is joined to a uniform solid right circular cylinder, of the same material, with base radius r and height l, so that the plane base of the cone coincides with a plane face of the cylinder. Find the distance of the centre of mass of the solid thus formed from the base.
Show that, if $6l^2 \geqslant h^2$, this solid can rest in equilibrium on a horizontal plane, with the curved surface of the cylinder touching the plane.

Given that $l = h$, show that the solid can rest in equilibrium with its conical surface touching the plane, provided that $r \geqslant \frac{1}{4}h\sqrt{5}$.

(JMB)

Chapter 18

Moment of inertia

18.1 The kinetic energy of a rigid body rotating about a fixed axis

In this chapter we return to the study of objects which are moving, i.e. dynamics. In particular we shall be looking at the motion of a rigid body when it rotates about a fixed axis. Most machines contain several parts which rotate, for example, pulleys, flywheels, or cogs. In this book we have frequently had examples featuring pulleys, but so far we have been obliged to regard them as 'light', in order to neglect their dynamic properties.

Firstly, we shall examine the kinetic energy of a rotating rigid body. In Chapter 10 we saw that the kinetic energy of a particle, of mass m, moving with speed v, is $\frac{1}{2}mv^2$. We have also seen (in Chapter 13) that if a point P is at a distance r from a fixed point O, where r is constant, and if OP is rotating with constant angular speed ω rad s^{-1}, then the linear speed of P is $r\omega$, (see Figure 18.1). So a particle of mass m, attached to point P, would have kinetic energy $\frac{1}{2}m(r\omega)^2$.

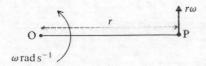

Figure 18.1

In the example which follows, we shall find the kinetic energy of a system of particles attached to a light rigid rod. Although this is a rather artificial situation, in that we shall neglect the dynamical effect of the rod itself, this will serve as an introduction to the method we shall use on bodies whose mass is distributed continuously. (A similar approach was used when we examined the centre of gravity of a rigid body.)

Kinetic energy of a rotating body

Example 1

A light rigid rod OX is 4 m long. A, B and C are three points on the
rod such that $OA = AB = BC = CX = 1$ metre. Small particles, of
mass m, $2m$, $4m$ and m kg are attached to A, B, C and X respectively
(see Figure 18.2). Find the kinetic energy of the system when the rod
rotates about an axis through O, at a constant angular speed of
ω rad s^{-1}.

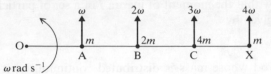

Figure 18.2

The system is shown in Figure 18.2, with the linear speed of each
particle indicated. The total kinetic energy is

$$[\tfrac{1}{2} \times m \times (\omega)^2] + [\tfrac{1}{2} \times 2m \times (2\omega)^2] + [\tfrac{1}{2} \times 4m \times (3\omega)^2] + [\tfrac{1}{2} \times m \times (4\omega)^2]$$
$$= \tfrac{1}{2}(m + 8m + 36m + 16m)\omega^2$$
$$= \tfrac{1}{2}(61m)\omega^2$$

The kinetic energy of the system is $\frac{61}{2} m \omega^2$ joules.

Qu. 1 Find the kinetic energy of the system of particles in Example 1, when the
rod rotates, at the same angular speed, about point B.

If, now, we generalise the situation described in Example 1, by considering particles of mass $m_1, m_2, m_3, \ldots, m_n$, fixed to points on the rigid rod at distances
$r_1, r_2, r_3, \ldots, r_n$, respectively, from O, the total kinetic energy would be

$$\tfrac{1}{2}m_1(r_1\omega)^2 + \tfrac{1}{2}m_2(r_2\omega)^2 + \tfrac{1}{2}m_3(r_3\omega)^2 + \ldots + \tfrac{1}{2}m_n(r_n\omega)^2$$
$$= \tfrac{1}{2}(m_1r_1^2 + m_2r_2^2 + m_3r_3^2 + \ldots + m_nr_n^2)\omega^2$$

The term in brackets is called the **moment of inertia** about O of the system of
particles; it is normally denoted by the symbol I. Using this, the kinetic energy of
the system can be written

$$\text{K.E.} = \tfrac{1}{2}I\omega^2$$

and using the 'sigma notation', we can write

$$I = \sum (m_i r_i^2)$$

Each term in the formula for I has the form (mass) × (length squared), so when
using SI units, moment of inertia is measured in kg m^2. The moment of inertia of
the system of particles in Example 1 was $61m$, but notice that the same system of
particles has moment of inertia $9m$ when the axis is at point B (see Qu. 1). This

illustrates an important feature of the moment of inertia of a rigid body, namely, that it depends on the position of the axis.

18.2 The moment of inertia of a uniform rod

We shall now consider the moment of inertia of a uniform rod, about an axis perpendicular to the rod, through one end. Let its mass be M and its length $2a$.

In the previous section we saw that the moment of inertia I of a set of particles arranged in a straight line is given by

$$I = \sum (m_i r_i^2)$$

In order to apply this to a rod whose mass is distributed continuously, it is necessary to use calculus. We shall write the mass per unit of length, as ρ. (Notice that, since the rod is 'uniform', ρ is constant and $M = 2a\rho$.)

Figure 18.3 shows the rod AB, and a typical point P at a distance x from end A. Consider the small element PQ, where $PQ = \delta x$.

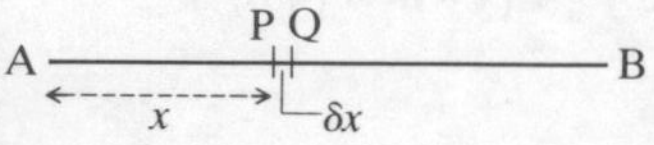

A ————————— P Q ————————— B

$\overset{\longleftarrow\!-\!-\!-\!-\!-\!-\!\longrightarrow}{\quad x \quad}$ $\overset{}{\underset{\delta x}{\rule{0pt}{0pt}}}$

Figure 18.3

The mass of the typical element PQ is then $\rho(\delta x)$ and its distance from A is x, so the moment of inertia is given by

$$I = \sum (x^2 \rho \, \delta x)$$

We now find the limit of this sum as δx tends to zero. As usual this means we evaluate the corresponding definite integral, i.e.

$$I = \rho \int_0^{2a} x^2 \, dx$$

$$= \rho \left[\tfrac{1}{3} x^3 \right]_0^{2a}$$

$$= \tfrac{8}{3} \rho a^3$$

But $2a\rho = M$, so $I = \tfrac{4}{3} M a^2$

> The moment of inertia of a uniform rod, of mass M and length $2a$, about an axis through one end is $\tfrac{4}{3} M a^2$.

Qu. 2 Write down the moment of inertia of a uniform rod, of mass m and length l, about an axis at one end. By regarding a uniform rod, of length $2l$ and mass M, as a pair of rods, each of length l and mass $m (= \tfrac{1}{2} M)$, rigidly joined together in a straight line, show that the moment of inertia of the composite rod about an axis through its midpoint is $M l^2$.

Exercise 18a

1 Four equal particles, each of mass m, are attached to points O, A, B and C, of a light rigid rod OC, where $OA = AB = BC = a$. Find the moment of inertia of this system about an axis through O. Hence write down its kinetic energy when it rotates about O at ω rad s^{-1}.

2 The rod OC in question 1 is freely pivoted at O, and initially it is held with OC horizontal. It is then released. Using the principle of conservation of energy, find, in terms of g and a, the angular velocity of the rod at the instant when C is vertically below O.

3 Show that the moment of inertia of the rod OC in question 1, about a point X on the rod, where $OX = x$, is $m(4x^2 - 12ax + 14a^2)$. Show that the moment of inertia is least when X is the midpoint of OC, i.e. when $x = \frac{3}{2}a$, and find this least value.

4 Prove, by integration, that the moment of inertia of a uniform rod of mass M and length $2a$, about an axis perpendicular to the rod and through its midpoint, is $\frac{1}{3}Ma^2$.

(Hint: Use the method of the preceding section; measure x from the midpoint of the rod and integrate from $x = -a$ to $x = +a$.)

5 Find the moment of inertia of the uniform rod in question 4 about an axis through a point X in the rod, at a distance c from one end. Show that the moment of inertia is least when X is at the midpoint of the rod.

18.3 The moment of inertia of a set of particles in two dimensions

Consider a set of four particles, each of mass m, attached to a light grid at points A, B, C and D whose coordinates are $(3a, 2a)$, $(7a, 2a)$, $(7a, 6a)$ and $(3a, 6a)$ respectively, as in Figure 18.4.

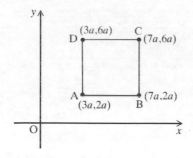

Figure 18.4

The moment of inertia, I_x, about the x-axis is given by

$$I_x = [m \times (2a)^2] + [m \times (2a)^2] + [m \times (6a)^2] + [m \times (6a)^2]$$
$$= 80ma^2$$

The moment of inertia, I_y, about the y-axis is given by

$$I_y = [m \times (3a)^2] + [m \times (7a)^2] + [m \times (7a)^2] + [m \times (3a)^2]$$
$$= 116\, ma^2$$

We can also find the moment of inertia about the axis through the origin which is *perpendicular* to the plane containing the particles. This is called the z-axis, and we shall write the moment of inertia about this axis as I_z. The distances of the particles from the z-axis are

$$OA = a\sqrt{13} \qquad OB = a\sqrt{53} \qquad OC = a\sqrt{85} \quad \text{and} \quad OD = a\sqrt{45}$$

Hence

$$I_z = [m \times 13a^2] + [m \times 53a^2] + [m \times 85a^2] + [m \times 45a^2]$$
$$= 196\, ma^2$$

(Notice that $I_z = I_x + I_y$. We shall see in the next section that this is significant.)

We can generalise this for a set of n particles, each of mass m, attached to fixed points in a plane. Let (x_i, y_i) be the coordinates of a typical particle, then, using the sigma notation,

$$I_x = \sum (my_i^2) \qquad I_y = \sum (mx_i^2) \quad \text{and} \quad I_z = \sum (mr_i^2)$$

where $r_i^2 = x_i^2 + y_i^2$.

18.4 The perpendicular axes theorem

From the last section, we know that

$$I_z = \sum (mr_i^2)$$

where $r_i^2 = x_i^2 + y_i^2$. So

$$I_z = \sum m(x_i^2 + y_i^2)$$
$$I_z = \sum (mx_i^2) + \sum (my_i^2)$$
$$= I_y + I_x$$

$$\therefore \quad \boxed{I_z = I_x + I_y}$$

In other words, the sum of the moments of inertia of a set of coplanar particles about two perpendicular axes *in the plane of the particles*, is equal to the moment of inertia about an axis which goes through the common point of the two axes and is perpendicular to them. This is known as the **perpendicular axes theorem**.

The same reasoning can be applied to a lamina, with a continuously distributed mass. For example, if a rectangular plate has moments of inertia I and J about axes through its centre and parallel to its sides, the moment of inertia about an axis through its centre and perpendicular to its plane is $(I + J)$.

18.5 The moment of inertia of a circular ring

Consider the ring shown in Figure 18.5. We take the magnitude of the cross-section to be negligible compared with the radius r (as in a child's hoop).

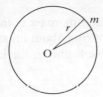

Figure 18.5

Divide the circumference into small segments, each of mass m. The moment of inertia, I, about an axis through the centre and perpendicular to the plane of the hoop, is given by

$$I = \sum (mr^2)$$

$$= r^2 \left(\sum m\right) \qquad \text{since } r \text{ is constant for all the small segments}$$

$$= Mr^2 \qquad \text{where } M \left(= \sum m\right) \text{ is the total mass}$$

> The moment of inertia of a circular ring, mass M and radius r, about an axis through its centre perpendicular to its plane is Mr^2.

Let the moment of inertia of the ring about a diameter be J: the moment of inertia about all diameters will also be J. Taking two perpendicular diameters and applying the perpendicular axes theorem, we have

$$J + J = Mr^2$$
$$\therefore \qquad J = \tfrac{1}{2}Mr^2$$

> The moment of inertia of a circular ring, mass M and radius r, about a diameter is $\tfrac{1}{2}Mr^2$.

18.6 The moment of inertia of a uniform circular disc

In order to find the moment of inertia of a uniform disc, mass M and radius r, about an axis through its centre and perpendicular to its plane, we divide the disc into concentric rings and we consider a typical one, bounded by circles of radius x and $(x + \delta x)$, as shown in Figure 18.6. We shall take the mass per unit of area, to be σ; the total mass can then be written $M = \pi r^2 \sigma$.

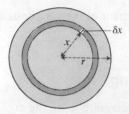

Figure 18.6

The area of the typical ring can be taken to be $2\pi x(\delta x)$ and its mass is then $2\pi x(\delta x)\sigma$. Using the formula for the moment of inertia of a ring about an axis perpendicular to its plane and through its centre, which we found in the last section, we have

moment of inertia of the typical ring $= 2\pi x(\delta x)\sigma x^2$
$$= 2\pi \sigma x^3(\delta x)$$

Summing for all such rings, and letting δx tend to zero gives

the moment of inertia of the disc $= \displaystyle\int_0^r 2\pi\sigma x^3 \, \mathrm{d}x$

$$= 2\pi\sigma\left[\tfrac{1}{4}x^4\right]_0^r$$

$$= \tfrac{1}{2}\pi\sigma r^4$$

But $\pi\sigma r^2 = M$, so:

> The moment of inertia of a uniform disc, mass M and radius r, about an axis through its centre and perpendicular to its plane is $\tfrac{1}{2}Mr^2$.

As in the previous section we can now apply the perpendicular axes theorem to this result to find the moment of inertia of a disc about a diameter, i.e.

> The moment of inertia of a uniform disc, mass M and radius r, about an axis along a diameter is $\tfrac{1}{4}Mr^2$.

18.7 The parallel axes theorem

We have already noted that the formula for the moment of inertia of a rigid body depends on the position of the axis, so for any one body there will be infinitely many different formulae. Fortunately the situation is not as complicated as it sounds: provided we can find the moment of inertia for an axis passing through the centre of gravity, we can modify the formula, using the parallel axes theorem, to find the moment of inertia about any parallel axis. The theorem, which we shall prove below, is fairly simple. If the axis is moved a distance a, we have only to add Ma^2, where M is the total mass, to the formula for the moment of inertia about an axis through the centre of gravity. If we write the moment of inertia about an axis through the centre of gravity as I_G, and that through point A as I_A, then

$$I_A = I_G + Ma^2$$

(where a is the distance between the two axes). For example, we have seen earlier that the moment of inertia of a uniform rod, mass M and length $2a$ about an axis through its centre is $\frac{1}{3}Ma^2$ (see Qu. 2). If now we move the axis to one end of the rod, that is a distance a, the formula for the moment of inertia becomes

$$\frac{1}{3}Ma^2 + Ma^2$$
$$= \frac{4}{3}Ma^2 \qquad \text{(as we saw in §18.2)}$$

Before we embark on the proof of this important theorem, there is a preliminary point to note. Suppose we have a set of particles, each of mass m, as in Figure 18.7 [one typical particle is shown, at point $P(x,y)$].

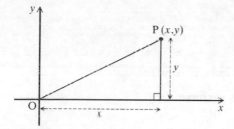

Figure 18.7

Proceeding as we did in Chapter 16, the position of the centre of mass $(\bar{x}, \bar{y})$ is given by

$$M\bar{x} = \sum(mx) \quad \text{and} \quad M\bar{y} = \sum(my)$$

where $M = \sum m$, the total mass.

If the centre of gravity is at the origin, then $\bar{x} = \bar{y} = 0$ and hence

$$\sum(mx) = \sum(my) = 0 \qquad\qquad \cdots \quad (1)$$

This will be a vital step in the proof which follows.

Let the moment of inertia of the rigid body about an axis through the centre of gravity G, and perpendicular to the plane of the axes shown in Figure 18.8, be I_G. We wish to find the moment of inertia about a parallel axis through a point A, (where the distance between the two axes is a). There is no loss of generality if we take G to be the origin, and A to be a point on the x-axis, making the coordinates of A $(a,0)$.

Consider a typical particle, mass m, situated at $P(x,y)$. Let $OP = r$, then

$$I_G = \sum (mr^2) \qquad \qquad \qquad \cdots \quad (2)$$

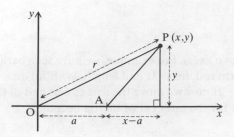

Figure 18.8

The moment of inertia, I_A, about a parallel axis through A, is given by

$$I_A = \sum (m \times AP^2) \qquad \qquad \cdots \quad (3)$$

Now

$$AP^2 = y^2 + (x - a)^2$$
$$= y^2 + x^2 - 2ax + a^2$$
$$= r^2 - 2ax + a^2 \qquad \text{(since } r^2 = x^2 + y^2\text{)}$$

Substituting for AP^2, in equation (3), we have

$$I_A = \sum m(r^2 - 2ax + a^2)$$
$$= \sum (mr^2) - 2a\left[\sum (mx)\right] + a^2\left[\sum m\right]$$

We know, from (2), that $\sum (mr^2) = I_G$, and, from (1), that $\sum (mx) = 0$. So writing $\sum m$ as M, the total mass, we have

$$I_A = I_G + Ma^2$$

as required.

Note

(a) The theorem requires that the original axis must pass through the centre of gravity.

(b) Since moving the axis from G *increases* the magnitude of the moment of inertia, it follows that the moment of inertia is least for an axis which passes through the centre of gravity.

(c) Unlike the 'perpendicular axes theorem', this result is not restricted to laminae.

Qu. 3 Find the moment of inertia of a uniform disc, mass M and radius r, about a tangent.

The reader is reminded that the reason for calculating the moment of inertia, I, about a given axis of a rigid body is that we can then write down the kinetic energy (K.E.) of the rigid body when it rotates about that axis with angular speed ω, i.e.

$$\text{K.E.} = \tfrac{1}{2}I\omega^2$$

(It is worth noting that this formula is very similar to the formula $\tfrac{1}{2}Mv^2$ for the kinetic energy of a particle of mass M moving with speed v. The moment of inertia, I, takes the place of the mass, M, and the angular speed, ω, replaces the speed v).

The next example shows how results from the preceding pages can be used to find the K.E. of a flywheel.

Example 2

A heavy flywheel consists of a uniform circular disc, of mass 20 kg and radius 0.5 m. Find its kinetic energy when it rotates at 30 r.p.m. about an axis perpendicular to its plane, passing through
(a) the centre of the disc;
(b) a point on the rim of the disc.

The angular velocity of the flywheel is 30 revolutions per minute
    $= 60\pi$ radians per minute
    $= \pi$ radians per second

The kinetic energy (K.E.) of a rigid body is $\tfrac{1}{2}I\omega^2$, where I is its moment of inertia about the given axis, and ω is its angular velocity in rad s^{-1}. In this example $\omega = \pi$, hence the K.E. of the flywheel is $\tfrac{1}{2}I\pi^2$ joules.

(a) If the axis of rotation passes through the centre of the disc, the moment of inertia is given by the formula $\tfrac{1}{2}Mr^2$, i.e.

$$I = \tfrac{1}{2} \times 20 \times 0.5^2 \text{ kg m}^2$$
$$= 2.5 \text{ kg m}^2$$

Hence the kinetic energy

$$= \tfrac{1}{2} \times 2.5 \times \pi^2 \text{ J}$$
$$= 12.3 \text{ J} \quad \text{(correct to three significant figures)}$$

(b) If the axis passes through a point on the rim
$$I = \tfrac{1}{2}Mr^2 + Mr^2 \quad \text{(parallel axes theorem)}$$
$$= \tfrac{3}{2}Mr^2$$
$$= \tfrac{3}{2} \times 20 \times 0.5^2 \text{ kg m}^2$$
$$= 7.5 \text{ kg m}^2$$

Hence the kinetic energy

$$= \tfrac{1}{2} \times 7.5 \times \pi^2 \text{ J}$$
$$= 37.0 \text{ J} \quad \text{(correct to three significant figures)}$$

18.8 The radius of gyration

Suppose that a certain rigid body has total mass M and that its moment of inertia about a given axis is I. Its **radius of gyration**, which is always denoted by k, about this axis is defined as follows

$$k = \sqrt{\left(\frac{I}{M}\right)}$$

or $I = Mk^2$

For example, we have already seen that the moment of inertia of a uniform rod, mass M and length $2a$ about its midpoint is $\frac{1}{3}Ma^2$, so its radius of gyration about the same axis is $\dfrac{a}{\sqrt{3}}$.

Summary

In the table below, M always represents the total mass; all the objects have 'uniform' mass:

Description	Dimensions	Position of axis	Moment of inertia	Radius of gyration
Rod	length $2a$	through centre, perpendicular to the rod	$\dfrac{Ma^2}{3}$	$\dfrac{a}{\sqrt{3}}$
Rod	length $2a$	through end, perpendicular to the rod	$\dfrac{4Ma^2}{3}$	$\dfrac{2a}{\sqrt{3}}$
Ring	radius r	through centre, perpendicular to plane	Mr^2	r
Ring	radius r	diameter	$\dfrac{Mr^2}{2}$	$\dfrac{r}{\sqrt{2}}$
Disc	radius r	through centre, perpendicular to plane	$\dfrac{Mr^2}{2}$	$\dfrac{r}{\sqrt{2}}$
Disc	radius r	diameter	$\dfrac{Mr^2}{4}$	$\dfrac{r}{2}$

Exercise 18b

1 Find the moment of inertia of a set of three particles, each of mass m, situated at the vertices of an equilateral triangle of side $2a$
  (a) about an axis along a median;
  (b) about an axis through the centroid and parallel to one of the sides;
  (c) about an axis through the centroid and perpendicular to the plane of the triangle.
  Check your answer to part (c) by using the perpendicular axes theorem.

2 Find the moment of inertia of a set of four particles, each of mass m, situated at the vertices of a square of side $2a$
   (a) about an axis along a diagonal;
   (b) about an axis along a line through the centre and parallel to one pair of sides;
   (c) about an axis through the centre and perpendicular to the plane of the square.
   Check your answer to part (c) by using the perpendicular axes theorem.

3 Find the moment of inertia of an equilateral triangle ABC, consisting of three uniform rods, AB, BC and CA, each of mass m and length $2a$, about an axis through its centroid and perpendicular to its plane.

4 Find the moment of inertia of a square ABCD, consisting of four uniform rods, AB, BC, CA and DA, each of mass m and length $2a$, about an axis through its centre and perpendicular to its plane.
   Find also the moment of inertia about an axis through the midpoints of a pair of opposite sides.

5 Find the moment of inertia of a set of four particles, each of mass m, situated at points whose coordinates are A (a,a), B $(5a,a)$, C $(6a,4a)$ and D $(2a,4a)$, about the each of the following axes:
   (a) the x-axis;
   (b) the y-axis;
   (c) the axis through the origin and perpendicular to the coordinate plane.

6 A heavy flywheel has mass $4\,kg$ and radius of gyration $0.5\,m$. Find its kinetic energy when it is rotating at 60 r.p.m.

7 A heavy flywheel is rotating at 120 r.p.m. It can be brought to rest in 10 revolutions by a torque of $20\,N\,m$. Find its moment of inertia about the axis of rotation.
   [Hint: work done = torque × angle turned through (in radians)]

8 A simple wheel, centre O, consists of a uniform circular ring, of mass M and radius r, and two uniform rods, each of mass M and length $2r$ intersecting at right-angles at O. Find the moment of inertia of the wheel, about an axis through O perpendicular to its plane, in terms of M and r. Show that when it is rotating at N r.p.m. its kinetic energy is $\frac{1}{1080} Mr^2 N^2 \pi^2$.

9 Show that the moment of inertia of a uniform lamina, of mass M, bounded by two concentric circles whose radii are r and $2r$ (i.e. like a washer), is $\frac{5}{2}Mr^2$, the axis being perpendicular to the plane of the lamina and through its centre.

10 Show that the moment of inertia of a uniform lamina, of mass M, in the form of an equilateral triangle of side $2a$, about an axis through a vertex and perpendicular to the plane of the triangle, is $\frac{5}{3}Ma^2$.

18.9 Moments of inertia–some extensions

We have already seen that the moment of a uniform circular lamina of mass m and radius r, about an axis through its centre and perpendicular to its plane, is $\frac{1}{2}mr^2$. Now suppose we stack a large number of these on top of each other, with their centres forming an axis perpendicular to their planes, making a uniform cylinder (rather like a stack of LP records) of radius r and mass $M (= \sum m)$. This is shown in Figure 18.9.

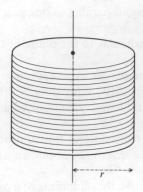

Figure 18.9

We can find the moment of inertia of the cylinder by summing the moments of inertia of the discs, i.e.

$$\text{moment of inertia of the cylinder} = \sum \left(\tfrac{1}{2}mr^2\right)$$
$$= \tfrac{1}{2}\left(\sum m\right)r^2$$

(Note that r is the same for all the discs, so we can make r^2 a factor of the expression.)

But $\sum m = M$, so:

> The moment of inertia of a solid uniform cylinder about its axis is $\tfrac{1}{2}Mr^2$.

Notice that the *form* of the moment of inertia formula is unchanged. A similar phenomenon will be observed whenever an object is 'extended' parallel to the axis; some examples of this appear in the following questions.

Qu. 4 Show that the moment of inertia of a uniform cylindrical shell, (i.e. a cylinder made from a rectangular lamina), of mass M and radius r, about its axis is Mr^2.

Qu. 5 Show that the moment of inertia of a uniform rectangular lamina, ABCD, mass M with sides of length $AB = 2a$ and $BC = 2b$, about the side AB is $\tfrac{4}{3}Mb^2$.

18.10 The moment of inertia of a solid of revolution

The moment of inertia of a solid of revolution about its axis of symmetry can be found by integration; the method is very similar to that we have already used when finding the centre of gravity (see Chapter 16).

Example 3

Find the moment of inertia of a uniform solid sphere, mass M radius r, about a diameter.

Let the density be ρ. The total mass M is then $\frac{4}{3}\rho\pi r^3$. As usual, we divide the solid into elements, in this case circular discs perpendicular to the axis. Consider a typical disc whose centre is at a distance x from O, the centre of the sphere. The radius of the disc is $\sqrt{(r^2 - x^2)}$ and its thickness is δx (see Figure 18.10).

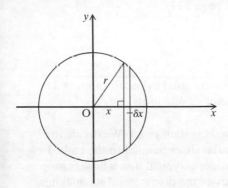

Figure 18.10

$$\text{mass of a typical disc} = \pi\rho(r^2 - x^2)(\delta x)$$

Using the formula for the moment of inertia of a uniform disc (see the Summary in §18.8, page 382), we have

$$\text{moment of inertia of typical disc} = \frac{1}{2} \times \pi\rho(r^2 - x^2)(\delta x) \times (r^2 - x^2)$$
$$= \frac{1}{2}\pi\rho(r^2 - x^2)^2 \,\delta x$$

Summing for all such elements (i.e. from $x = -r$ to $x = +r$) and letting δx tend to zero, we have

$$\text{moment of inertia of the sphere} = \int_{-r}^{+r} \frac{1}{2}\pi\rho(r^2 - x^2)^2 \,dx$$

$$= \frac{1}{2}\pi\rho \int_{-r}^{+r} (r^4 - 2r^2x^2 + x^4)\,dx$$

$$= \frac{1}{2}\pi\rho \left[r^4x - \frac{2}{3}r^2x^3 + \frac{1}{5}x^5 \right]_{-r}^{+r}$$

$$= 2 \times \frac{1}{2}\pi\rho[r^5 - \frac{2}{3}r^5 + \frac{1}{5}r^5]$$

$$= \pi\rho[1 - \frac{2}{3} + \frac{1}{5}]r^5$$

$$= \pi\rho\left(\frac{15 - 10 + 3}{15}\right)r^5$$

$$= \pi\rho(\tfrac{8}{15})r^5$$
$$= \tfrac{4}{3}\pi\rho r^3(\tfrac{2}{5}r^2)$$

But $M = \tfrac{4}{3}\pi\rho r^3$, so:

> The moment of inertia of a uniform solid sphere, of mass M and radius r, about a diameter, is $\tfrac{2}{5}Mr^2$.

Qu. 6 Show that the moment of inertia of a thin hollow sphere, mass M radius r, about a diameter is $\tfrac{2}{3}Mr^2$.
(Hint: compare Example 6, Chapter 16, page 339).

Example 4

Find the moment of inertia of a uniform solid cylinder, of mass M, length l and radius r, about a diameter of one of its ends.

Let the density be ρ. The total mass M is then $\rho\pi r^2 l$. We divide the solid into elements, in this case circular discs parallel to the ends. Let the axis of rotation be Oy, and consider a typical disc whose centre is at a distance x from O. The radius of the disc is r and its thickness is δx (see Figure 18.11), so its mass is $\rho\pi r^2(\delta x)$.

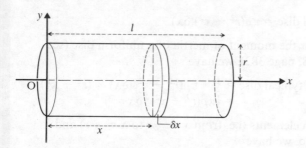

Figure 18.11

The formula for the moment of inertia of a disc about its diameter is $\tfrac{1}{4}mr^2$, (see §18.6, page 378), so

> moment of inertia of typical disc, about its own diameter
> $= \tfrac{1}{4}\rho\pi r^2(\delta x) \times r^2$
> $= \tfrac{1}{4}\rho\pi r^4 \delta x$

Using the parallel axes theorem,

> the moment of inertia of this disc about the y-axis
> $= \tfrac{1}{4}\rho\pi r^4 \delta x + (\rho\pi r^2 \delta x)x^2$
> $= (\tfrac{1}{4}\rho\pi r^4 + \rho\pi r^2 x^2)\delta x$

Summing for all such elements (i.e. from $x = 0$ to $x = l$) and letting δx tend to zero, gives

$$\text{moment of inertia of cylinder} = \int_0^l (\tfrac{1}{4}\rho\pi r^4 + \rho\pi r^2 x^2)\,dx$$

$$= \rho\pi r^2 \int_0^l (\tfrac{1}{4}r^2 + x^2)\,dx$$

$$= \rho\pi r^2 \left[\tfrac{1}{4}r^2 x + \tfrac{1}{3}x^3 \right]_0^l$$

$$= \tfrac{1}{12}\rho\pi r^2 [3r^2 l + 4l^3]$$

$$= \tfrac{1}{12}\rho\pi r^2 l [3r^2 + 4l^2]$$

But $M = \rho\pi r^2 l$, so:

> The moment of inertia of a uniform solid cylinder, of mass M, length l and radius r, about a diameter of one of its ends is $\tfrac{1}{12}M(3r^2 + 4l^2)$.

Qu. 7 Verify the result of the last example by checking that it reduces to the moment of inertia of a thin disc, radius r, about a diameter, when l tends to zero, and to the moment of inertia of a rod, length l, about one end, when r tends to zero.

Exercise 18c

In the questions below, the objects described are uniform, and in each case M represents the total mass.

1 Find the moment of inertia of a rectangular lamina, with sides of length $2a$ and $2b$, about an axis through its centre and perpendicular to its plane.
2 Find the moment of inertia of a solid cylinder, radius r and length $2l$, about an axis through its centre and perpendicular to its axis of symmetry.
3 Find the moment of inertia of a solid cone, height h and base radius r, about its axis of symmetry.
4 Find the moment of inertia of a solid cone, height h and base radius r, about an axis which is a diameter of its base.
5 The formula for the moment of inertia of a thin spherical shell, mass m and radius r, about a diameter, is $\tfrac{2}{3}mr^2$. Use this to prove that the moment of inertia of a solid sphere, mass M and radius R, about a diameter is $\tfrac{2}{5}MR^2$.

18.11 Moment of inertia and kinetic energy

The idea of the moment of inertia of a rigid body, I, was introduced in order to write down the kinetic energy of a rigid body rotating with angular velocity ω: in §18.1 we saw that

$$\text{K.E.} = \tfrac{1}{2}I\omega^2$$

We shall now apply this, together with the various formulae for moments of inertia which we have at our disposal, to problems involving rotating rigid bodies.

Example 5

A heavy flywheel has moment of inertia of $10 \, \text{kg m}^2$ and radius $0.5 \, \text{m}$. It is mounted on frictionless bearings, with its axis horizontal. A light inextensible string, which has one end attached to the rim of the flywheel, is wound around its circumference.

A block of mass $10 \, \text{kg}$ is attached to the free end of the string, which hangs vertically. The system is released from rest. Find the angular velocity of the flywheel when the block has fallen $2 \, \text{m}$.

Since there is no friction, the potential energy (P.E.) lost by the block, will all be converted into kinetic energy (K.E.).

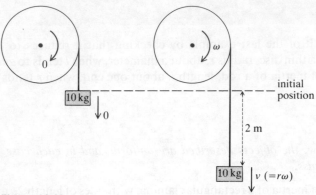

Figure 18.12

P.E. lost by the block, when it has fallen 2 metres,
$$= mg \times 2 \text{ joules}$$

where m is the mass of the block; in this case $m = 10$

∴   P.E. lost by the block $= 20g$ joules   . . . (1)

K.E. gained by the flywheel $= \frac{1}{2}I\omega^2$

where I is its moment of inertia ($= 10$, in this example) and ω is its angular velocity in rad s^{-1}

∴   K.E. gained by the flywheel $= 5\omega^2$ joules   . . . (2)

However, the block has also gained kinetic energy:

K.E. gained by the block $= \frac{1}{2}mv^2$

where v is the linear speed of the block in m s^{-1}

Since the string is inextensible, the speed v of the block is equal to the linear speed of a point on the rim of the flywheel, i.e. $r\omega$ m s^{-1}, where r is its radius; in this case $v = 0.5\omega$ m s^{-1}. Hence

$$\text{K.E. gained by the block} = \tfrac{1}{2} \times 10 \times (0.5\omega)^2 \text{ joules}$$
$$= 1.25\omega^2 \text{ joules} \qquad \ldots \quad (3)$$

Hence, adding (2) and (3), the total K.E. gained is $6.25\omega^2$. But the P.E. lost (see (1)) is $20g$ joules, so using the principle of conservation of energy

$$6.25\omega^2 = 20g$$
$$\therefore \qquad \omega^2 = 3.2g$$

Taking g to be 9.81 m s^{-2}, gives $\omega = 5.6$, correct to two significant figures.

When the block has fallen 2 m, the angular velocity of the flywheel is 5.6 rad s^{-1}.

Example 6

A uniform rod AB, of mass M and length l, is freely pivoted at A. Initially the rod is horizontal, then it is released. Find its angular velocity when B is vertically below A.

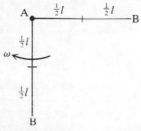

Figure 18.13

The moment of inertia of the rod about A is

$$\tfrac{4}{3}M(\tfrac{1}{2}l)^2 = \tfrac{1}{3}Ml^2$$

Let the angular velocity when B is vertically below A be ω. The K.E. at this point is then

$$\tfrac{1}{2} \times \tfrac{1}{3}Ml^2 \times \omega^2 = \tfrac{1}{6}Ml^2\omega^2$$

The centre of gravity of the rod has fallen a vertical distance $\tfrac{1}{2}l$, so the P.E. lost is $\tfrac{1}{2}Mgl$. Hence, using the principle of conservation of energy,

$$\tfrac{1}{6}Ml^2\omega^2 = \tfrac{1}{2}Mgl$$
$$\therefore \qquad \omega^2 = 3g/l$$

Hence the angular velocity of the rod when B is vertically below A is $\sqrt{(3g/l)}$.

In the next example, we shall use an important piece of calculus technique involving the chain rule, namely:

$$\frac{d}{dt}(v^2) = \frac{d}{dv}(v^2) \times \frac{dv}{dt}$$

Hence

$$\frac{d}{dt}(v^2) = 2v\frac{dv}{dt}$$

Example 7

A heavy circular pulley of radius r and mass M, has a perfectly rough cable passing over it (i.e. the cable does not slip when it is in contact with the pulley). The pulley is freely pivoted about an axis through its centre, and the cable may be taken to be light and inextensible. The free ends of the cable hang vertically and particles A and B are attached to its ends. The mass of A is m_1 and the mass of B is m_2, where $m_1 > m_2$.

The system is released from rest. Find the acceleration of the particles and the tension acting on A.

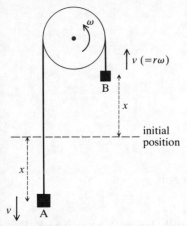

Figure 18.14

Consider the situation when A has descended a distance x. The P.E. lost by A is m_1gx and that gained by B is m_2gx. Hence

the net loss of P.E. $= m_1gx - m_2gx$
$$= (m_1 - m_2)gx$$

The moment of inertia of the pulley is $\frac{1}{2}Mr^2$. Let its angular velocity when particle A has descended a distance x be ω. The K.E. gained by the pulley is then $\frac{1}{4}Mr^2\omega^2$. The particles have also gained K.E., which we can write as $(\frac{1}{2}m_1v^2 + \frac{1}{2}m_2v^2)$; notice that they have the same speed v because the cable is 'inextensible'. Also, since the cable does not slip, $v = r\omega$. Hence

$$
\begin{aligned}
\text{the total K.E. gained} &= \tfrac{1}{4}Mr^2\omega^2 + \tfrac{1}{2}m_1v^2 + \tfrac{1}{2}m_2v^2 \\
&= \tfrac{1}{4}Mv^2 + \tfrac{1}{2}m_1v^2 + \tfrac{1}{2}m_2v^2 \\
&= \tfrac{1}{4}(M + 2m_1 + 2m_2)v^2
\end{aligned}
$$

Using the principle of conservation of energy, we have

$$
\tfrac{1}{4}(M + 2m_1 + 2m_2)v^2 = (m_1 - m_2)gx
$$
$$
(M + 2m_1 + 2m_2)v^2 = 4(m_1 - m_2)gx
$$

Differentiating with respect to t, the time, gives

$$
(M + 2m_1 + 2m_2)\,2v\frac{\mathrm{d}v}{\mathrm{d}t} = 4(m_1 - m_2)\,g\,\frac{\mathrm{d}x}{\mathrm{d}t}
$$

But $\dfrac{\mathrm{d}x}{\mathrm{d}t} = v$, so

$$
(M + 2m_1 + 2m_2)\,2v\frac{\mathrm{d}v}{\mathrm{d}t} = 4(m_1 - m_2)\,gv
$$

Hence, cancelling $2v$,

$$
(M + 2m_1 + 2m_2)\frac{\mathrm{d}v}{\mathrm{d}t} = 2(m_1 - m_2)g
$$

$$
\therefore \quad \frac{\mathrm{d}v}{\mathrm{d}t} = \frac{2(m_1 - m_2)g}{(M + 2m_1 + 2m_2)}
$$

The acceleration of the particles is

$$
\frac{2(m_1 - m_2)g}{(M + 2m_1 + 2m_2)}
$$

Figure 18.15

To find the tension T_1 in the cable acting on A, we use Newton's second law of motion i.e.

$$
m_1g - T_1 = m_1 \times \frac{2(m_1 - m_2)g}{(M + 2m_1 + 2m_2)}
$$

$$
\therefore \quad T_1 = m_1g - \frac{2m_1(m_1 - m_2)g}{(M + 2m_1 + 2m_2)}
$$

$$= \frac{m_1g(M + 2m_1 + 2m_2) - 2m_1(m_1 - m_2)g}{(M + 2m_1 + 2m_2)}$$

$$= \frac{[(M + 2m_1 + 2m_2) - 2(m_1 - m_2)]m_1g}{(M + 2m_1 + 2m_2)}$$

$$\therefore \qquad T_1 = \frac{(M + 4m_2)m_1g}{(M + 2m_1 + 2m_2)}$$

Hence the tension acting on A is

$$\frac{(M + 4m_2)m_1g}{(M + 2m_1 + 2m_2)}$$

Qu. 8 Prove that the tension acting on B in Example 7 is

$$\frac{(M + 4m_1)m_2g}{(M + 2m_1 + 2m_2)}$$

18.12 The compound pendulum

In §12.8 we considered the motion of a simple pendulum, that is, a pendulum consisting of a heavy particle supported by a light inextensible string. With our knowledge of moments of inertia, we can now consider the effect of a more complicated pendulum, one consisting of a rigid body which is free to oscillate in a vertical plane about a smooth pivot. Such a pendulum is called a **compound pendulum**.

Figure 18.16a represents a rigid body, of mass M, whose centre of gravity is at point G. It can rotate freely about an axis through point O, and the distance of the centre of gravity from the axis is h. The line OG swings to and fro in a vertical plane; as it does so, the point G moves along the arc of a circle. Figure 18.16b shows the pendulum at time t; OY is the downward vertical and angle YOG is θ radians.

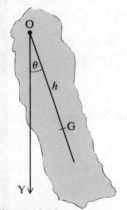

Figure 18.16a

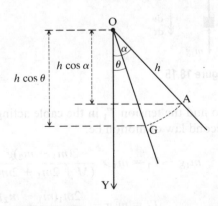

Figure 18.16b

When $t = 0$, the angle θ is equal to α, and the pendulum is released from this position (i.e. when G is at A in Figure 18.16b).

At time t the angular velocity is $\dfrac{d\theta}{dt}$ and we write this as $\dot{\theta}$. When $t = 0$, $\dot{\theta} = 0$.

We shall apply the principle of conservation of energy to the motion of the pendulum. The kinetic energy gained by the pendulum is $\frac{1}{2}I\dot{\theta}^2$, where I is the moment of inertia of the pendulum about an axis through O.

Initially G is at A, and in this position its depth below the fixed point O is $h\cos\alpha$. At time t, the depth of G below O is $h\cos\theta$, and consequently the vertical displacement of the centre of gravity in this time interval is $(h\cos\theta - h\cos\alpha)$. Hence, the P.E. lost is

$$Mgh(\cos\theta - \cos\alpha)$$

Equating the K.E. gained and the P.E. lost, we have

$$\tfrac{1}{2}I\dot{\theta}^2 = Mgh(\cos\theta - \cos\alpha) \qquad \qquad \text{. . . . (1)}$$

When we differentiate with respect to t, $\dot{\theta}^2$ becomes $2\dot{\theta}\ddot{\theta}$, and $\cos\theta$ becomes $-(\sin\theta)\,\dot{\theta}$, and hence differentiating equation (1) with respect to time gives

$$I\dot{\theta}\ddot{\theta} = -Mgh(\sin\theta)\,\dot{\theta}$$

Dividing both sides by $\dot{\theta}$ reduces this to

$$I\ddot{\theta} = -Mgh\sin\theta \qquad \qquad \text{. . . . (2)}$$

As in §12.8, this equation is difficult to solve, but if the oscillations are small in magnitude, we can use the approximation

$$\sin\theta \approx \theta \quad \text{where } \theta \text{ is small}$$

Consequently, for small oscillations equation (2) can be written

$$I\ddot{\theta} = -(Mgh)\theta$$

or $$\ddot{\theta} = -\left(\frac{Mgh}{I}\right)\theta \qquad \qquad \text{. . . . (3)}$$

Now equation (3) has the form of SHM equation (see §12.5, page 211), with $n^2 = \dfrac{Mgh}{I}$, and its period of oscillation is

$$T = \frac{2\pi}{n}$$

$$= 2\pi\sqrt{\left(\frac{I}{Mgh}\right)}$$

Qu. 9 Find the period of a compound pendulum, consisting of a uniform rod, of mass M and length $2a$, when it makes small oscillations about a frictionless pivot through one end.

Exercise 18d (Miscellaneous)

1 Show, by integration, that the moment of inertia of a uniform circular disc, of mass m and radius a, about an axis through the centre and perpendicular to the plane of the disc is $\frac{1}{2}ma^2$.

Deduce the moment of inertia of the disc about a diameter.

Show that the moment of inertia of a uniform right circular cone, of height r, base radius r and mass M, about an axis through its vertex and parallel to a diameter of the base is $\frac{3}{4}Mr^2$.

The above cone is free to turn about a fixed smooth pivot at its vertex and is released from rest with its axis horizontal. Find the angular speed of the cone when its axis is vertical.

(L)

2 Show, by integration, that the moment of inertia of a uniform circular disc, of mass M and radius a, about an axis which passes through its centre and is perpendicular to its plane is $\frac{1}{2}Ma^2$.

Without further integration, deduce the moment of inertia of the disc
(a) about an axis perpendicular to its plane and passing through a point on its circumference;
(b) about a diameter.

A uniform disc, of mass M and radius a, is suspended from a smooth pivot at a point on its circumference and rests in equilibrium. Calculate the period of small oscillations when the centre of the disc is slightly displaced
(c) in the plane of the disc;
(d) perpendicular to the plane of the disc.

(L)

3 Show, by integration, that the moment of inertia of a uniform equilateral triangular lamina, of side $2a$ and mass m, about an axis through a vertex perpendicular to its plane is $\frac{5}{3}ma^2$.

Deduce that the moment of inertia of a uniform regular hexagonal lamina, of side $2a$ and mass M, about an axis through a vertex perpendicular to the plane of the lamina is $\frac{17}{3}Ma^2$.

A compound pendulum consists of a uniform regular hexagonal lamina ABCDEF, of side $2a$ and mass M, with a particle of mass $\frac{1}{2}M$ attached to the vertex D. The pendulum oscillates about a smooth horizontal axis which passes through the vertex A and is perpendicular to the plane of the lamina.

Show that the period of small oscillations is $\pi\sqrt{\left(\dfrac{41a}{3g}\right)}$.

(L)

4 Prove, by integration, that the moment of inertia of a uniform rod of mass m and length l, about an axis through its midpoint and perpendicular to its length, is $\frac{1}{12}ml^2$.

Three such rods AB, BC, CD are rigidly joined so as to form three sides of a square ABCD, and the rods are suspended under gravity from a smooth horizontal axis through D which is perpendicular to the plane of the square. Show that

(a) the moment of inertia of the system about the axis through D is $3ml^2$;

(b) the distance of the centre of gravity of the system from D is $\frac{5}{6}l$.

The system is released from rest with DC horizontal, and A vertically below D. Find the greatest value of its angular velocity in the subsequent motion.

Find also the period of small oscillations about the equilibrium position.

(JMB)

5 A pulley, of radius a, which can rotate smoothly about its fixed horizontal axis, has moment of inertia I about this axis. A light inelastic string passing over the pulley carries masses m_1 and m_2 ($m_2 < m_1$) at its ends. There is no slip between the string and the pulley and the hanging parts of the string are vertical. Show that the acceleration of the mass m_1 is

$$\frac{(m_1 - m_2)ga^2}{I + (m_1 + m_2)a^2}$$

Find the tensions T_1 and T_2 in the hanging parts of the string and show that

$$T_1 + T_2 < (m_1 + m_2)g$$

(JMB)

6 A thin uniform plate P is formed by removing a square of side a from a square plate of side $3a$, one side of the smaller square being part of a side AB of the larger square. Show that

(a) the distance of the centre of mass of the plate P from AB is $\frac{13}{8}a$;

(b) the radius of gyration of the plate P about AB is $a\sqrt{(\frac{10}{3})}$.

The plate P, which can rotate freely about a fixed horizontal axis along AD, is released from rest from the position in which it is horizontal. Calculate the angular velocity and the angular acceleration when the plate has turned through 30°.

If when the plate is horizontal it is given an initial angular velocity greater than $\sqrt{\left(\frac{39g}{40a}\right)}$, prove that it will make complete revolutions.

(O & C)

7 A triangular frame is made of three uniform rods OA, OB, AB each of mass m and with lengths $2a$, $2a$, $2a\sqrt{2}$ respectively. The frame is free to rotate in a vertical plane about a fixed smooth horizontal axis passing through O. Find the distance of the centre of gravity of the frame from O and the moment of inertia of the frame about the axis.

The frame is at rest with AB horizontal and below O. It is then set in motion by being given an angular velocity ω so that B begins to rise. When the frame has turned through an angle θ, prove that

$$\left(\frac{d\theta}{dt}\right)^2 = \omega^2 - \frac{3g\sqrt{2}}{4a}(1 - \cos\theta)$$

Hence show that, if $\sqrt{2} - 1 < \dfrac{4a\omega^2}{3g} < 2\sqrt{2}$, B will rise above the level of O, but the frame will not make complete revolutions.

If $\omega^2 = \dfrac{3g\sqrt{2}}{4a}$, find the greatest height above O reached by B.

(O & C)

8 Write down the moment of inertia of a uniform circular disc, of mass m and radius r, about an axis L in its plane, given that L is at a distance kr from the centre of the disc, where $0 < k \leqslant 1$.

The disc is held with its plane at a small angle to the vertical, with the axis L horizontal and the centre of the disc below the level of L. The disc is then released from rest and makes small oscillations about L. Prove that

$$\frac{d^2\theta}{dt^2} = -\frac{4kg}{(1 + 4k^2)r}\theta$$

where θ is the angle that the plane of the disc makes with the vertical at time t and powers of θ above the second are neglected.

Obtain the period of the oscillations, and find the value of k for which the period is least.

(C)

9 A uniform straight rod AB, of length $2l$ and mass m, is free to swing in a vertical plane about the fixed point A. A particle of mass $6m$ is attached to the rod at a distance x from A. Write down the moment of inertia of the system about an axis through A perpendicular to the rod.

The system makes small oscillations in the vertical plane about the position of stable equilibrium. Show that these oscillations have period

$$2\pi\sqrt{\left[\frac{4l^2 + 18x^2}{3g(l + 6x)}\right]}$$

Deduce that, for the period to be as small as possible, $x = \frac{1}{3}l$.

The particle is attached to the rod at a distance $\frac{1}{3}l$ from A. The rod is held with B vertically above A, slightly displaced and released from rest. Show that the greatest speed attained by B during the subsequent motion is $(24gl)^{\frac{1}{2}}$.

(C)

10 Two uniform circular discs are fixed at their centres to the ends A and B of a uniform straight rod of length $8a$ and mass $3m$. The disc at A has mass $2m$ and radius a. The disc at B has mass $4m$ and radius a. The discs and the rod are coplanar. The body is pivoted on a smooth horizontal axis which is perpendicular to the plane of the discs and which passes through a point O of the rod, where $AO = 3a$. Show that the moment of inertia of the body about this axis is $140ma^2$.

Initially the body is resting in equilibrium with B vertically below O. The body is set in motion by giving **B** a horizontal velocity of magnitude V. In the subsequent motion the body first comes to instantaneous rest when AB is horizontal. Find V in terms of a and g, where g is the acceleration due to gravity.

(C)

Answers

Chapter 1

Exercise 1

2	Greatest	Least
(a)	10.2	9.8
(b)	0.4	0
(c)	26	24
(d)	1.08	1
(e)	1	0.923
(f)	0.102	0.098
(g)	2.5	∞

Chapter 2

Qu. 1 19.995 m/s
Qu. 2 $(20 - 5h)$ m/s
Qu. 3 1.67 m/s^2
Qu. 4 0.67 m/s^2
Qu. 5 (a) 8 m/s^2 (b) 1.6 m/s^2

Exercise 2a

(Where results have been obtained from a graph, it may not be possible to achieve the degree of accuracy shown here.)

1 33 min, 4.4 km; 67 min, 8.9 km
3 10 m s^{-1}, -20 m s^{-1}
5 (a) 0 s, 3.14 s, 6.28 s (b) 0.41 s, 2.73 s, 6.69 s (c) 3.55 s, 5.87 s
(d) 2.70 m s^{-1} (e) -3.27 m s^{-1}

6 30 m s^{-1}
7 (a) 30.5 m s^{-1} (b) 30.05 m s^{-1} (c) $(30 + 5h) \text{ m s}^{-1}$
8 $(10 - 5h) \text{ m s}^{-1}, 10 \text{ m s}^{-1}$
9 -20 m s^{-1}
10 (a) 8.41 m s^{-1} (b) 9.59 m s^{-1} (c) 9.98 m s^{-1}
    (d) 10.00 m s^{-1} velocity is 10 m s^{-1}

Exercise 2b

1 (a) 2 m s^{-2} (b) 0.204 m s^{-2} (c) 9.75 m s^{-2}
2 (a) 0.463 m s^{-2} (b) 1.67 m s^{-2} (c) -2.78 m s^{-2}
3 36 km/h
4 (a) 3.47 s (b) 6.94 s (c) 27.8 s
5 (a) 48.2 m (b) 96.5 m (c) 386 m
6 $a = 2$, 20 m s^{-1}
7 1188 m
8 (a) 480 m (b) 500 m
9 (a) 4.8 m s^{-2} (b) 1.9 m s^{-2} (c) 0.85 m s^{-2}
10 18 m

Exercise 2c (Miscellaneous)

1 (a) 13.9 m s^{-1} (b) 12.2 m s^{-1} (c) 26.8 m s^{-1} (d) 44.7 m s^{-1}
2 72 km/h
3 $102 \text{ min}, 20.4 \text{ km}$
4 (a) 19.5 m s^{-1} (b) 24 m s^{-1} (c) 30 m s^{-1}
5 (a) $2 \text{ s}, 5 \text{ s}$ (b) 8 s
6 (a) $0 \text{ s}, 10 \text{ s}$ (b) $2 \text{ s}, 8 \text{ s}$ (c) 12 s (d) 6 m s^{-1}
    (e) -2 m s^{-1} (f) -10 m s^{-1}
7 (a) 20 (b) 20 (c) 16.4; $16 + 4h$
8 $8T + 4h, 8T$
9 (a) 1.67 m/s^2 (b) -3.33 m/s^2 (c) -3 m/s^2 (d) 4 m/s^2
10 (a) 20 m s^{-1} (b) 10 m s^{-1} (c) 15 m s^{-1} (d) -45 m s^{-1}
11 100 m
12 $8 \text{ s}, 7.5 \text{ m s}^{-2}$

Chapter 3

Exercise 3a

1   4 2 4 3 ± 10 4 75

5   2 6 $\dfrac{v - u}{t}$ 7 $\dfrac{2s}{u + v}$ 8 $\dfrac{2(s - ut)}{t^2}$

9   $\dfrac{v^2 - u^2}{2a}$ 10 $\dfrac{2s - at^2}{2t}$

11 $1.875 \, \text{m s}^{-2}$, $40 \, \text{s}$ **12** (a) $80 \, \text{m}$ (b) $320 \, \text{m}$
13 $40 \, \text{m s}^{-1}$ **14** $10 \, \text{m s}^{-1}$
16 (a) $30 \, \text{m s}^{-1}$ (b) $2 \, \text{m s}^{-2}$ (c) $1.5 \, \text{m s}^{-2}$
17 (a) $4 \, \text{s}$, $16 \, \text{s}$ (b) $6 \, \text{s}$, $14 \, \text{s}$
18 $1878 \, \text{m}$
19 $4 \, \text{s}$
20 $25 \, \text{s}$, $600 \, \text{m}$

Exercise 3b

1 $1.41 \, \text{s}$, $14.1 \, \text{m s}^{-1}$ **2** $12 \, \text{m s}^{-1}$
3 $2.5 \, \text{s}$ **4** $1 \, \text{s}$
5 $2 \, \text{s}$ **6** $5.58 \, \text{s}$
7 $5.51 \, \text{m}$ **10** $\dfrac{2u}{3g}, \dfrac{4u}{3g}$

Exercise 3c (Miscellaneous)

1 $2.5 \, \text{m s}^{-2}$ **2** $164 \, \text{m}$, $76.8 \, \text{km h}^{-1}$ **3** $u = 65$; $211 \, \text{m}$
4 $700 \, \text{m}$
6 (a) $16 \, \text{m}$, $12 \, \text{m}$ (b) $40 \, \text{m}$, $48 \, \text{m}$ (c) $72 \, \text{m}$, $60 \, \text{m}$
8 (a) $83 \, \text{m}$ (b) $8.3 \, \text{m s}^{-1}$ (c) $1.9 \, \text{m s}^{-2}$
9 $2 \, \text{s}$, $300 \, \text{m}$ **10** $100 \, \text{s}$
11 (a) $9 \, \text{m}$ (b) $4 \, \text{m s}^{-1}$ **12** $320 \, \text{m}$, $v = 55$ **13** $8\frac{1}{2} \, \text{min}$
14 (a) $10 \, \text{m s}^{-1}$ (b) (i) $5.77 \, \text{s}$ (ii) $8.66 \, \text{m s}^{-1}$
15 $10/u$, $14/u$, $6/u$ (a) $u = 5$ (b) $x = 2\frac{1}{2}$, $y = 4\frac{1}{6}$

Chapter 4

Exercise 4a

1 (a) $10 \, \text{m s}^{-1}$ (b) $0 \, \text{m s}^{-1}$ (c) $-10 \, \text{m s}^{-1}$
2 (a) $30 \, \text{m s}^{-1}$ (b) $43.2 \, \text{m s}^{-1}$; $6.6 \, \text{m s}^{-2}$
3 $t = 2$, $x = 20$
4 (a) $0 \, \text{m s}^{-2}$ (b) $12 \, \text{m s}^{-2}$ (c) $0 \, \text{m s}^{-2}$
5 $-3 \, \text{m s}^{-2}$, $+3 \, \text{m s}^{-2}$
6 $20 \, \text{m s}^{-1}$
7 $14 \, \text{m s}^{-1}$, $12 \, \text{m s}^{-2}$
8 (a) $3 \, \text{m s}^{-1}$, $0 \, \text{m s}^{-2}$ (b) $0 \, \text{m s}^{-1}$, $-9 \, \text{m s}^{-2}$
  (c) $3 \cos 6 \approx 2.88 \, \text{m s}^{-1}$, $-9 \sin 6 \approx 2.51 \, \text{m s}^{-2}$
9 $0.1 \, \text{m s}^{-1}$, $0.002 \, \text{m s}^{-2}$
10 (a) $\frac{1}{4}\pi \, \text{s}$, $\frac{3}{4}\pi \, \text{s}$; $\sqrt{2} \, \text{m}$, $0 \, \text{m}$

Exercise 4b

1 $x = 3t^2 + t^3 + 10$, $122 \, \text{m}$ **2** $x = 40$

3 (a) 14 m   (b) 134 m   (c) $(6n^2 - 6n + 14)$ m
4 1.25 m   5 $v = 3t^2 + 10$; $x = t^3 + 10t$
6 $x = 8t^{\frac{3}{2}} - 40t + 4$ 7 $v = 2x + 5$, $x = 10$
8 $v = \sqrt{(2x^4 + 4)}$, $x = 2$ 9 $v = \pm\sqrt{(100 - 4x^2)}$, $\pm 8 \, \text{m s}^{-1}$
10 $v = \sqrt{(9 + 16/x)}$

Exercise 4c (Miscellaneous)

1 (a) $x = 12t^2 - 2t^3 - 32$ (b) 18 m   (c) $(24 - 12t)\,\text{m s}^{-2}$ (d) $24\,\text{m s}^{-1}$
2 $v = 2\cos 2t - 2\sin 2t$, $\frac{1}{8}\pi$
3 (a) $83\frac{1}{3}$ m   (b) $208\frac{1}{3}$ m
4 40 300 km h^{-1}
5 $\frac{1}{2}$s, $5\frac{1}{2}$s; $-20\,\text{m s}^{-2}$, $+20\,\text{m s}^{-2}$; 39 m
6 (a) 15 m   (b) 2.31 s   (d) 49.3 m
7 (a) $1\frac{1}{3}$ h, $53\frac{1}{3}$ km   (b) 180 km h^{-2}, -180 km h^{-2} (c) 60 km h^{-1}

8 $x = \dfrac{k}{\omega}\left(t - \dfrac{1}{\omega}\sin \omega t\right) + ut$

9 (a) $\dfrac{1 - \cos 2\pi t}{2\pi} + \frac{1}{2}t$ (b) $\frac{7}{12} < t < \frac{11}{12}$

10 320 m

Chapter 5

Qu. 1 (a) 6.40, 51.3°   (b) 3.61, 146.3°   (c) 8.49, $-45°$ (d) 6, 180°
Qu. 2 (a) 1, 36.9°   (b) 1, 143.1°   (c) 1, $-36.9°$ (d) 5, 90°
       (e) 3, 180°
Qu. 3 (a) 1, α (b) 1, $(\frac{1}{2}\pi - \alpha)$ (c) 1, $(\pi - \alpha)$ (d) 1, $-\alpha$ or $(2\pi - \alpha)$
Qu. 4 (a) $5\mathbf{i} - 2\mathbf{j}$ (b) $-\mathbf{i} + 10\mathbf{j}$ (c) $16\mathbf{i} + 8\mathbf{j}$ (d) $4\mathbf{j}$
Qu. 5 $\dfrac{d\mathbf{F}}{dt} = -(10\sin t)\mathbf{i} + (10\cos t)\mathbf{j}$
Qu. 6 $\mathbf{V} = (t^2 + 5)\mathbf{i} + (1.5t^2 + 3)\mathbf{j}$
Qu. 7 $\mathbf{a} . \mathbf{b} = -3$

Exercise 5a

1 (a) $6\mathbf{i} - 2\mathbf{j}$ (b) $-2\mathbf{i} + 8\mathbf{j}$ (c) $16\mathbf{i} - 9\mathbf{j}$ (d) $2\mathbf{i} + 25\mathbf{j}$
2 (a) $\sqrt{40}$, $-18.4°$ (b) $\sqrt{68}$, 104.0°   (c) $\sqrt{337}$, $-29.4°$
   (d) $\sqrt{629}$, 85.4°
3 $\mathbf{F} = 8.66\mathbf{i} + 5\mathbf{j}$
4 $\mathbf{V} = \begin{pmatrix} -10 \\ 17.3 \end{pmatrix}$
5 $\mathbf{a} = 4.33\mathbf{i} + 2.5\mathbf{j}$, $\mathbf{b} = -4.33\mathbf{i} + 7.5\mathbf{j}$
6 $m = 5$, $n = -2$
7 $\{1, 3, -2\}$, or any multiple
8 $\overrightarrow{AC} = \begin{pmatrix} 62.9 \\ -38.3 \end{pmatrix}$, $AC = 73.7$ km,   bearing = 121.3°

9 18.4°
10 (a) $5i + 6j$ (b) $26i + 15j$ (c) $21i + 9j$

Exercise 5b

2 $\frac{16}{65}$
3 $m = 10$
4 $u = (\cos \alpha)i + (\sin \alpha)j$, $v = (-\sin \alpha)i + (\cos \alpha)j$
5 $\dfrac{du}{dt} = (-\sin t)i + (\cos t)j$

6 $\dfrac{dr}{dt} = 2ti + 3t^2j$

8 $p = 2t^2i + 2t^3j + c$; $p = 2t^2i + (2t^3 + 10)j$
9 $r = 5t^2j + at + b$

Exercise 5c (Miscellaneous)

1 $p = 5, q = 25, p.q = 75$; $\frac{3}{5}$
2 $u = 5, v = 5, c = \sqrt{50}, d = \sqrt{50}$; $u.v = 0, c.d = 0$; a square
3 $\begin{pmatrix} 10.7 \\ 8.46 \end{pmatrix}$, 13.6

4 $\pm[(\cos \alpha)i - (\sin \alpha)j)]$
5 $\frac{4}{5}, \frac{12}{25}, \frac{9}{25}$
6 $t = 0, 16$; $20i + 80j, 20i - 80j$
7 $10ti - 5t^2j$
8 A circle
9 0.6, 0.9
11 $\begin{pmatrix} 4 \\ 1 \\ 3 \end{pmatrix}$, $\begin{pmatrix} 2 \\ 5 \\ 0 \end{pmatrix}$;   (a) 13   (b) $\sqrt{26}$ (c) $\sqrt{29}$;   12.1 square units

12 (a) (i) b (ii) $\frac{1}{2}b$ (iii) $a + \frac{1}{2}b$ (iv) $a - b$ (b) $b + x(a - b)$
   (c) $y(a + \frac{1}{2}b)$
13 (a) $\pm\begin{pmatrix} 0.6 \\ -0.8 \end{pmatrix}$ (b) $m = 5, n = 1$; -0.179
14 (a) -3 (b) $\frac{4}{3}$ (c) $10 \ or \ -0.4$
15 $(1, -1), \begin{pmatrix} 2 \\ 4 \end{pmatrix}$; N_1 and N_2 intersect at $(7, 11)$.

   $(1, 2, 4),$ $\begin{pmatrix} -1 \\ 0 \\ 1 \end{pmatrix}$; $(1, 0, 1),$ $\begin{pmatrix} 2 \\ 2 \\ 3 \end{pmatrix}$;

The lines do not meet and are not parallel.

Revision Exercise 1

1 78 m s^{-1}, 82 m s^{-1}
2 (a) 3.7 m s^{-2} (b) 1.4 m s^{-2} (c) 0.5 m s^{-2}
3 50 m
4 $\dfrac{v^2 - u^2}{2a}$
6 7at, 1.75a, 56.5at^2
7 $\dfrac{4x}{3t^2}, \dfrac{x}{3t}$
8 $\dfrac{15u^2}{32g}$
9 (a) $b + \frac{1}{2}ft^2 - ut$ (b) $vt + b - ut - v^2/(2f)$;
least distance $= (2bf - u^2)/(2f)$
10 $u(2k^2 - 1)^{\frac{1}{2}}$; $(n - 4)k^2 + 4k - n = 0$
11 $2 - 2t + \frac{1}{2}(f - 4)t^2$; $\frac{400}{9}$ m
12 10 m s^{-1}; $25 + 15T - \frac{13}{16}T^2$; $T = 20$
14 (a) $v = e^{-4t} - \frac{1}{2}$ (b) $-\frac{1}{2}$ (c) $x = \frac{1}{4}(\frac{1}{2} - v) + \frac{1}{8}\ln(v + \frac{1}{2})$
15 8.2 m
16 T_1, t_0 ; $2Vt_0/\sqrt{5}$
17 $(-4, 11)$
18 (a) $27/\sqrt{850}$ (c) volume $= 9$, height $= \frac{18}{11}$
19 (a) 1 (b) .82° (c) $\frac{9}{13}$ (d) $\frac{5}{9}$
20 $x = b - \dfrac{\lambda}{a^2}\left(\dfrac{a \cdot b}{1 + \lambda}\right)a$

Chapter 6

Qu. 1 speed $= 2.5$ m s^{-1}, bearing $= 127°$
Qu. 3 (a) $-j$ (b) $15i + 7.5j$ (c) $\sqrt{45}$

Exercise 6a

1 (a) 7 m s^{-1}, N (b) 3 m s^{-1}, N (c) 6.40 m s^{-1}, 039°
 (d) 8.32 m s^{-1}, 020°
2 53.9 m s^{-1} at 22° to the direction of the car
3 192 miles
4 $x = \dfrac{T(V^2 - w^2)}{2V}$
5 (a) 10 mph (b) 70 mph (c) 50 mph
6 (a) $3j$ km h^{-1} (b) $(i + j)$ km h^{-1} (c) $(t - 5)i + (3 - 2t)j$ km
 (d) $(i - 2j)$ km h^{-1}
7 298 knots, bearing 191°
8 64.0 mph, bearing 141°
9 9.50 m s^{-1} at 11.4° to the downward vertical
10 $V = 2\sqrt{2}$; $\sqrt{2}i - \sqrt{2}j$

Exercise 6b

1 (a) $5\mathbf{i} + 11\mathbf{j}$ (b) $14\mathbf{i} + 22\mathbf{j}$ (c) $24\mathbf{i} + 22\mathbf{j}$
2 $50\sqrt{3}\mathbf{i} + 50\mathbf{j}$; $-10\mathbf{j}$, $t = 10$, $500\sqrt{3}$ m
3 $\sqrt{(2 + 2\cos t)}$, $t = \pi$ or 3π
4 Towards O
5 Towards (1,0)

Exercise 6c (Miscellaneous)

1 1.38 knots, 141°
2 430 mph
3 $-2\mathbf{i} + 4\mathbf{j}$, 2.24 km
4 $\mathbf{v} = (25 - 8t)\mathbf{i} - 4t\mathbf{j}$, $\mathbf{a} = -8\mathbf{i} - 4\mathbf{j}$
5 (a) 1 m
6 $\mathbf{v} = (2 \cos t)\mathbf{i} + (2 \sin 2t)\mathbf{j}$;
  $t = \frac{1}{4}\pi, \frac{3}{4}\pi, \frac{5}{4}\pi, \frac{7}{4}\pi, \ldots$; $t = \frac{1}{2}\pi, \pi, \frac{3}{2}\pi, \ldots$
7 $\mathbf{r} = (ut)\mathbf{i} - (5t^2)\mathbf{j}$, 23.8 m s^{-1}
8 6.34 knots along BA; 1.42 knots along AB
9 (a) $\mathbf{v} = 3\mathbf{i} + 4\mathbf{j} + 10t\mathbf{k}$ (b) 323° (d) $10\sqrt{5}$ m (e) $t = 4$
10 $\frac{1}{6}(\mathbf{i} + 900\mathbf{j}) \times 10^{-6}$ m, $t = 6$
11 $(10 - 30t, 0)$, $(0, 10 - 40t)$, $PQ = \sqrt{(200 - 1400t + 2500t^2)}$;
   16 min 48 s after noon; 50 km h^{-1}, bearing 143°
12 $(-1.2 + 0.2t)\mathbf{i} + (3.2 - 0.2t)\mathbf{j} + \mathbf{k}$; $t = 16$
13 5.6 km, 12.44 hours
14 62 km, 7 h after noon
15 $-3\mathbf{j}, 4\mathbf{i}$; $4\mathbf{i} + 3\mathbf{j}$; $\mathbf{i} - 4\mathbf{j}$; 0.6

Chapter 7

Qu. 2 99.0 m s^{-1}
Qu. 3 15°, 75°

Exercise 7a

1 5 m, 35 m
2 15 m, 35 m
3 5.10 m, 35.3 m
4 15.3 m, 35.3 m
5 $\dfrac{u^2}{8g}, \dfrac{3u^2}{8g}$
6 21%
7 29.1 m s^{-1}
8 1.54%
9 23.4 m s^{-1}
10 1.9%
11 18.9°, 71.1°
12 12.6 m s^{-1}
13 (a) $2v/g$ (b) $2uv/g$
14 1930 m
15 74.3°, 27.0°

Exercise 7b

3 $\frac{1}{4}R, \frac{3}{4}R$

10 $(\pm U^2/g, 0), (0, \frac{1}{2}U^2/g);$ 21.5 m

Exercise 7c (Miscellaneous)

1 (a) 12 m s^{-1} (b) 17.4 m   (c) falling

2 $\tan \alpha = \frac{1}{2}$ or 7

3 From 2 s to 10 s (approximately)

4 $2\frac{2}{3}$ s

5 15 m, 2 s

6 $U = \frac{2}{5}V\sqrt{10}$

7 $V = 90 \text{ m s}^{-1}$, 50.7°;   0.8

9 $40\sqrt{5}$ m

10 $\sqrt{2000} \text{ m s}^{-1}$, 120 m, 90 m

11 Answers are standard projectile formulae. Angle of elevation $= 30°$

12 1.5 s

13 (a) $U \cos \alpha$, $U \sin \alpha - gt$

   (b) $x = (U \cos \alpha)t$, $y = (U \sin \alpha)t - \frac{1}{2}gt^2$

14 45°, 30 m s^{-1}

15 25 m s^{-1}

Chapter 8

Qu. 1 10j

Qu. 2 (a) 45.3 N   (b) 21.1 N

Qu. 3 (a) 342 N   (b) 940 N

Exercise 8 (Miscellaneous)

1 225 m

2 4690 N

3 28.1 m

4 11 000 N, 9810 N, 8310 N

5 $g(\sin \alpha - \sin \beta)$

6 $\left(\dfrac{Pm}{M + m}\right)$N

7 (a) 1.50 m s^{-2} (b) 2.22 N

8 662 m

9 21 400 N

10 $\begin{pmatrix} -4 \\ 3 \end{pmatrix}$

11 5.55 m s^{-2}, 35.9° to the x-axis

12 25 000 N, 70 000 N, 35 000 N

13 1.41 N, $-45°$ to the positive x-axis

15 $-\mathbf{j}, +\mathbf{j}$

Chapter 9

Exercise 9a

1 20 N; 2.5 m s^{-2}, 37.5 N
2 3.5 m s^{-2}, 26 N
3 2 m s^{-2}, 12 N, 24 N
4 3$\frac{1}{3}$ m s^{-2}, 26$\frac{2}{3}$ N; $\frac{1}{3}$ m
5 2.5 m s^{-2}, 5 m s^{-2}, 250 N

6 $\left(\dfrac{M-m}{M+m}\right)g,$ $\dfrac{2mMg}{M+m}$

7 $\dfrac{mg-F}{m+M},$ $\dfrac{mMg+mF}{m+M}$

8 $\dfrac{Mg-mg\sin\alpha-F}{(M+m)},$ $\dfrac{Mmg(1+\sin\alpha)+MF}{(M+m)}$

9 $\left(\dfrac{4m+M}{3m+M}\right)mg,$ $2\left(\dfrac{2m+M}{3m+M}\right)mg$

10 $\dfrac{mMg(\sin\beta+\sin\alpha)}{(m+M)}$

Exercise 9b

1 39.6 m s^{-1} **2** 4.11 m s^{-2} **3** 1.46 m s^{-2}
4 5.56 m s^{-2} **5** 1.73 m s^{-2} **6** 0.159
7 0.466   **8** 0.070; 23.3 m s^{-1}

Exercise 9c (Miscellaneous)

1 652 m   **2** 10.7 m s^{-1} **3** 2.38 s
4 22.4 N, 2.35 m s^{-2} **5** 12.7 N, 3.60 m s^{-2} **6** 1.21 s
10 $\frac{1}{2}g$, $3mg$ **11** (a) 1 m s^{-2} (b) 99 N   (c) 55 N
12 1$\frac{2}{3}$ m s^{-2}; 5 m   **13** 50 N
14 (a) 2 m s^{-2}, 40 N, 36 N   (b) 1 m s^{-2}, 45 N, 33 N
15 (a) $\frac{9}{25}g$, $\frac{48}{25}mg$ (b) $\frac{7}{25}g$, $\frac{54}{25}mg$

Revision Exercise 2

1 (a) $-2\mathbf{i} - 4\mathbf{j}$ (b) $\sqrt{26}$; $a = 8, b = 4$

2 $v = \frac{1}{20}u, w = \frac{3}{4}u$; $\frac{1}{20}(-17\mathbf{i} + 16\mathbf{j})$

3 $301°$, 1.66 h; 755 km, $034°$

4 8.89 km h^{-1}, $257°$; 12.3 km h^{-1}, $097°$

5 (a) $\frac{4}{3}$ (b) $\frac{3}{4}$; $\cos\theta = \frac{10}{11}$

6 $\sqrt{(3ag)}$; $2\sqrt{(a/g)}$, $\sqrt{(ag)}$, $30°$ below the horizontal

7 $V = 149$ m s^{-1}, $\tan\alpha = 2.8$, $OA = 500$ m;

119 m s^{-1}, $\frac{81}{70}$

8 (a) $\frac{9}{14}gT^2, \frac{9}{7}T$ (b) $\frac{1}{7}(gT\sqrt{37})$

9 $\tan\alpha = 1$; (a) 5.83 s (b) $47.7°$ below the horizontal (c) 71.2 m

10 $(1 + \sqrt{2})V, (\sqrt{2} - 1)V^2/g$

11 $p = 4, q = -9$; $y = 26, t = 5$

12 3920 N; $x = 8$

13 0.9 m s^{-2}; $\frac{4}{3}$ s, 30 m

14 $k\sin(nt),\quad u + \dfrac{k}{n} - \dfrac{k}{n}\cos(nt),\quad ut + \dfrac{kt}{n} - \left(\dfrac{k}{n^2}\right)\sin(nt)$

15 A circle, centre O, radius 4; 1000 N; $-(\cos 5t)\mathbf{i} - (\sin 5t)\mathbf{j}$;

$(\pi + 5t)$ radians; 9 m

16 $\sqrt{(3l/g)}, mg$

17 $\sqrt{(\frac{3}{5}gl)}$

18 $\frac{1}{4}g$

20 $\frac{2}{5}g$ m s$^{-2}, \frac{21}{5}g$ N, $\frac{4}{7}$ s

Chapter 10

Qu. 1 $750g$ J $= 7360$ J

Qu. 2 1000 J

Qu. 3 5 J

Qu. 4 1200 kJ

Qu. 5 20 kW

Exercise 10a

1 (a) 200 J (b) 50 J (c) 6 J

2 2520 J

3 7.36×10^8 J

4 (a) 1.25×10^6 J (b) 1.25×10^6 J (c) 1.25×10^6 J

5 25 J

6 4000 J, approx

7 14.7 J

8 (a) 0.245 J (b) 0.0613 J (c) 0.229 J

9 $mgr(1 + \cos\alpha)$

10 $\frac{1}{2}x[\mu(w\sqrt{3} + W) + W\sqrt{3} - w]$

Exercise 10b

1 (a) $4.43 \, \mathrm{m \, s^{-1}}$ (b) $6.26 \, \mathrm{m \, s^{-1}}$ (c) $5.79 \, \mathrm{m \, s^{-1}}$ (d) $3.13 \, \mathrm{m \, s^{-1}}$
2 $31.3 \, \mathrm{m \, s^{-1}}$ **3** $\frac{4}{3} \mathrm{m}$ **4** $2.94 \times 10^{10} \, \mathrm{J}; \, 7.62 \times 10^5 \, \mathrm{N}$
5 $0.4 \, \mathrm{J}$ **6** $60 \, \mathrm{kN}$ **7** $20.4 \, \mathrm{cm}, \, 79.4°$
8 $9.80 \, \mathrm{m \, s^{-1}}$ **9** $0.0981 \, \mathrm{J}$ **10** (a) $36.6 \, \mathrm{m \, s^{-1}}$ (b) $24.6 \, \mathrm{m \, s^{-1}}$
11 $232 \, \mathrm{kJ}; \, 164 \, \mathrm{kJ}$
14 $209 \, \mathrm{kJ}$

Exercise 10c

1 $65.4 \, \mathrm{W}$ **2** $58.9 \, \mathrm{kW}$ **3** $1.66 \, \mathrm{kW}$ **4** $33.8 \, \mathrm{kW}$
5 $53.1 \, \mathrm{kW}; \, 1.10 \, \mathrm{m \, s^{-2}}$ **6** $4.08 \, \mathrm{m}$ **7** $30 \, \mathrm{m \, s^{-1}}$
8 $a = 150, \, b = 0.4; \, 31.6 \, \mathrm{kW}$ **9** $\frac{4}{3} u \, \mathrm{km \, h^{-1}}$

Exercise 10d (Miscellaneous)

1 (a) $50 \, \mathrm{J}$ (b) $-50 \, \mathrm{J}$ (c) $25 \, \mathrm{J}$ (d) $26 \, \mathrm{J}$ (e) $125 \, \mathrm{J}$
2 $1.6 \, \mathrm{J}$ **3** $120 \, \mathrm{W}$ **4** $2.5 mgx; \, \frac{1}{4} mgx; \, \frac{3}{4} \sqrt{(2gx)}$
5 (a) mgd (b) $\sqrt{(2gd)}$ (c) $0.4mgd$ (d) $3mgd$
6 (a) $20 \, \mathrm{J}$ (b) $48 \, \mathrm{J}$ (c) $28 \, \mathrm{J}$ (d) $2\frac{1}{3} \mathrm{N}; \quad 5 \, \mathrm{m}$
7 $-200\mathbf{i} - 50\mathbf{j} \, \mathrm{N}, \, 11 \, \mathrm{J}$
9 $10^4 \, \mathrm{N}; \, 12.5 \, \mathrm{m \, s^{-1}}$ **10** $500 \, \mathrm{N}, \, 15\,000 \, \mathrm{W}$ **11** $7 \, \mathrm{m \, s^{-1}}$
12 $800 \, \mathrm{N}$ (a) $16 \, \mathrm{m \, s^{-1}}$ (b) $\frac{25}{16} \mathrm{m \, s^{-2}}$ **13** $22.7, \, 1.37 \, \mathrm{m \, s^{-2}}$
14 $475 \, \mathrm{N}; \, 34.4 \, \mathrm{kW}$ **15** $32.5 \, \mathrm{kW}; \, 29.3 \, \mathrm{km \, h^{-1}}$

Chapter 11

Qu.3 0.6

Exercise 11a

1 (a) $50 \, \mathrm{N \, s}$ (b) $150 \, \mathrm{N \, s}$ (c) $50 \, \mathrm{N \, s}$
2 (a) $5 \, \mathrm{m \, s^{-1}}$ (b) $(3\mathbf{i} + 4\mathbf{j}) \, \mathrm{m \, s^{-1}}$ (c) $(5\mathbf{i} - 5\mathbf{j}) \, \mathrm{m \, s^{-1}}$
3 (a) $17\,400 \, \mathrm{N \, s}$ at $100°$ to the original direction   (b) $1740 \, \mathrm{N}$
4 (a) $\sqrt{261} \, (\simeq 16.2) \, \mathrm{N \, s}$ (b) $17.5 \, \mathrm{m \, s^{-1}}$
5 $25 \, \mathrm{s}$
6 $6 \, \mathrm{N \, s}, \, 60 \, \mathrm{N}$
7 $10 \, \mathrm{N \, s}, \, 10 \, \mathrm{N}$
8 $100 \, \mathrm{N \, s}, \, 100 \, \mathrm{N}$
9 $0.569 \, \mathrm{N \, s}, \, 1.48 \, \mathrm{m \, s^{-1}}$
10 (a) $17.5 \, \mathrm{N \, s}$
    (b) $\sqrt{200} \, (\simeq 14.1) \, \mathrm{N \, s}$, at $135°$ to the original direction
    (c) $7.37 \, \mathrm{N \, s}$, at $151°$ to the original direction

Exercise 11b

1 (a) $4.5 \, \mathrm{m \, s^{-1}}$ (b) $3 \, \mathrm{m \, s^{-1}}$ **2** $2.5 \, \mathrm{m \, s^{-1}}$

3 (a) 2.5 m s^{-1}, $502\,500$ J   **4** $\frac{1}{3}(3v - 2u)$
5 $-2u, 3u$;   60%
6 10 m s^{-1}; 5 m s^{-1}; $\cos^{-1} 0.75$ $(\simeq 41.4°)$
7 $55\frac{5}{9}\%$
8 $\sqrt{125}$ $(\simeq 11.2)$ mph;   N $\tan^{-1} 0.5$ E (i.e. bearing $026.6°$)
9 (a) $\dfrac{mu \sin \alpha}{(M + m)}$ (b) $mu \cos \alpha$
10 $-u, 3u$

Exercise 11c

1 (a) $2 \times 0.75^2 \text{ m}$ (b) $2 \times 0.75^4 \text{ m}$ (c) $2 \times 0.75^{2n} \text{ m}$
2 (a) 0.958 s   (b) 3.83 s
5 $-1 \text{ km h}^{-1}, 3 \text{ km h}^{-1}$
6 $u, \frac{3}{2}u$

Exercise 11d (Miscellaneous)

1 $1 \text{ m s}^{-1}, \frac{1}{8}$ **2** $0.24, 0.16$
3 (a) $\frac{5}{4}$ (b) $M = 10, m = 8$ **4** $401 \text{ m s}^{-1}, 2.25°$
5 (a) 5 m s^{-1} (b) 25 m s^{-1} (c) $5\sqrt{7}$ $(\simeq 13.2) \text{ m s}^{-1}$
6 (a) $20e^6 \text{ m}$ (b) $4e$ s   (c) $\frac{1}{3}$ (d) 25 m
7 $\frac{1}{2}; 0 \text{ m s}^{-1}, 8 \text{ m s}^{-1}$ **8** $-\frac{1}{2}U, \frac{1}{12}U$
9 $\frac{1}{2}(1 - e)u$, $\frac{1}{4}(1 - e^2)u$, $\frac{1}{4}(1 + e)^2 u$
10 (a) $\frac{1}{3}(1 - 2e)V$, $\frac{1}{3}(1 + e)V$; $e = \frac{1}{2}$
11 $(-200\mathbf{i} - 50\mathbf{j})$ N, 11 J   **12** $1/\sqrt{e}$
13 $\frac{1}{3}(7 - 8e)u$, $\frac{1}{3}(16e + 7)u$; $e = \frac{1}{8}$; $3u/(1 + k)$, $k > \frac{1}{2}$
14 $\frac{4}{9}u, \frac{1}{9}u$; yes   **15** $\frac{1}{5}(3 - 2e)u$, $\frac{2}{5}(1 + e)^2 u$

Chapter 12

Qu. 1 (a) 2.5 N   (b) 56 mm   (c) $T = \frac{1}{16}x$
Qu. 2 $T = 25x$;   (a) 50 N   (b) 12 cm
Qu. 6 0.994 cm

Exercise 12a

1 (a) 125 N   (b) 8 cm   (c) 87.5 N
2 $4W$ newtons; $\frac{1}{4}kl$ metres
3 $\dfrac{Wl}{2\lambda}; \dfrac{2Wl}{\lambda}$
4 1.25 J
5 0.4 m, 4 J
6 (a) 0.2 m s^{-2} (b) 0 m s^{-2} (c) $0.04x \text{ m s}^{-2}$ (d) 0.1 m s^{-1}
9 235 N
10 $a = \dfrac{\lambda}{l}$, $b = 0$, $c = -\dfrac{\lambda}{3l^3}$

Exercise 12b

1 (a) $\frac{1}{8}\pi^2\,\mathrm{m\,s^{-2}}$ (b) $\frac{1}{4}\pi\,\mathrm{m\,s^{-1}}$ (c) $\pm\frac{1}{5}\pi\,\mathrm{m\,s^{-1}}$

2 $8\pi\,\mathrm{s}$, $\frac{1}{5}\sin\left(\frac{1}{4}t\right)$ 3 $1/(80\pi^2)\,\mathrm{m}$

4 $(0.02 - 8x^2)\,\mathrm{J}$ 5 $0.8\,\mathrm{N}$

6 $4\pi\,\mathrm{s}$,   (a) $(5, 0)$ (b) $(2, 0)$ $(8, 0)$

7 $4\,\mathrm{m\,s^{-1}}$

8 $\pi\,\mathrm{s}$, $5\,\mathrm{m}$, $10\,\mathrm{m\,s^{-1}}$

9 $5\,\mathrm{m\,s^{-1}}$, $(0.05\pi + 0.2)\,\mathrm{s}$

10 $\dfrac{u}{n}, \dfrac{2\pi}{n}$; $l + \left(\dfrac{u}{n}\right)\sin(nt)$, where $n^2 = \dfrac{\lambda}{ml}$

Exercise 12c

1 $\frac{1}{5}\pi\sqrt{5}\,(\simeq 1.40)\,\mathrm{s}$, $0.2\,\mathrm{m}$; $\sqrt{0.6}\,(\simeq 0.775)\,\mathrm{m\,s^{-1}}$

2 $0.2\pi^2\,(\simeq 1.97)\,\mathrm{N\,m^{-1}}$

3 $0.2\,\mathrm{m}$, $\frac{1}{5}\pi\sqrt{2}\,(\simeq 0.889)\,\mathrm{s}$; $\sqrt{2}\,(\simeq 1.41)\,\mathrm{m\,s^{-1}}$

4 $40/\pi^2\,(\simeq 4.05)\,\mathrm{kg}$

5 (a) $0.4\pi\,(\simeq 1.26)\,\mathrm{s}$ (b) $0.2\pi\,(\simeq 0.628)\,\mathrm{s}$

6 $\frac{1}{10}\pi\sqrt{10}\,(\simeq 0.993)\,\mathrm{s}$; $\sqrt{7.5}\,(\simeq 2.74)\,\mathrm{m\,s^{-1}}$

7 $M = 4m$

8 (a) $2\pi\sqrt{\left(\dfrac{m(k_1 + k_2)}{k_1 k_2}\right)}$ (b) $2\pi\sqrt{\left(\dfrac{m}{k_1 + k_2}\right)}$

9 $5mg$;   (a) $\dfrac{l}{5}$,   (b) $2\pi\sqrt{\left(\dfrac{l}{5g}\right)}$,   (c) $\sqrt{\left(\dfrac{gl}{5}\right)}$

10 $v^2 = \dfrac{g(l^2 - 25x^2)}{5l}$

Exercise 12d (Miscellaneous)

1 $3, 4$; ± 5 2 $\sqrt{l}$ against T, or l against T^2

3 $\frac{2}{3}\,\mathrm{m}, \frac{2}{3}\,\mathrm{J}$ 4 $\frac{1}{2}\sqrt{3}\,(\simeq 0.866)\,\mathrm{m\,s^{-1}}$

5 $\frac{5}{2}a$ 7 $k = 3$; $\frac{3}{4}l$

8 $\frac{1}{2}mg, \frac{3}{2}mg$;   (a) $\frac{3}{2}l$,   (b) $\pi\sqrt{\left(\dfrac{2l}{g}\right)}$ 9 $\frac{5}{18}\pi^2\,\mathrm{J}$; $2\,\mathrm{s}$

10 $\sqrt{(gl)}$; $-(3gx/l + g)$; $H = l/3$, $K = l/\sqrt{3}$

11 $\frac{5}{3}l$ 12 $\frac{1}{2}l$

13 $\dfrac{\pi}{6n}$

14 (a) $12\,\mathrm{s}, 24/\pi\,(\simeq 7.64)\,\mathrm{m}$ (b) $5\,\mathrm{s}$; $12\sqrt{2}\,(\simeq 17.0)\,\mathrm{s}, \sqrt{2}\,(\simeq 1.41)\,\mathrm{m\,s^{-1}}$

15 (a) $0.08\,\mathrm{m}$ (b) $0.6\sqrt{5}\,(\simeq 1.34)\,\mathrm{m\,s^{-1}}, 0.04\,\mathrm{m}$

Chapter 13

Qu. 1 (c) towards the centre of the circle

Qu. 2 $\frac{1}{18}\pi$ ($\simeq 0.175$) mm s^{-1}

Qu. 3 $-\frac{1}{1800}\pi$ ($\simeq -0.001\,75$) rad s^{-1}

Qu. 4 $300/\pi$ ($\simeq 95.5$) r.p.m., 20 m s^{-1}

Exercise 13a

1 3.49 rad s^{-1} (a) 48.9 cm s^{-1} (b) 20.9 cm s^{-1}

2 10.5 rad s^{-1}, 1.57 m s^{-1}

3 23.1 rad s^{-1} 4 95.5 r.p.m.

5 (a) 24 rad s^{-1} (b) 229 r.p.m.   6 313 km h^{-1}, 86.9 km h^{-1}

7 3.49 rad s^{-2}, 1.75 rad s^{-2}, 18 800 rad

8 308 rad

9 0.514 rad s^{-2} 10 $\frac{1}{2}\omega t$

Exercise 13b

1 (a) 8 m s^{-2} (b) 50 m s^{-2} (c) 3.16 m s^{-2}

2 8.23 m s^{-2} 3 216 km h^{-1}

4 309 m   5 1.97 m s^{-2}

6 0.0338 m s^{-2} 7 2.29 km

8 13.5 r.p.m.

10 (a) $\mathbf{r} = 5t^2\mathbf{u}$ (b) $\mathbf{v} = (10t)\mathbf{u} + (10t^2)\mathbf{n}$ (c) $\mathbf{a} = (10 - 20t^2)\mathbf{u} + (40t)\mathbf{n}$

     (i) 20   (ii) $20\sqrt{5}$ ($\simeq 44.7$)   (iii) $10\sqrt{113}$ ($\simeq 106$)

Exercise 13c

1 47.3 r.p.m.   2 36 km h^{-1}

5 1090 N   7 36 000 km

8 6670 km, 7760 m s^{-1} 10 48.9°

11 300 N, 78.5°   14 15.8°; 40 100 N

Exercise 13d

1 5 m s^{-2} at $\tan^{-1}(\frac{4}{3})$ to the horizontal

2 $\sqrt{0.26}$ ($\simeq 0.51$) m s^{-2}

4 19.8 m s^{-1}

5 29.9 r.p.m.

7 $\frac{3}{2}a$

Exercise 13e (Miscellaneous)

1 51.2 N; 17.3 rad s^{-1} 2 $2mg$; $2mg$, $\sqrt{3l}$, $\sqrt{(g/l)}$

3 (a) 21.7 N   (b) 15.6 m; 4   5 $2\pi\sqrt{(ma/F)}$; 1.60 hours

6 (a) $\frac{1}{2}g$ (b) $g(2 - \sqrt{3})$

7 (a) (i) $m(l\omega^2 - g)$ (ii) $m(l\omega^2 + g)$ (b) $\dfrac{g}{4\pi^2}$

8 $4mg/\sqrt{3}$ 9 $\frac{3}{5}ml\omega^2$

10 (b) $\frac{1}{2}\sqrt{3}m(2g - l\omega^2)$, $\frac{1}{2}m(3l\omega^2 - 2g)$

Revision Exercise 3

1 1800 N; 804 N   2 10 m s^{-1}; 0.2 m s^{-2}

3 $a = 0.2$, $b = 600$; 0.5 m s^{-2} 4 0.683 m s^{-2}

5 $14Mu \sin \alpha$; 7.47 m s^{-1}

6 (a) 1 m s^{-1}, 5 m s^{-1}; 45 N s (b) 67.5 J

7 $\frac{2}{3}mV$ 8 $\dfrac{n(n-1)d}{2u}$, $\dfrac{mu^2(n-1)}{2n}$

9 three; $\frac{1}{18}mu^2$, $\frac{2}{3}mu$ 10 $4V$, $45 \, mV^2$, $-V$, $14V$

11 (b) $2\sqrt{(ga)}$, $2a$ 12 (a) 0.44 m   (b) 0.03 m

13 0.23%

14 $\dfrac{40 \, mu^2 + 45mga}{72a}$, $\dfrac{40mu^2 - 45mga}{72a}$

15 $\dfrac{ga^2}{r^2}$; $\dfrac{2\pi}{a}\sqrt{\left[\dfrac{(a+h)^3}{g}\right]}$

16 (b) $\frac{1}{2}ag$, $\sqrt{(\frac{1}{2}ag)}$ 17 mg

18 $\theta = \frac{1}{6}\pi$, $v = \sqrt{(3ag)}$

Chapter 14

Qu. 2 $10\sqrt{2}$, $y = x$; $(2,2)$; $10\sqrt{5}$, $2y = x + 2$

Exercise 14a

1 100 N at 36.9° to the x-axis   2 79.0°

3 26.6°, 224 N   4 (a) 32.8°   (b) 92.3 N

5 60 N, parallel to BA, through C   6 21.8°, 53.9 N

7 (a) 100 N vertically upwards

   (b) 57.7 N   (c) 100 N at 30° to the horizontal

8 68.4 N, 97.4 N   9 250 N

Exercise 14b

1 25 N, 053°

2 (a) $18\mathbf{i} + 5\mathbf{j}$ (b) $66.5\mathbf{i} + 21.9\mathbf{j}$ (c) $44.3\mathbf{i} + 83.3\mathbf{j}$ (d) $12.5\mathbf{i} + 3.33\mathbf{j}$

3 120°, 17.3 N   4 13 N; 76.7°, 72.1°, 22.6°

5 53.1°, 19.5 N   6 11.5°, 6.63 kN

7 93.9 kN, 230 kN   8 60 cm; 5 N, 4 N

9 5.12 N   10 (b) 66.1 N   (c) 99.7 N

Exercise 14c

1 0.115
3 86.6 N; 0.433
5 $3\mu_1 < 2\mu_2$
7 0.125; 1.4W

2 0.151
4 (a) 115 N   (b) 360 N
6 between 9.6 and 10.9
8 $\theta = \lambda$

Exercise 14d (Miscellaneous)

1 24 N, 20.8 N
3 $\frac{8}{27}$
6 (i) 7.30 units   (ii) 1.93 units,   7.55 units
8 $\theta = \lambda$
11 24

2 5 N, 037°
5 (b) $P = \frac{1}{2}W(\sin\alpha + \mu\cos\alpha)$

10 $m = 2$, $\mu = 0.225$
13 $2 - \sqrt{3}$

Chapter 15

Qu. 1 1.5 m
Qu. 2 (a) 40 N m   (b) 34.6 N m
Qu. 3 48.4 N at 55.7° to the x-axis
Qu. 4 $(\frac{3}{4}, 0)$
Qu. 5 $c = 8$

Exercise 15a

1 (a) $+30$ N m   (b) $+20$ N m   (c) -32 N m   (d) -26 N m
  (e) -14 N m
2 1200 N m   **3** 7 m
4 50 N, (13, 0)   **5** $+12$ N m
6 $x = 7$ **7** (0, 5)
8 The force undergoes a translation bi
9 25 N; $4y = 3x - 10$ **10** 5 N $y = 9$

Exercise 15b

1 53.3 cm, 1450 N   **2** 91.9 cm
3 xW/L, $W(1 - x/L)$ **4** 3.5 N, 7.5 N
5 $W = 11$ **6** 13 N
7 80 N; 60 N   **8** 0.2 N
9 400 kN   **10** 57.2°
11 544 kN, 156 kN
12 62.5 N, 210 N at 17.4° with the vertical
13 $\dfrac{aW + HT}{4a}$, $\dfrac{aW - HT}{4a}$
14 57.7 N, 208 N   **15** 1250 N
16 (a) none   (b) 93.2 N
17 2000 N; 1630 N at 79.4° with the downward vertical
20 32.5°

Exercise 15c (Miscellaneous)

1 0.4 m   **2** (a) $X = 3$ (b) 100 N, 50 N

3 $P = 6\frac{2}{3}$, $Q = 10$, $R = 6\frac{2}{3}$ **4** (b) 34° $2W$

5 (a) $100 \cos \theta$ (c) 94.3   (d) 32°

6 (a) $P = 7$, $Q = 5$, 36 N m;   (b) (i) $2\sqrt{2}$ N   **7** 2.7

8 $\dfrac{(6 - y)w}{(12 - x - y)}$, $\dfrac{(6 - x)w}{(12 - x - y)}$ (b) 3

9 (a) 26 N, 22.6°

10 (a) 19.5, 16.2, 31.9   (b) 1.96

11 $\frac{2}{3}W \sec \theta$, vertical; $W(1 - \frac{2}{3} \sec \theta)$, vertical

12 $\frac{1}{2}W\sqrt{3}$;   (a) $\frac{1}{2}W\sqrt{39}$ (b) 73.9°

13 (b) $\frac{10}{19}$ **14** $\frac{1}{2}W\sqrt{5}$, $\frac{1}{2}W$, 0   **15** 5 m

Chapter 16

Qu. 1 1.5 N

Qu. 2 $5.6\pi g$

Qu. 3 (2,3)

Exercise 16a

1 Preston   **2** Birmingham

3 4 cm, 3 cm   **4** 5 cm

5 3.5 cm

Exercise 16b

1 66 cm   **2** 41 cm

3 $(a, \frac{7}{10}a)$ **5** $(\frac{1}{3}(c + a), \frac{1}{3}h)$

6 $3\mathbf{i} + 3\mathbf{j}$ **7** $4.4\mathbf{i} + 2.8\mathbf{j}$

8 $\frac{1}{3}(\mathbf{a} + \mathbf{b} + \mathbf{c})$

Exercise 16c

1 6   **2** $\frac{81}{19}$

3 $8\frac{1}{3}$ **4** $2\frac{2}{3}$

5 $1\frac{1}{2}$ **6** $1\frac{2}{7}$

7 $2 \ln 2$ **8** $\left(\dfrac{e^2 + 1}{2(e^2 - 1)}\right)$

9 $\dfrac{3h}{4}$ **10** $\dfrac{3r}{8}$

Exercise 16d

1 $(\frac{2}{3}h, 0)$

2 $(1.2, 0)$

3 $\left(\dfrac{40}{3\pi}, 0\right)$

4 $(0, 2.4)$

5 $(0, \frac{1}{8}\pi)$

6 $(\frac{1}{3}b, \frac{1}{3}h)$

7 $(\frac{9}{16}, \frac{7}{10})$

8 $(\frac{4}{5}a, \frac{2}{7}a)$

9 $\left(\dfrac{4r}{3\pi}, \dfrac{4r}{3\pi}\right)$

10 $\left(\dfrac{1}{e-1}, \dfrac{e+1}{4}\right)$

Exercise 16e (Miscellaneous)

1 $2\mathbf{i} + \mathbf{j} + 2\mathbf{k}$

2 $\frac{1}{4}(\mathbf{a} + \mathbf{b} + \mathbf{c} + \mathbf{d})$

3 $(\frac{2}{3}h, 0)$

5 $\left(0, \dfrac{2a^2}{5b}\right)$

6 (a) 51.1   (b) 2.82

7 8.41 cm

Chapter 17

Qu. 1 103 cm

Qu. 2 $\frac{3}{20}r$

Qu. 3 the intersection of the lines of action of W and μR

Exercise 17a

1 43 cm

2 54.2 cm

3 1.94 m

4 $\frac{1}{3}r$

5 $\frac{15}{14}l$

6 (0, 5.5)

7 (2.5, 3.5)

8 (3.36, 8.35)

9 $\frac{1}{13}(14 + 3\sqrt{3})a$

10 $\dfrac{28r}{9\pi}$

Exercise 17b

3 21°

5 $9h/13$

Exercise 17c (Miscellaneous)

1 (a) $10g, 70g$ (b) $\frac{10}{3}$ (c) 120

2 1.8 kg, 2 m

3 39.8°; 34.8°

5 36°

6 $\dfrac{h^2 - 3a^2}{4(2a + h)}$;   (a) 17°   (b) $h = a\sqrt{3}$

7 $(\frac{7}{18}a, \frac{4}{9}a)$

8 $\frac{1}{4}W\sqrt{7}$

10 $(\frac{3}{5}a, \frac{3}{4}a)$; $\frac{5}{4}$

Revision Exercise 4

1 $4\mathbf{j} - 8\mathbf{k}$ **2** (a) $P/(mg)$

3 (b) $\dfrac{mg(\mu\cos\alpha + \sin\alpha)}{(\cos\alpha - \mu\sin\alpha)}$

4 (a) $\sqrt{3}$, N 30° E; $\sqrt{3}$, S 30° W   (b) $k = 5$

5 $\lambda(2\sec\theta - 1)$, $\lambda\tan\theta(2 - \cos\theta)$, $W + \lambda(2 - \cos\theta)$

6 $k = \frac{35}{6}$; 43.2 m

7 $R = \frac{3}{4}W$, $F = \frac{1}{4}W\tan\alpha$; $\frac{1}{3}\sqrt{3}$

8 (a) $10P$ (b) $\frac{4}{3}$ (c) $\frac{9}{4}a$ (e) $20Pa$

    (f) $S = 6P$, $T = 3.5P$, $U = 4.5P$

9 (a) $\frac{5}{6}mg$ (b) $\frac{1}{6}mg\sqrt{13}$; $\frac{2}{3}$; $\frac{7}{10}mg$

10 $\tan\theta = \frac{3}{4}$

11 $2a/\pi$, from O   **12** $\frac{45}{112}$

13 $\frac{5}{6}mg$ **14** $\dfrac{H + h}{3H - h}$

15 $\frac{24}{65}$ **16** $\frac{12}{5}a$; $\frac{1}{9}$

17 $\dfrac{3(1 - \lambda^4)h}{4(1 - \lambda^3)}$ **18** $\frac{10}{9}a$; $\frac{33}{16}$

19 (a) $P < 2\mu W$ (b) $P < \mu W$ (c) $P < \dfrac{Wa}{a + h}$ (d) $P < \dfrac{Wa}{2h}$

Upper cube slides. Lower cube topples.

20 $\dfrac{6l^2 + 4hl + h^2}{4(3l + h)}$

Chapter 18

Qu. 1 $4.5m\omega^2$

Qu. 2 $\frac{1}{3}ml^2$

Qu. 3 $\frac{5}{4}Mr^2$

Qu. 9 $4\pi\sqrt{\left(\dfrac{a}{3g}\right)}$

Exercise 18a

1 $14ma^2$; $7ma^2\omega^2$ **2** $\sqrt{\left(\dfrac{6g}{7a}\right)}$

3 $5ma^2$ **5** $\frac{1}{3}M(3c^2 - 6ac + 4a^2)$

Exercise 18b

1 (a) $2Ma^2$ (b) $2Ma^2$ (c) $4Ma^2$

2 (a) $4Ma^2$ (b) $4Ma^2$ (c) $8Ma^2$

3 $2Ma^2$

4 $\frac{16}{3}Ma^2$; $\frac{8}{3}Ma^2$

5 (a) $34Ma^2$ (b) $66Ma^2$ (c) $100Ma^2$

6 19.7 J

7 15.9 kg m^2

8 $\frac{5}{3}Mr^2$

Exercise 18c

1 $\frac{1}{3}M(a^2 + b^2)$ **2** $\frac{1}{12}M(3r^2 + 4l^2)$

3 $\frac{3}{10}Mr^2$ **4** $\frac{1}{20}M(3r^2 + 2h^2)$

Exercise 18d (Miscellaneous)

1 $\frac{1}{4}ma^2$; $\sqrt{\left(\dfrac{2g}{r}\right)}$

2 (a) $\frac{3}{2}Ma^2$ (b) $\frac{1}{4}Ma^2$ (c) $\pi\sqrt{\left(\dfrac{6a}{g}\right)}$ (d) $\pi\sqrt{\left(\dfrac{5a}{g}\right)}$

4 $\sqrt{\left(\dfrac{2g}{3l}\right)}$, $2\pi\sqrt{\left(\dfrac{6l}{5g}\right)}$

5 $T_1 = \dfrac{m_2g(I + m_1a^2)}{I + (m_1 + m_2)a^2}$, $T_2 = \dfrac{m_1g(I + m_2a^2)}{I + (m_1 + m_2)a^2}$

6 $\sqrt{\left(\dfrac{39g}{80a}\right)}$, $\dfrac{39g\sqrt{3}}{160a}$

7 $\frac{2}{3}a\sqrt{2}$, $\frac{16}{3}ma^2$; $a\sqrt{2}$

8 $\frac{1}{4}mr^2(1 + 4k^2)$, $\pi\sqrt{\left[\dfrac{r(1 + 4k^2)}{kg}\right]}$, $k = \frac{1}{2}$

9 $\frac{4}{3}ml^2 + 6mx^2$

10 $\sqrt{\left(\frac{85}{14}ga\right)}$

Index

419